Foreword

Most of us are familiar with the Automotive Market—of course. Here we think immediately about car manufactures, big brands like Mercedes Benz, BMW, Jaguar Land Rover, etc. It is well known, highly relevant from an economic point of view as the sector makes a significant contribution to the gross domestic product and the global environment.

But in the automotive industry there is another, hidden but highly systemically relevant market that is not so obvious and does not immediately catch our eye. And yet, as consumers, we experience and rely on it throughout the entire lifetime of our vehicles. It's the Automotive Aftermarket.

The fact that this market grows in the shadows may also be because it is not talked about often. My appreciation therefore goes to Maria Elo, Fotios Katsardis, and all the authors out of the industry and academia of this volume to highlight the Automotive Aftermarket with its diversity, its strength, its contribution to our society, and the attractiveness in working in this sector of the automotive industry. Ask yourself the question: What would happen to our vehicles without the Automotive Aftermarket?

This book is showing and explaining the market itself, the various players, and the diversity of this industry and stretches the bow from small to big, from workshops to industry. Always on top of technologies and technological trends, this volume outlines the tremendous challenges with the upcoming electrification also offering multiple opportunities for people joining this industry.

Given the enormous challenge that our society faces today, the book also shows the relevant contribution that our industry makes and can make over the lifetime of a vehicle when it comes to sustainability. Let us think, for example, of approaches like reman, re-use, refurbish, or recycling.

Part I of this volume gives a good insight into this unique market and the importance of the sector for our society. Part II describes what is necessary to deal with the challenges the various players are facing in this industry, but also how they position themselves to make the best possible use of the new opportunities that arise in parallel. Part III is showing us that we have similar requirements but internationally the industry is dealing differently with these requirements. Part IV concludes in showing the innovation power, the business ecosystem as well as global challenges this sector is facing.

As one who has worked in leading positions in the Automotive Aftermarket for over 30 years, I'm very pleased having the opportunity to write this foreword for the book as I personally believe in this industry—its opportunities, innovation power, and high relevance for each of us in our daily mobile lives.

GROUPAUTO International
Neuilly-sur-Seine, France

Guenter Weber

Foreword

The automotive industry is among the most important ones in many countries, if not economically then at the level of media attention and discussions about its past and continued current contribution regarding greenhouse gas emissions as well as the future potential regarding the reduction thereof. Given this dominance, it might surprise that the discourse does not represent the entire value chain. In fact, like in many other industries of durable goods, a lot of value creation happens over the lifecycle of a good, i.e. after the actual sales. While in other industries such as construction of real estate or mass-transportation vehicles examples such as facility management or maintenance have become well-established as parts of the analysis, the automotive industry keeps being dominated by the focus on the development, production, and sales of cars, rather than what happens after the point-of-sales—the so-called "Automotive Aftermarket" (AA). This biased perspective is surprising because the transformation of the automotive industry—both feasible due to technological innovation and necessary due to ecological challenges—should affect this aftermarket no less than the earlier stages of the value chain, if not even more. Conversely, there might also be a shift of value creation away from OEMs to the aftermarket, given that internal combustion engines become less relevant over the next decade or so, thereby re-iterating the value migration theses rebutted for the case of challenges from software firms entering the car cockpit (Jacobides et al., 2013) and put under a broader lens by several others (cf. Schulze et al., 2015).

This volume emerges as a pioneering academic venture that seeks to unravel the complexities, nuances, and underappreciated facets of the aftermarket. Through a careful examination of existing literature, this collection of essays presents a holistic, multidisciplinary exploration of the AA, shedding light on its profound economic, societal, and environmental impacts.

The introductory chapters set the stage by navigating the contours of existing literature, revealing a conspicuous void in scholarly attention toward the AA. While primary manufacturing has garnered extensive scrutiny, the aftermarket surfaces as an intricate tapestry, woven with distinct managerial challenges and societal reverberations. The editors' keen observations on the dearth of comprehensive representations of the AA serve as the driving force behind this book, compelling an exploration of various disciplinary perspectives to construct a nuanced understanding of the aftermarket.

This volume is not merely an academic exercise but a collective effort of editors and authors deeply entrenched in the automotive aftermarket. Their experiential knowledge, gleaned from both managerial and educational roles, permeating this book, illuminates challenges in disseminating sector-specific information to diverse audiences. The editors' dual roles as business experts and educators resonate with the overarching theme of the book—bridging the gap between academia, industry, and broader societal contexts.

A distinctive feature of this book lies in its embracing of diverse disciplinary viewpoints, acknowledging the interconnected nature of the automotive aftermarket. The stark contrast in scholarly attention between the broader automotive industry and the aftermarket, vividly illustrated through search engine results, becomes the impetus for this comprehensive exploration. By weaving together these disciplinary threads, the book aims to offer a mosaic of perspectives that goes beyond traditional hierarchical representations.

The thematic structure of the book unfolds seamlessly, providing a comprehensive framework for delving into various facets of the AA. From a foundational exploration of the aftermarket as a system-relevant sector to nuanced discussions on knowledge, education, talent development, regional dynamics, market intricacies, internationalization challenges, and the confluence of innovation, business ecosystems, and sustainability concerns—each part contributes to a holistic understanding.

From a German perspective, this book is particularly important as the automotive industry is one of the largest contributors to GDP and employment. Working with SMEs, which form a large part of this industry, complementing the household names of German OEMs the network of RKW Kompetenzzentrum and regional RKWs together with its partners is keen on supporting firms in this sector. Providing them an academically and practically well-founded basis for their transformational efforts is a very timely initiative.

In conclusion, the book is a transformative journey through the intricate landscape of the AA as an industry in its own right. By embracing diverse viewpoints and challenging hierarchical representations, this book paves the way for a nuanced understanding of the automotive aftermarket's role in shaping the trajectory of the automotive industry. As we embark on this exploration, we anticipate a collective shift in scholarly discourse, fostering a more comprehensive and interconnected understanding of the automotive ecosystem and the often-overlooked aftermarket.

Head of Entrepreneurship, RKW Competence Center Gründung, RKW Kompetenzzentrum
Eschborn, Germany

Florian Täube

References

Jacobides, M. G., MacDuffie, J. P., & Tae, C. J. (2013). How agency and structure shaped value stasis in the automobile ecosystem. Strategic Management Journal, 37(9), 1942–1967.

Schulze, A., Paul MacDuffie, J., & Täube, F. A. (2015). Introduction: knowledge generation and innovation diffusion in the global automotive industry—change and stability during turbulent times. Industrial and corporate change, 24(3), 603–611.

Contents

Contributors

Raushan Aman Matti Ahtisaari Institute, University of Oulu, Oulu, Finland

David Breitenbach PatentRenewal.com ApS, Copenhagen, Denmark

Álvaro de la Cruz Tomás FIGIEFA, Brussels, Belgium

Erik Deitersen africon GmbH Management Consultancy, Bremen, Germany

Ralph Dreher Chair of Technical Vocational Didactics (TVD), Universität Siegen, Lehrstuhl für Technikdidaktik am Berufskolleg, Siegen, Germany

Maria Elo University of Southern Denmark (SDU), Odense, Denmark

Belt and Road Institute of International Business, School of Economics, Shanghai University, Shanghai, China

Abo Akademi University, Turku, Finland

Turku school of Economics, University of Turku, Turku, Finland

Institute Arbeit und Technik, Gelsenkirchen, Germany

Monika Hajdas Uniwersytet Ekonomiczny we Wrocławiu, Wrocław, Poland

Tilo Halaszovich Constructor University, School of Business, Social and Decision Sciences, Constructor University, Bremen, Germany

Andreja Jaklič Faculty of Social Sciences, University of Ljubljana, Ljubljana, Slovenia

Jens Jüngst Chair of Technical Vocational Didactics (TVD), Universität Siegen, Lehrstuhl für Technikdidaktik am Berufskolleg, Siegen, Germany

Amanda Katsardis University of Aberdeen, Aberdeen, UK

Fotios Katsardis Temot International Autoparts GmbH, Ratingen, Germany

Hasmeet Kaur Roland Berger, Munich, Germany

Gyula Kimpán TEMOT International Autoparts GmbH, Ratingen, Germany

Daniel C. F. Köhler Business Unit Truck and Trailer Components, BPW Bergische Achsen KG, Wiehl, Germany

Iris Koleša Faculty of Social Sciences, University of Ljubljana, Ljubljana, Slovenia

Fynn Komp Chair of Technical Vocational Didactics (TVD), Universität Siegen, Lehrstuhl für Technikdidaktik am Berufskolleg, Siegen, Germany

Zoran Nikolic Wolk After Sales Experts GmbH, Bergisch Gladbach, Germany

Claire Croucher Petersen Department of Technology and Innovation, University of Southern Denmark, Odense M, Denmark

Susanne Sandberg School of Business and Economics, Linnaeus University, Kalmar, Sweden

Felix Sterk Robert Bosch GmbH, Automotive Aftermarket, Karlsruhe Institute of Technology, Institute of Information Systems and Marketing, Institute of Information Systems and Marketing (IISM), Karlsruhe, Germany

Anders Leth Villumsen Department of Technology and Innovation, University of Southern Denmark, Odense M, Denmark

Preface: Global Perspectives and Disciplinary Angles in Understanding the Automotive Aftermarket

Maria Elo and Fotios Katsardis

Abstract

Despite rich literature and research on the primary automotive industry and business, there is very limited literature focusing on the automotive aftermarket and the related business actors and ecosystems. Automotive aftermarket can be addressed by multiple disciplinary angles, e.g. engineering, marketing, supply chain, innovation, strategy, and other management disciplines. The complex and intertwined nature of the automotive aftermarket as a structure may partly shadow its importance as a societally, technologically, economically, environmentally, and politically relevant business sector. In general, there is lack of interest and awareness of the automotive aftermarket regardless of its effects on economy, employment, sustainability, and mobility. This gap extends from education to institutional regulation. The automotive aftermarket is not "just" aftersales repair, in fact, it represents a fabric of diverse business contributing to mobility solutions, data management and sustainability development. This book explores inherent themes, challenges, and potentials of the automotive aftermarket from different global, thematic, and disciplinary angles written by a range of

M. Elo (✉)
University of Southern Denmark (SDU), Odense, Denmark

Belt and Road Institute of International Business, School of Economics, Shanghai University, Shanghai, China

Abo Akademi University, Turku, Finland

Turku school of Economics, University of Turku, Turku, Finland

Institute Arbeit und Technik, Gelsenkirchen, Germany
e-mail: melo@sam.sdu.dk

F. Katsardis
Temot International Autoparts GmbH, Ratingen, Germany
e-mail: f.katsardis@temot.com

M. Elo, F. Katsardis (eds.), *Automotive Aftermarket*, Management for Professionals, https://doi.org/10.1007/978-3-031-62419-3_1

authors and experts. It contributes to advancing the automotive aftermarket related knowledge and awareness.

1 Introduction

There is a range of literature and research addressing the automotive business, especially the primary manufacturing business, but little extant literature focusing on the automotive aftermarket. In general, the literature focuses largely on technology, engineering and manufacturing, and respective disciplines, but also on marketing, supply chain, innovation, strategy, and other management disciplines. This variety of disciplinary angles illustrates that the automotive sector in total offers numerous perspectives that are distinct and have stand-alone features regarding research, management, and business. The complex and intertwined nature of the automotive aftermarket as well as its importance as a societally, technologically, economically, environmentally, and politically relevant business sector tends to be approached often as a processual subset of the automotive sector. However, this is a misleading approach as there are multiple activities taking place that are not "just" some sort of aftersales maintenance and repair issues. In this sense, the extant literature is missing out on its representation of the automotive aftermarket sector, its actors, operations, ecosystems, and management.

A simple comparison shows the neglected nature of the automotive aftermarket as a focus. Regarding research literature, Google Scholar is a tool that easily visualizes the amount of existing work on a topic. In October 2023, "automotive aftermarket" search word brought 21,300 results, "independent automotive aftermarket" showed 11,300 results, while "automotive industry" brought 2,410,000 results in Google Scholar. In a similar vein, Google search engine illustrates respective discrepancies across general information available in the Internet. "Automotive aftermarket" in Google search brings 93,200,000 results, "independent automotive aftermarket" shows 6,890,000, and again "automotive industry" illustrates a considerable magnitude of interest with 637,000,000 results.

This missing interest and awareness are interesting, even a paradox, as the automotive industry—and the inherent automotive aftermarket sector—represents a sector of one of the world's largest industries with notable effects on economy, employment, sustainability, and mobility to mention some of the key aspects. Hence, there is a need to delve into this industry and update the lenses that are used to address the "aftermarket" and "aftersales" part of this landscape. The rationale of the book stems from this need to understand the gap that the automotive aftermarket represents and introduce different perspectives for examining and discussing it. There are several elements that call for more managerial, research, and policy attention underlining the relevance of AA economically, societally, and environmentally. The automotive aftermarket is that side of the business that has the most direct and long-term effects with car owners, users, and drivers maintaining their mobility function, but also with other mobility, including heavy vehicles, like buses, trucks, and other highly relevant types of machines that allow our economy and society to operate. Interestingly, it seems to be partly forgotten in many public debates that all these vehicles, including public transport, need to be maintained and serviced. While in research servitization has gained notable attention as a form of value creation (Verstrepen et al., 1999). Many discussions tend to focus on technological and future trends, such as electrification, autonomous driving, connected vehicles, and shared or diverse mobility instead (Grebe et al., 2021). Yet, beyond the future and such potential, the older and extant vehicles also generate legitimate concerns to be discussed. For example, the functioning in the post-sales life cycle is relevant from agricultural vehicles enabling food production to ambulances providing safe medical patient transport. These concerns may be particularly relevant in developing countries that have different challenges often related to earlier technologies in use. In short, there is a multitude of particular aftermarket subthemes and elements that together form a systemic function, a system-relevant sector for a society. This function is enabled and managed by a range of different AA business actors. For some reason, this side of the automotive business has remained largely ignored, for example, in media, research, and education.

The book editors have a long relationship with automotive aftermarket in different roles. In management and education activities, it becomes clear that there is not only a lack of research but also a lack of awareness and interest in this sector among different audiences. We as editors have experienced this gap in our everyday work. As a business expert, the role and meaning of AA are often seen as secondary importance or even ignored; moreover, in all sorts of contexts, there is a continuous need to explain the sector to audiences who are not "insiders." This lack of sectoral information and understanding seems to be a rather global problem. More AA information and knowledge production is required to produce an updated understanding of AA and "a big picture" of it for all stakeholders and audiences.

Also as an educator, it has been interesting to see that automotive management students, even engineering students, do not really perceive or know this sector, its functions, and roles, nor its importance in facilitating sustainable mobility or circular economy. Especially, service-related business tends to receive less attention compared to manufacturing business among students. In a study carried out at the University of Southern Denmark, we also found that the majority of the respondents, whether engineering or business students, had very incorrect perceptions of the sector. Additionally, while looking for teaching cases, journal articles, or other materials explaining specific aspects or some insights into this sector, it became evident that there is a very limited offering of such sources—and the large majority is actually commercial material or just annual reports. Commercial material is not peer-reviewed or edited in a similar way to academic research material, and any organization can produce or share what they see as information (Tewell, 2015). However, in the era where information literacy and critical thinking are becoming increasingly important, the shortage of validated, legitimate information, and research literature leaves room for misinterpretations and alternative perceptions of all sorts.

When we think of the management challenges and opportunities in the AA business, we also need to prepare for the future AA talent and their need to capture this business appropriately. For managerial decision-making, both strategic information and the ability to use such market intelligence for that process are crucial elements

(Novicevic et al., 2004). Once we turn to the actual business market, there is a multitude of professionals who possess extensive experience, knowledge, and capabilities related to AA and who have the information literacy required to critically examine, interpret, and present perspectives of AA. Therefore, the need for a book that would offer a starting point for general understanding, such as what this sector is, how it operates, where it is active, what institutional structures and organization settings it has, and what kind of challenges it faces, became apparent. This knowledge needs to be bridged and disseminated and new avenues need to be opened. Due to the nature of AA having multiple sides and audiences, it is clear that we need to bring diverse expert voices and insights together, beyond academic research or commercial discussions.

For crafting a selection of viewpoints that come from different perspectives, we have invited for this book a range of stakeholders, such as industry experts, managers, entrepreneurs, educators, representatives of associations, and junior and senior scholars, who bring particular lenses and topics to this book. These are very different authors who altogether generate a multidisciplinary puzzle of the automotive aftermarket with the chapters they contribute as puzzle pieces. Naturally, it is not realistic to create an exhaustive or holistic book covering all topics. Instead, the aim of the book is to start this discussion on institutionalizing the automotive aftermarket as a stand-alone discussion and sector with a distinct identity and role in the global value network, research, and socioeconomic landscape. The overarching lens of the book is the managerial element as all these discussions relate to management in some form.

The book's objectives are (a) to bring together multiple disciplinary perspectives addressing the automotive aftermarket and respective knowledge, (b) provide insights into specific relevant aspects of the automotive aftermarket, (c) increase awareness of the sector as a whole and its characteristics, and (d) trigger a discussion on the future of automotive aftermarket showcasing some of its challenges and opportunities.

Multiple authors bring their perspectives from business, research, and institutional sectors to contribute to these objectives. The book is organized into four parts. Each part has its own perspectives and foci.

2 Part I: Introduction to the Automotive Aftermarket—A System-Relevant Sector

The book starts with a section that presents several overviews of AA and its characteristics as a sector spanning global environments and engaging in diverse operations. This shared understanding is necessary to produce before going deeper into specific other perspectives on the AA. As Laborda and Moral (2020)[1] note, "there is one economic sector related to the automobile industry that has received little attention, both in theoretical and empirical economic analyses, that is growing in relevance and importance. We refer to aftermarket automotive services whose

[1] See more in https://www.mdpi.com/2071-1050/12/18/7817.

profitability has made car manufacturers themselves consider automotive post–sales as an additional business unit." Hence, the raison d'etre and business logic per se of the automotive aftermarket deserve attention. Beyond the considerations of the business models or roles in the value chain, understanding the context and its framework conditions is required as even within the EU there are differences (David & Terstriep, 2015). These are not just market-related differences but also institutional and nonmarket differences influencing business strategies (Corredoira & McDermott, 2014). Moreover, the global landscape of the automotive sector moves along with the developmental patterns that change the frameworks, such as innovation and technology (Schulze et al., 2015).

The book starts with the chapter "Automotive aftermarket – introduction to a global business," authored by Fotios Katsardis, Temot International Autoparts, and TALENTS4AA. This chapter compiles multiple viewpoints that together generate an overview of the development of the sector into a global industry. It introduces relevant elements from different angles, which shape the industry and its pathways into different markets. Further, the chapter analyzes the heterogeneity of the sector in its geographies and brings insights into different areas and their developments. It presents perspectives related to the supply chain, distribution, and marketing, among others. Yet, the underlying red thread of service business is inherent in the functions and markets presented. The chapter contributes a sectoral overview that combines management, trade, and service lenses.

The next chapter "Understanding the automotive aftermarket" by Hasmeet Kaur, partner, global managing director, Roland Berger, offers a viewpoint from international consulting to the characteristics of the sector that are relevant for analysis and comparison. This chapter discusses the sector as a business market and provides key thematic elements that managers need to capture in their decision-making process when trying to understand the global automotive aftermarket, such as its structures and channels, organizational forms, actor types as archetypes, and contextual differences. Beyond reflecting on these aspects, the discussion illuminates also the driving forces of the automotive aftermarket, its main challenges and opportunities, and its market trends. The chapter contributes to understanding this market and its management concerns.

After these introductory contributions, there are two important vignettes that bring distinct perspectives from the angle of top managers with extensive industry expertise from the United States and Europe.

The first viewpoint "An expert vignette from the American perspective—John R. Washbish, president and CEO" compiled by Elo and Katsardis provides an interview offering a perspective representing both the American entrepreneurial field with independent family businesses and the top management and alliance level. Mr. Washbish identifies several elements of relevance for AA from consumer behavior and instant gratification to consolidation and private equity-generated changes. The need for education and training is highlighted as an important success factor, similar to the family business dynamics. In addition, several market potentials are pointed out, locally and internationally, even in one's "backyard."

The second viewpoint "'The Biggest Hidden Champion': a vignette from Michael Söding, industry veteran from Schaeffler Group" compiled by Elo and

Katsardis offers a more European perspective to the global automotive aftermarket. This vignette provides a German case of extensive expertise that reflects experiences from renowned supplier companies, particularly from the Schaeffler Group. Mr. Söding reflects on his career and the industry's progress, highlighting several opportunities linked to issues such as learning, career, solutions, and ecosystems. He also explicates challenges, for example, access to data, standardization, talent development, German-centric views, and lack of sectoral visibility. Yet, as the title suggests, he underlines the hidden dimension of the sector with a positive connotation.

Beyond the societal role of the automotive aftermarket sector, both these vignettes point also to the different institutional and regulatory frameworks that shape the evolution of the sector, its challenges, and opportunities.

The next contribution from the perspective of automotive aftermarket distributors FIGIEFA named "Right to repair in the automotive industry," authored by Álvaro de la Cruz Tomás shows the importance of institutions and how those enable or inhibit free market and respective competition. The meaning of legislation is a panacea for shaping the structures and rules governing this sector. The chapter provides insights into the challenges that the regulative framework generates for free and independent aftermarket actors while presenting their origins and mechanisms. These institutional and regulatory issues around the data and its role in the automotive service business contribute to many elements of this sector. This becomes important to understand as access to data appears inherently as an enabling factor for multiple activities in many of the other chapters too.

3 Part II: Perspectives on Knowledge, Education, and Talent Development

The part of the book has collected chapters that bring very different but yet relevant perspectives to the discussion on the automotive aftermarket together. It compiles views that illustrate how AA is developing as a business in a knowledge economy that constantly innovates products, services, learning, and other processes (Bouncken et al., 2021). AA is also addressed as a proactive field needing and developing solutions for education that anticipates changes in AA and technology.

Knowledge sharing is not only relevant at the management level but also at other organizational levels within the workshops and across employees (Alam et al., 2022; Schmidt et al., 2021). Beyond tacit knowledge, codified knowledge, information, and quality of data play a major role in management (Smith et al., 2023). Among experts, there has been notable criticism related to the roles of OEM companies and their monopolistic advantages that relate to vehicle data ownership, access, and costs. This criticism continues also in the training and education context as large multinational corporations may rely on self-trained personnel and run inhouse training units, while the many independent aftermarket actors, typically smaller family businesses (e.g., workshops), become reduced in the depth and scope of operations focusing more on sales and installation of parts. Many processual elements remain blocked for them, hence, also the knowledge transfer and learning become delimited and inequal in the same market. This is a concern for sectoral capacity building.

Furthermore, in this part, the automotive sector as a whole is reflected as a sector that needs to address inequalities related to gender and diversity. When talent is scarce, the need to understand bottlenecks and glass ceilings hindering organizational development and competitiveness becomes even more crucial (Hornbuckle, 2023). Simultaneously, AA is called to advance awareness of it and its attractiveness as a career option rethinking ways to organize and execute employer branding activities (Gupta et al., 2018).

In this part, the chapter starting the discussion is "Marketing information systems in the automotive aftersales market," authored by Zoran Nikolic from Wolk After Sales Experts GmbH, Germany. This chapter presents a view of the managerial decision-making-related information that reflects the particular characteristics of the automotive aftermarket. It points out the complexity of the information that an aftersales market has due to its nature, and the need to capture and assess information that supports managerial decision-making on strategy and diverse market operations. He notes that the AA is becoming more emancipated in terms of the sector and its origin as the secondary market. This refers to a distinct field of data flows forming the base for a marketing information system that is AA specific. The chapter offers an overview of marketing information systems and their use; additionally, it offers several insights into the specific cases and dashboards illustrating the function. It highlights the need to understand the data needs and their use professionally, for example, via marketing information systems, to maintain and advance the competitive advantage.

Next, the chapter "Vocational education training (VET) for electrical-driven cars: development of a training concept for the aftersales market (Project DIAKOM-E)" by Ralph Dreher, Jens Jüngst, and Fynn Komp from the University of Siegen brings up the importance of training and education in the evolving AA. As the electrification of mobility increases, also the knowledge and skills needed to serve the vehicles evolve. Appropriate training and access to the vehicle data function as prerequisites for repair and service. The German experience offering dual vocational training concepts provides a particular setting for understanding these needs and practices, and Project DIAKOM-E shows the development of a real-life training project. The chapter refers to the need to advance manufacturer-independent training that builds sectoral capacity, while reducing the OEM dominance. There is currently a lack of both manufacturer-dependent and manufacturer-independent training; hence, DIAKOM-E has set itself the goal of offering manufacturer-independent training in the diagnosis and repair of BEV vehicles, which can be made available to all workshops (brand-related or independent) for employee training. Also here, data and information link to the core issue of vocational training and its functioning for all. The project introduces a learning situation matrix that facilitates learning and inclusion of heterogeneous groups of learners. Additionally, the chapter provides a range of conclusions reflecting the particular characteristics of AA versus OEM.

Not only information and skills are necessary, but also the people per se. The next chapter "Gender equality at work: tackling gender equality in senior leadership in the automotive aftermarket" by Amanda Katsardis, University of Aberdeen, and Raushan Aman, University of Oulu, introduces the perspective of gender and gender equality in the automotive aftermarket. This chapter shifts the attention from the workshop to the management level and senior leadership positions. The chapter illustrates the gap in the SDG 5 targets on gender equality through the lens of the female managers and leaders in AA. This research project finds that the sector still suffers from gender bias that impedes women's career advancement, although it also shows that several companies are investing in gender and diversity policies and mentoring programs. The study underlines that especially leadership positions have remained limited for women and suggests multiple recommendations for action from increasing flexibility and family-friendliness to fairness and equitable mentoring programs. It points out that the sector faces a dual challenge of reducing bias and increasing its attractiveness and suitability for women in recruiting, developing careers, and retaining female talent strategies.

The following chapter "Acquiring talent for the automotive aftermarket: how sustainability is becoming a key driver in talent acquisition and development for the sector" by Gyula Kimpan, Temot International GmbH, and Maria Elo, University of Southern Denmark, continues to address the human capital side of the sector focusing on the talent attraction. This chapter introduces the issue of how to attract and acquire new and young talent, especially students, to the automotive aftermarket sector. It provides a perspective from the students, their perceptions of the AA, and its attractiveness. The chapter presents these preliminary findings from a research project in Denmark that took place among engineering and business students who represent a potential talent source for AA. The results show a lack of awareness of the sector and its opportunities, as well as elements such as sustainability and circularity that pull students to perceive a sector as attractive and interesting. Recommendations for updating AA communication strategies are provided.

4 Part III: Perspectives on Regions, Markets, and Internationalization

The automotive aftermarket is not a homogeneous industry or sector that has global markets similar to some other industries. Different regions are hosting different activities, and markets have distinct features related to these activities as well as the type of actors, car parc, mobility patterns, culture, and many other aspects (Subramoniam et al., 2009). Interestingly, there are only limited studies addressing the geographies of the AA, such as service networks, business clusters, or comparative business development patterns (Aboltins, 2022; Marshall, 1989). Instead, issues such as supply chain and logistics have gained more research attention (Li, 2015), together with production and operations management (Durugbo, 2020).

AA markets studied as business markets are often addressed in a rather siloed or fragmented manner, with focus on particular products and services (Durugbo, 2020;

Huber & Bauer, 2009), and less as markets operating in systemic competition (Hawker, 2011). Regional approaches and differences in marketing practices and strategies form another aspect of interest as marketing practices differ even within categories and countries (Al Fayad, 2014). This highlights the complex nature of the AA that moves at multiple speeds and processes in different contexts while competing within a multilayered business. These diverse markets, whether regions or product-related markets, represent a plethora of different opportunities and challenges. Responding to international opportunities, market-related arbitrage, and advantage seeking, such as location advantages, but also to supplier ecosystems and next-shoring with proximity to demand and innovation, many businesses continue to develop their international expansion and integration (Frigant & Layan, 2009; George et al., 2014; Ramamurti, 2009). As Ramamurti (2009) notes, there are several main strategies related to internationalization, particularly in the emerging markets, and the respective roles; natural resource vertical integrator, local optimizer, low-cost partner, global consolidator, and global first mover. AA companies can be found reflecting many of such strategies and patterns across the world.

In this part, the chapter "Automotive aftermarket in the Western Balkans: opportunities for the region and the European automotive industry" by Iris Koleša and Andreja Jaklić, University of Ljubljana, illuminates the context of Western Balkans and explores views from AA firms, clusters, and support system representatives on regional similarities and industry development. Regions matter for sectoral and industrial development. The chapter offers insights into local strengths and weaknesses. It presents specific features of Western Balkan countries that have undergone numerous transformations and compares these markets and sectoral developments. The countries are Croatia, Slovenia, Bosnia-Herzegovina, North Macedonia, Serbia, Montenegro, Kosova, and Albania. The chapter provides recommendations particularly for policymaking and calls for modernization of the higher education system and employee development, advancing public–private partnerships and networking, as well as faster development of internationalization and investments in the region, for example, in digitalization, automation, and technology.

The chapter "Service design, marketing, and automotive aftermarket: perspectives from a Polish study" by Monika Hajdas, University of Wroclaw, presents findings from a local study related to the marketing of service business in AA. The chapter discusses service business and its branding and value propositions from the perspective of customers, that is, user insights. The study elaborates user-centered design as an instrument to address customer needs, in this case female car drivers. Developing personas to reflect a particular target group, its needs and wishes are introduced for successful updating of service marketing approaches. These reflections on the challenges of customer interaction take place at the workshop level, and hence, provide food for thought for service businesses but also their marketing partners. The chapter calls for addressing the female customer experiences with updated marketing instruments to incorporate this target group's potential.

The following chapter "The relevance and perception of African business potentials: German perspectives and strategies on internationalization to Africa" by Tilo

Halaszovich, Constructor University, and Erik Deitersen, Africon GmbH Management Consultancy, focuses on the next big market, Africa, and its role as a target market for German suppliers. The African automotive aftermarket is increasing and becoming an interesting opportunity for growth and expansion. This chapter illustrates and discusses the hurdles and approaches that German firms are having for entering local markets in Africa. These may be explained by their liabilities of foreignness. Different patterns and strategies are presented, while exports represent a dominant entry strategy. The study finds that the local partners and their particular knowledge are instrumental in a successful entry process.

5 Part IV: Perspectives on Innovation, Business Ecosystems, and Global Challenges

Interestingly, while the primary production in the automotive industry compiles and produces the vehicle, it is not necessarily the innovator behind novelties and innovations against common perceptions. Instead, the automotive aftermarket serves as the source of many innovations. Major supplier companies, such as Bosch Group, Schaeffler Group, ZF Friedrichshafen AG, and Denso Corporation, are central actors in generating patents and innovations. However, the focus of interest in related research tends to relate mainly to service-related innovations (Dombrowski & Engel, 2014; Wang & Kimble, 2016). Yet, innovations from the automotive aftermarket are not limited to those. For example, Mahmoud-Jouini et al. (2007) point out that suppliers can actually leap-jump their direct customers in the value network and generate breakthrough usage innovations. Such cases are theoretically and practically fascinating and relevant to understanding the creation of innovation in a sectoral context with diverse power constellations and business flows (von Hippel, 2006).

The effects of the automotive and mobility sector are not limited to technological innovations or economic impact, and there are multiple forms of effects that address also environmental and social concerns and impacts (Williams & Blyth, 2023). The AA sector has a notable employment and entrepreneurial impact too. There are multiple stakeholders and business actors creating the web of operations that create and maintain automotive mobility. It is an interconnected sector where changes and critical events are connected, not confined, to nature and may hence generate domino effects and complicated capture processes (MacKay & Chia, 2013). The automotive aftermarket is an integral part of the societal and economic tissue of a place, participating in and creating business and entrepreneurial ecosystems (Prakash & Manchanda, 2022). The entrepreneurial resource base is changing with aging populations. The AA ecosystems enabling lower entry barriers and more inclusive entry points for diverse entrepreneurs, differently than the dealer networks, provide societally relevant value for different people and regions.

However, there are very limited studies on the ecosystems surrounding the automotive sector, and no extant research on the automotive aftermarket per se as an ecosystem. An OECD study by Dechezleprêtre et al. (2023) takes a country-level view for a holistic analysis and combines inter-country input–output tables, patent data, mergers and acquisitions transactions, cross-country micro-distributed data, and firm-level balance sheet data. By doing so, they identify ecosystemic boundaries, connected sectors, both upstream and downstream, knowledge and technology providers that represent, for example, universities, and assess the nature of the ecosystem's microstructure that continuously evolves. Dechezleprêtre et al. (2023) call for effective policies that would focus on innovation and competition and would facilitate the growth of young firms.

Beyond such industry concerns on innovation and technology, the green and digital transitions call for novel perspectives to develop the automotive ecosystem as a whole as well as the automotive aftermarket. As a result, sustainability is gaining attention in management and research, but sectoral AA research attention remains scarce.

Sustainability in the automotive industry in general is often enacted due to external pressures; those lead companies to address the need, for example, for sustainable supply chain management practices (Mathivathanan et al., 2018; Szász et al., 2021). Another important part of the discussion is the circularity that is deeply linked to the AA operations. Circular economy and recycling were inherent elements in the AA already before the public attention started pointing out them, but such life cycle management and related marketing issues have also diverse challenges (Diener & Tillman, 2016; Elo & Kareila, 2014). There is a range of supply management issues, product life cycle strategies, and regulations and business considerations related to issues such as remanufacturing, and these also have contextual characteristics (Subramoniam et al., 2009). These aspects underline the complexity, internationality, and intertwined processual nature of the AA business that managers need to cope with when advancing sustainability and related innovations in the AA ecosystem. The AA is facing also a plethora of aspects to consider regarding the overall development of the sector and the technological and structural evolution (Waas et al., 2021; World Economic Forum, 2022).

The chapter "Innovation capabilities in the automotive aftermarket: case study from Automechanika Frankfurt 2022" by Anders Leth Villumsen and Claire Croucher Petersen from the University of Southern Denmark, Department of Technology and Innovation, brings a perspective from young experts who participated in person at the Automechanika Frankfurt 2022 as student talent ambassadors and collected material for analysis. They address innovation theory and showcase five particular innovations from different types of automotive aftermarket actors. These cases illustrate how different AA companies develop and market their innovations to their users and customers while advancing their innovation capabilities continuously and increasingly toward service-centric innovation. The authors identify two main elements shaping and innovation: sustainability and legislation. The sustainability elements per se referred mainly to economic and environmental sustainability.

The next chapter "How can the Nordic automotive aftermarket provide opportunities counteracting the challenges of disadvantaged entrepreneurs?" is written by David Breitenbach, Leibniz University Hannover, Susanne Sandberg, Linnaeus University, and Maria Elo, University of Southern Denmark. This chapter explores more the social sustainability side of the Nordic AA and how it may operate as a gateway for capacity development of entrepreneurial potential regarding disadvantaged entrepreneurs, such as immigrants and refugees. These groups face impediments and entry barriers in both employment and entrepreneurship, but the study finds that the AA has suitable characteristics with low entry barriers and demand for labor and entrepreneurial succession. The AA industry and its network structures may facilitate entrepreneurial entry and inclusion, despite language problems, providing both opportunities and support structures. The chapter offers a view underlining the development potential of entrepreneurial entry processes as a sector.

The following chapter "Automotive remanufacturing: the ultimate form of circular economy" is written by Daniel Köhler, BPW Bergische Achsen KG, who is also the chairman of the international nonprofit association Automotive Parts Remanufactures Association Europe AISBL (APRA Europe). This chapter takes a deeper and more explicit look at the sustainability discussion. It provides an overview of remanufacturing and circularity in the context of automotive aftermarket. It presents diverse benefits related to remanufacturing and explains the process of remanufacturing in this sector while offering a management view to it, hence departing from the dominant engineering studies. It considers critical aspects, such as identification of suitable products, and management of cores and their flows across the process. Additionally, it showcases examples of specific parts. The chapter finds remanufacturing as the ultimate form of recycling and circular economy and underlines the AA's role in reducing carbon footprint by engaging in remanufacturing strategies.

The final chapter "Leveraging car connectivity in the automotive aftermarket and beyond" by Felix Sterk, Karlsruhe Institute of Technology, connects many of the previously discussed dots of the AA sector and provides a study that offers a viewpoint to the platform ecosystems and connected cars. The chapter discusses how AA companies respond to the changing business and technology landscape by updating and designing new business models and solutions. It highlights the specific constellation in which AA companies operate by using alternative technical gateways to access vehicle data. This data is central to the generation of archetypal business models that the study presents. This chapter interconnects the original equipment manufacturer-relevant Android Automotive and the future perspectives of the automotive aftermarket. It contributes to understanding the paradigm shift that the technologies introduce and reflects the potential embedded in these.

Acknowledgments We would like to acknowledge all the support from industry, TALENTS4AA, APRA, AUTOMECHANIKA FRANKFURT, diverse professionals, and experts, as well as academic scholars who have helped us assess potential authors and extant projects to be invited. The lead editor expresses her gratitude to the University of Southern Denmark for its support and the Nordic Research Council's NOS HS project 329071 for its contribution in organizing workshops related to the automotive aftermarket and respective talents. We cordially thank all the contributors for their outstanding efforts and valuable insights. Without these contributors, we would not have been able to generate such a rich and broad discussion on the automotive aftermarket. Finally, we would like to thank all anonymous reviewers across the world who have kindly contributed as experts and scholarly audiences advancing the quality and relevance of the contributions. Thank you! We also acknowledge the importance of students and young people as readers and co-creators of sustainable future mobility and wish that there will be other efforts to continue and complement the discussions that our book project stimulates.

Disclaimer The editors and authors are not responsible for the integrity or validity of the secondary data sources. They have used data as provided for research purposes. Any concerns regarding such secondary data must be taken up with the original data provider. Further, we would like to note that reused illustrations, figures, and tables are subject to intellectual data protection and are reused with permission of the owner/source, all rights reserved. The sources are visible in individual chapters.

References

Aboltins, K. (2022). The car aftersales market in Europe and Latvian regions. *Regional Formation and Development Studies, 11*(3), 6–17.

Al Fayad, F. S. (2014). Examining the broken sales-cycle in the Saudi Arabia automotive market. *International Journal of Business and Management, 9*(8), 41.

Alam, M. Z., Kousar, S., Shafqat, N., & Shabbir, A. (2022). Drivers and challenges of tacit knowledge sharing in automotive workshop employees. *VINE Journal of Information and Knowledge Management Systems, 52*(1), 71–86.

Bouncken, R. B., Kraus, S., & Roig-Tierno, N. (2021). Knowledge-and innovation-based business models for future growth: Digitalized business models and portfolio considerations. *Review of Managerial Science, 15*(1), 1–14.

Corredoira, R. A., & McDermott, G. A. (2014). Adaptation, bridging and firm upgrading: How non-market institutions and MNCs facilitate knowledge recombination in emerging markets. *Journal of International Business Studies, 45*, 699–722.

David, A., & Terstriep, J. (2015). Regional framework conditions for future careers related to transport and mobility. *Journal of Strategy and Management, 8*(3), 211–230.

Dechezleprêtre, A., Díaz, L., Fadic, M., & Lalanne, G. (2023). How the green and digital transitions are reshaping the automotive ecosystem. In *OECD science, technology, and industry policy papers, No. 144*. OECD Publishing. https://doi.org/10.1787/f1874cab-en

Diener, D. L., & Tillman, A. M. (2016). Scrapping steel components for recycling—Isn't that good enough? Seeking improvements in automotive component end-of-life. *Resources, Conservation and Recycling, 110*, 48–60.

Dombrowski, U., & Engel, C. (2014). Impact of electric mobility on the after sales service in the automotive industry. *Procedia CIRP, 16*, 152–157.

Durugbo, C. M. (2020). After-sales services and aftermarket support: a systematic review, theory and future research directions. *International Journal of Production Research, 58*(6), 1857–1892.

Elo, M., & Kareila, T. (2014). Management of remanufacturing–strategic challenges from intellectual property rights. *International Journal of Manufacturing Technology and Management, 28*(4–6), 306–335.

Frigant, V., & Layan, J. B. (2009). Modular production and the new division of labour within Europe: The perspective of French automotive parts suppliers. *European Urban and Regional Studies, 16*(1), 11–25.

George, K., Ramaswamy, S., & Rassey, L. (2014). Next-shoring: A CEO's guide. *McKinsey Quarterly, 1*, 26–39.

Grebe, U. D., Hick, H., Rothbart, M., von Helmolt, R., Armengaud, E., Bajzek, M., & Kranabitl, P. (2021). Challenges for future automotive mobility. *Systems Engineering for Automotive Powertrain Development*, 3–30.

Gupta, R., Sahoo, S. K., & Sahoo, T. R. (2018). Employer branding: a tool for employee retention. *International Journal of Business and Management Invention (IJBMI) ISSN*.

Hawker, N. W. (2011). Automotive aftermarkets: A case study in systems competition. *The Antitrust Bulletin, 56*(1), 57–79.

Hornbuckle, R. (2023). *Organizational leadership culture and gender inequality in the C-Suite*. Doctoral dissertation, Walden University.

Huber, F., & Bauer, S. K. (2009). High performance in after-sales support services in the automotive industry. *International Journal of Automotive Technology and Management, 9*(3), 334–347.

Laborda, J., & Moral, M. J. (2020). Automotive aftermarket forecast in a changing world: The stakeholders' perceptions boost! *Sustainability, 12*(18), 7817.

Li, F. (2015, June). Study on the spare part logistics management model facing the automotive aftermarket. In *2015 International conference on management, education, information and control* (pp. 1488–1495). Atlantis Press.

MacKay, R. B., & Chia, R. (2013). Choice, chance, and unintended consequences in strategic change: A process understanding of the rise and fall of NorthCo Automotive. *Academy of Management Journal, 56*(1), 208–230.

Mahmoud-Jouini, S. B., Forero, J., & Midler, C. (2007, July). Aftermarket as premarket for breakthrough usage innovation. In *2007 IEEE international engineering management conference* (pp. 207–213). IEEE.

Marshall, J. N. (1989). Corporate reorganization and the geography of services: Evidence from the motor vehicle aftermarket in the West Midlands region of the UK. *Regional Studies, 23*(2), 139–150.

Mathivathanan, D., Kannan, D., & Haq, A. N. (2018). Sustainable supply chain management practices in Indian automotive industry: A multi-stakeholder view. *Resources, Conservation and Recycling, 128*, 284–305.

Novicevic, M. M., Harvey, M., Autry, C. W., & Bond, E. U., III. (2004). Dual-perspective SWOT: A synthesis of marketing intelligence and planning. *Marketing Intelligence & Planning, 22*(1), 84–94.

Prakash, D., & Manchanda, P. (2022). GoMechanic: Building India's largest automotive aftermarket ecosystem. *Asian Case Research Journal, 26*(03), 247–261.

Ramamurti, R. (2009). What have we learned about emerging—Market MNEs? In R. Ramamurti & J. Singh (Eds.), *Emerging multinationals in emerging markets* (p. 399). Cambridge University Press.

Schmidt, R., Bell, R., & Warren, V. (2021). Keeping the wheels of the automotive industry turning: The use of tacit knowledge by product development workers in a multinational automotive manufacturer. *Journal of Intellectual Capital, 22*(6), 1106–1125.

Schulze, A., Paul MacDuffie, J., & Täube, F. A. (2015). Introduction: Knowledge generation and innovation diffusion in the global automotive industry—Change and stability during turbulent times. *Industrial and Corporate Change, 24*(3), 603–611.

Smith, D., Elliot, M., & Sakshaug, J. W. (2023). To link or synthesize? An approach to data quality comparison. *ACM Journal of Data and Information Quality, 15*(2), 1–20.

Subramoniam, R., Huisingh, D., & Chinnam, R. B. (2009). Remanufacturing for the automotive aftermarket-strategic factors: literature review and future research needs. *Journal of Cleaner Production, 17*(13), 1163–1174.

Szász, L., Csíki, O., & Rácz, B. G. (2021). Sustainability management in the global automotive industry: A theoretical model and survey study. *International Journal of Production Economics, 235*, 108085.

Tewell, E. (2015). A decade of critical information literacy: A review of the literature. *Communications in Information Literacy, 9*(1), 2.

Verstrepen, S., Deschoolmeester, D., & Van den Berg, R. J. (1999). Servitization in the automotive sector: creating value and competitive advantage through service after sales. In *Global production management: IFIP WG5. 7 International conference on advances in production management systems, September 6–10, 1999, Berlin, Germany* (pp. 538–545). Springer.

von Hippel, E. (2006). *Democratizing innovation* (Vol. 1). The MIT Press.

Waas, A., Beck, M., Herzberg, R., Hauser, J., Schlehuber, F., Wolk, A. & Nikolic, Z. (2021) *At the crossroads: The European aftermarket in 2030*. BCG, CLEPA, Wolk After Sales Experts, BCG.com

Wang, H., & Kimble, C. (2016). How external factors influence business model innovation: A study of the Bosch group and the Chinese automotive aftermarket. *Global Business and Organizational Excellence, 35*(6), 53–64.

Williams, I. D., & Blyth, M. (2023). Autogeddon or autoheaven: Environmental and social effects of the automotive industry from launch to present. *Science of the Total Environment, 858*, 159987.

World Economic Forum. (2022). Automotive. Retrieved December 9, 2022, from https://www.weforum.org/communities/automotive

Maria Elo, full professor of international business and entrepreneurship at the University of Southern Denmark, professor at the BRIIB Institute of International Business at Shanghai University, Senior Research Fellow, University of Turku, Adjunct professor at Åbo Akademi University and Research Fellow at Institut für Arbeit und Technik works in the area of international business, -entrepreneurship and -migration with topics such as sustainability, business ecosystems, policy frameworks, networks, internationalization, mobility forms, human stickiness, migrant entrepreneurship, transnational and family businesses, diaspora investment and remittances. She has published books and articles, for example, in Journal of World Business, Journal of International Management, Journal of International Business Policy, Industrial Marketing Management, Regional Studies, Journal of International Entrepreneurship and International Journal of Entrepreneurship and Small Business. Maria is one of the founders of Talents4AA and a Talents4AA ambassador leading the collaboration with the education sector.

Fotios Katsardis is an experienced professional with extensive international experience in the automotive aftermarket. He holds a Bachelor of Arts (BA) degree in Economics, a Master of Arts (MA) in European Economic Studies, and an MBA in International Management, and was a Robert Schuman Scholar by the European Parliament. His working experience gained in different countries relates to multinational enterprises, strategic alliances, purchasing groups, SMEs, and family businesses. He has worked in different international positions for a renowned global Original Equipment Supplier and for the past 19 years as the CEO and President of a global automotive aftermarket alliance that operates across 110 countries. Fotios combines industry expertise, academic qualifications, and a passion for the automotive aftermarket. He is an acknowledged speaker and panellist of the sector. He is also the Chairman of the initiative Talents4AA, initiated with his sector peers, aiming at raising awareness of the automotive aftermarket and attracting new talents to the sector.

Part I

Introduction to Automotive Aftermarket: A System Relevant Sector

Automotive Aftermarket: Introduction to a Global Business

Fotios Katsardis

Abstract

The automotive aftermarket represents both secondary and tertiary sectors in the automotive industry, focusing on vehicle repair and maintenance once leaving the production line. It has a vast scope, encompassing the entire lifespan of a vehicle after the first sale. This sector plays a fundamental role in the transformation of raw materials into essential parts, accessories, equipment, and chemicals, which are then distributed and retailed through a variety of channels.

The automotive aftermarket brings together a diverse range of stakeholders. In fact, part of this sophisticated network are vehicle manufacturers, spare parts producers, garage and testing equipment companies, warehouse distributors, retailers, jobbers, e-commerce platforms, intermediaries such as insurances, leasing companies and fleets, repair and service garages, body shops and providers of essential data, and information for the overall aftermarket's operations.

The automotive aftermarket, by actively promoting the repair and maintenance of vehicles over the often more wasteful option of replacement, aligns with the broader sustainability goals. For vehicle owners, this sector ensures they are not short of choices when it comes to servicing, customizing, or maintaining their vehicles, tailored to their specific preferences and needs. At its core, the automotive aftermarket is dedicated to ensuring not just freedom of movement, but safe and sustainable mobility for all.

F. Katsardis (✉)
Temot International Autoparts GmbH, Ratingen, Germany
e-mail: F.katsardis@temot.com

M. Elo, F. Katsardis (eds.), *Automotive Aftermarket*, Management for Professionals, https://doi.org/10.1007/978-3-031-62419-3_2

1 The Global Automotive Aftermarket

1.1 An Overview

The automotive aftermarket is a crucial secondary and tertiary sector[1] focusing on the service, maintenance, and enhancement of vehicles following their production and initial sale. It ensures the comprehensive care and enhancement of vehicles throughout their entire life cycle and includes a great array of activities starting from manufacturing, remanufacturing, and distribution at all levels from wholesale to retail and the sale and installation of various vehicle replacement and maintenance parts, chemicals, lubricants, equipment, and accessories.

Key participants in the automotive aftermarket include spare parts manufacturers, data suppliers, repair and service garages, and vehicle body shops, among others. These stakeholders collaboratively ensure the production, replenishment, availability, and distribution of essential automotive parts and services required for vehicle repair, maintenance, and improvement. The market is practically divided into the original equipment service with parts and services provided by the vehicle manufacturers or their authorized dealers and the independent aftermarket (IAM) that are normally servicing all brands and all makes.

The automotive aftermarket deals with two categories of vehicles: (a) passenger cars and (b) commercial vehicles. The latter includes a wide variety of vehicles from light to heavy commercial and trucks, buses, agricultural, earth-moving machinery, and special applications vehicles. As a separate subsegment, this has not been examined carefully yet but is growing in several regions (Asia, LATAM), and has been the main mobility solution, including two and three wheelers (vehicles such as bicycles, motorcycles, trucks, and auto-rickshaws).

The sector can be segmented into further categories such as replacement part types, distribution, and service channels. It deals with a vast array of product categories, including but not limited to batteries, brakes, transmissions, filters, steering and suspension components, ignition systems, tires, lubricants, chemicals, and accessories. These parts and services are essential for the ongoing repair, maintenance, and improvement of operational vehicles. Understanding and anticipating product failure rates and maintenance needs is a complex aspect of the automotive aftermarket. In engineering literature, techniques and strategies like the bathtub curve[2] are utilized to anticipate failure rates, plan maintenance schedules, and manage warranties and recalls, aiming to enhance the reliability and longevity of auto-

[1]The three-sector theory is an economic theory, developed by Fisher (1935), Clark (1940) and Fourastié (1949). They divided the economy into primary, which deals with the raw material and their generation including farming, mining, and fisheries; secondary, with manufacturing producing products and finished goods; whereas it also constitutes the so-called tertiary sector that mainly deals with the services provided from one business to another and, in general, to the final customers—consumers.

[2]The "bathtub curve" in risk and reliability engineering is a widely used concept to illustrate the failure rates of products, including automotive parts, over time (see more in Jiang, 2013). The

motive parts and minimize defects and failures. In the distribution of parts, different quantitative methods are utilized to determine the optimum stock management and serve in the best way possible the market maintaining the optimum possible level of stock, which is a heavy cost to bear.

The definition of the distribution channels in the aftermarket literature follows a particular jargon with regional variations (European, American, etc.) The American Autocare Association has made a systematic approach and provided a good defini tion even though terms are used sometimes in a separate way from continent to continent.[3] The distribution as such has different channels that start from original equipment manufacturers (OEMs), which are the vehicle manufacturers producing the cars, being the source of everything. The original equipment suppliers (OESs) provide the products as well as the independent aftermarket suppliers (IAM) to the wholesalers (also known as warehouse distributors [WDs]) or the big retailers, and from there to the retailers, jobbers, e-tailers, etc. (Ballou, 2004; Kornafel, 2004).[4] At the end of the distribution channels are the service channels, which include OEM and dealerships garages and service stations, automotive centers, and independent workshops, either franchised or own brand.

Depending on the flow of the products from the manufacturer to the final consumer, the aftermarket distribution process can be three, two, or even one step, that is, the product arrives to the final consumer, that is, the driver or the installer, with or without the usage of the warehouse distributor and/or the jobber. A simple visualization of the aftermarket distribution flows of the European aftermarket is shown in Fig. 1.

This is admittedly a rather simplistic view and not valid in other regions; in reality, every single country follows only grosso modo these patterns. So, in Europe, products cannot be sold without comprehensive services, training is indispensable, and delivery five times a day to the garages is almost a triviality, whereas in contrast

---

curve is divided into three distinct phases, resembling the shape of a bathtub. Here is a breakdown of each phase:

1. Infant mortality phase (early failures): In this initial phase, the failure rate is high because of manufacturing defects, poor materials, or design flaws. Products that fail in this stage do so relatively quickly, and these are often identified during initial testing or shortly after the product goes into use.
2. Normal life phase (random failures): In this middle phase, the failure rate is low and constant. The product operates reliably, and only a few parts fail due to unforeseen circumstances or random events.
3. Automotive parts in this phase have passed the initial "break-in" period and operate as expected, with only occasional failures.
4. Wear-out phase (end-of-life failures): Toward the end of the product's useful life, the failure rate begins to increase as components wear out and reach the end of their operational life. For automotive parts, this is when wear and tear take their toll, and the components need to be replaced due to aging and extended use.

[3] See more at https://www.autocare.org/ and in "Data and Information", n.d.

[4] Distribution in logistics and supply chain management often involves various models of transportation to facilitate the delivery of goods or services from producers to consumers. See more in Ballou, R.H. (2004). *Business logistics/supply management: planning, organizing, and controlling the supply chain* (5th ed.). Prentice Hall Inc.; Kornafel, P. (2004). *Inventory Management and Purchasing: Tales and Techniques from the Automotive Aftermarket*, 1st Books.

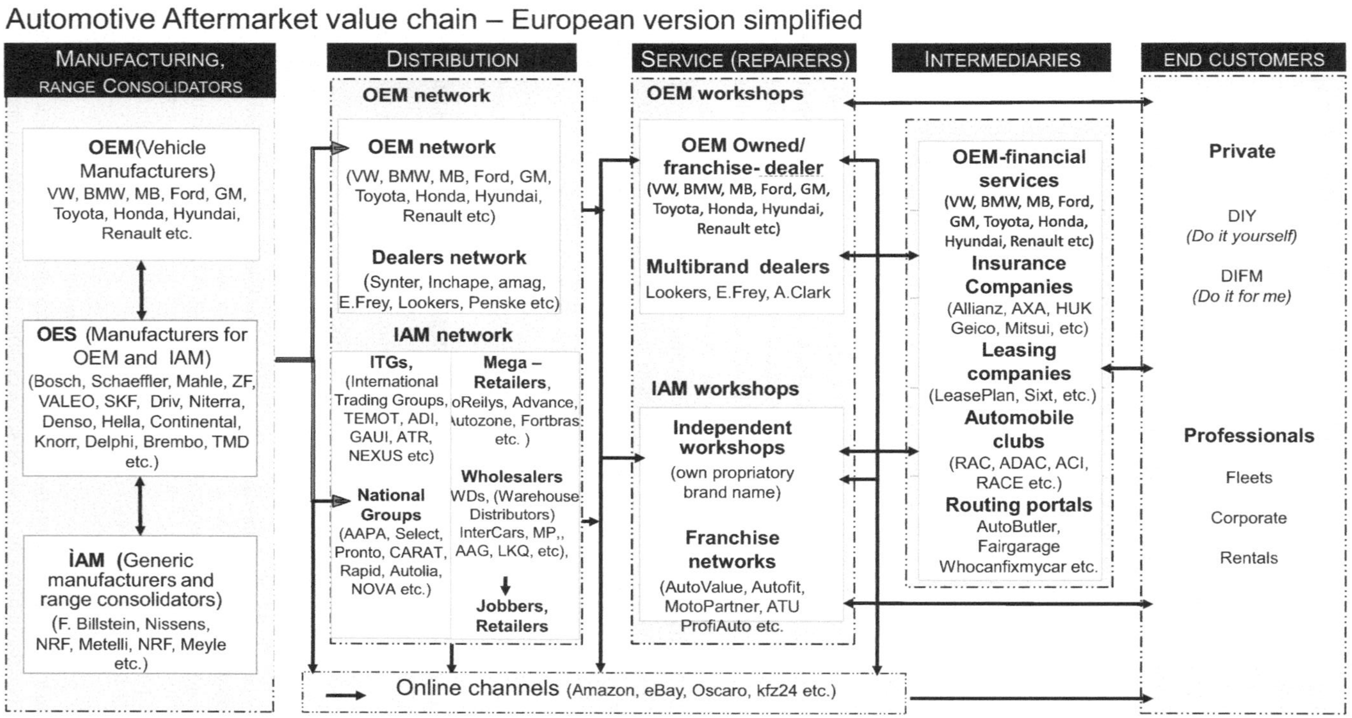

Fig. 1 Automotive aftermarket, European value chain. (Source: TEMOT International Autoparts GmbH, based on Roland Berger material for TEMOT 2021, used with permission)

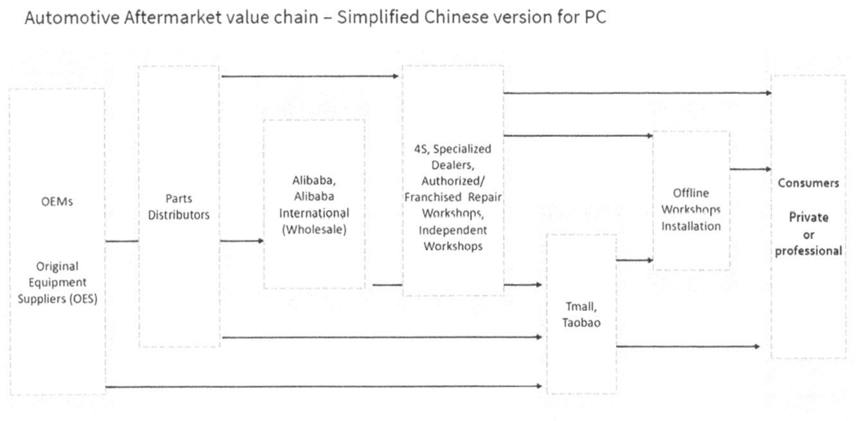

Fig. 2 Chinese automotive aftermarket value chain (simplified) (Source: Temot International Autoparts GmbH, 2023, used with permission)

in Sub-Saharan Africa, the issue of how to find any available products whatsoever and purchasing power to afford it is dominant. This makes fake and cheap second-hand products dominate the market. The South European aftermarkets are very fragmented; the Asian aftermarkets are still in transition, with big differences in the level of the participants; and the Russian, African, and South American aftermarkets are also still characterized by illicit activities. Huge wholesalers prevail in North America and expand to Western Europe; brands dominate the European markets but private brands are the main source of income for the big boxes in the United States. In China, big Internet platforms have built their own ecosystem, with the key distribution players (suppliers, wholesalers, and repairers) and utilization of very sophisticated technology to the final consumer. Figure 2 shows the respective value chain with examples of the Internet giant Alibaba [see section China in Fig. 2, Chinese Automotive Aftermarket value chain (simplified)" (Temot International Autoparts GmbH, 2023)].

2 Structure of the Automotive Aftermarket

The automotive aftermarket is composed of various entities involved in the entire process from manufacturing, distribution, retailing, installation, and servicing of vehicle parts, accessories, and equipment (Wikipedia, n.d.-a, see more at https://www.autocare.org/ and https://talents4aa.com/#manifesto).

The main types of entities involved in the automotive aftermarket are

- Original equipment manufacturers (OEMs): They are the vehicle manufacturers (VMs). OEMs are primarily involved in manufacturing the "original" parts of their vehicles, but they also play a role in the aftermarket industry by selling these spare and replacement parts, often tagged as "genuine" parts. Examples are Ford, Toyota, Volkswagen, and Hyundai.

- Original equipment suppliers (OES): They are the manufacturers and suppliers of equivalent OE products. These entities produce aftermarket parts and accessories or source from different parties and complement their program in the product categories they compete in. They specialize in various areas such as mechanical parts, electronic components, body parts, or accessories. Examples are Bosch, Continental, Denso, ZF, Valeo, Driv, Schaeffler, NGK, Febi, and Mahle (BCAR Auto Parts, n.d.).
- Remanufacturers: These entities specialize in rebuilding and remanufacturing certain vehicle parts such as engines and transmissions, steering and suspension, etc., contributing to the aftermarket supply chain.[5] Examples are Jasper Engines & Transmissions, Remy/BBB, Cardone Industries, ZF, and Bosch (APRA Remanufacturing | APRA Europe).
- Wholesaler distributors: They purchase products in bulk from manufacturers and distribute them to various entities in the supply chain such as retailers and repair shops. Distributors play a crucial role in making products available across different markets. In the United States, they are like warehouse distributors (WDs), which are entities that store vast quantities of products from the manufacturers and supply them to the retailers and repair shops. They often have expansive networks that allow products to be distributed over a wide geographical area. Examples are AutoZone, Advance Auto Parts, and NAPA Auto Parts.
- Retailers: They sell aftermarket parts and accessories directly to the consumers through their own outlets or franchises. They also operate through various channels, including physical stores and online platforms. Examples are O'Reilly Auto Parts, Pep Boys, Amazon, and eBay (Grand View Research, 2020).

---

[5] The Automotive Parts Remanufacturing Association (APRA) defines the remanufacturing process as follows

- "Remanufacturing is a standardized industrial process by which cores are returned to same-as-new, or better, condition and performance. The process is in line with specific technical specifications, including engineering, quality, and testing standards. The process yields fully warranted products.
- An industrial process is an established process, which is fully documented, and capable to fulfil the requirements established by the remanufacturer.
- A core is a previously sold, worn or non-functional product or part, intended for the remanufacturing process. During reverse logistics, a core is protected, handled and identified for remanufacturing to avoid damage and to preserve its value. A core is not waste or scrap and is not intended to be reused before remanufacturing.

REMANUFACTURED PART:

- A remanufactured part fulfils a function which is at least equivalent compared to the original part. It is restored from an existing part (CORE), using standardized industrial processes in line with specific technical specifications. A remanufactured part is given the same warranty as a new part and it clearly identifies the part as a remanufactured part and states the remanufacturer." Source: Remanufacturing | APRA Europe, https://www.apraeurope.org/remanufacturing.

- Jobbers: They are smaller retailers or distributors that sell automotive parts and accessories to professional installers and do-it-yourself customers. Examples are Carquest, Federated Auto Parts, and Auto Value (*Merriam-Webster* defines jobbers as a wholesaler who operates on a small scale or who sells only to retailers and institutions; www.merriam-webster.com/dictionary/jobber).
- Service providers, that is, repair shops: Service providers include workshops and garages that install and replace parts, accessories, perform maintenance, improvements, and repair vehicles. They interact directly with the consumers being the drivers or professionals like fleets, insurance companies, leasing firms, etc. They are normally small family-owned companies with great expertise and pride in their own capabilities. They invest a lot in tools, equipment, and continuous training. They can be independent (their own name is normally their brand name) or part of a franchise chain. They normally procure their parts from WDs, jobbers, or other distributors. Examples are AutoValue, Midas, Firestone, Monro, and Bumper to Bumper in the United States; Autofit, ATOY Huolto, and Bosch Car Service in Europe.
- e-Commerce or e-tailing is used as a synonym for sales of products and services through an electronic (Internet) platform. When products are sold at the wholesale level from professionals to professionals, it is called business-to-business (B2B); and when products are sold to the final customer and consumer, it is called business-to-consumer (B2C) (Deloitte, 2022). Electronic cataloging and ordering systems as well as stock management began in the aftermarket long before the 1990s when the Internet started expanding, and with the creation of virtual Internet marketplaces and platforms they contributed to the big expansion. From the end of the 1990s to the beginning of 2000, this big expansion made people believe that e-commerce would abolish the traditional distribution in a short period of time. (This proved to be a wrong assessment as, by 2023, e-commerce had taken on average a 20% market share depending on the country of course.) The success of the Internet, however, meant that today more than 80% of all people dealing with the aftermarket use the Internet at least to browse and get information. e-commerce, in general, created a big commercial upheaval with huge price transparency and obliged the companies to adjust business models and practices to defend their position or get as much as they could from Internet sales, creating their own relevant activities or selling directly to the platforms. They were forced to reconsider their distribution channels, transform their warehousing and stocks to Internet webpages, and deal with shipping and transport centers. As is well known from the other sectors, successful e-commerce requires huge investment in strong brands, technologically advanced web solutions with continuously changing sites, and an expertise in shipping equal to nothing. The biggest asset is data and subsequently an area of continuous investments in customer profiling and transactional data analytics (Deloitte, 2019). There are also numerous online resellers or retailers on the Internet that provide a huge program of all kinds of products for consumers to browse, find, and purchase aftermarket parts and accessories. Examples are RockAuto, PartsGeek,

AutoAnything, Oscaro, eBay, Amazon, Mercado Libre, Alibaba, and Tuhu (Grand View Research, 2020; Hayes, 2021).

- Intermediaries: Companies like insurances, fleets, leasing companies, and other third parties that directly influence the choice of parts and service providers for vehicle owners act as intermediaries, facilitating services and transactions, and playing a significant role in the demand for aftermarket products. They can qualify and choose repairers, offer packages, and connect the professional and private drivers with repair providers. Examples are State Farm, Geico, Progressive, Allianz, HUK, RBS, whocanfixmycar, and Autobutler.
- Collaborative entities, alliances, and international trading groups: Almost all distributors, wholesalers, and retailers have some kind of alliances or trading groups to leverage collective bargaining, resource sharing, and extending their market reach. Since the 1990s, the so-called (note: a contested term) international trading groups (ITGs) emerged and influenced the international networking massively. The ITGs are organizations that enable automotive aftermarket distributors and suppliers across various countries to collaborate and strengthen their position in the market. These groups, among other benefits, facilitate in theory bulk purchasing, coordinate and make supply chain more efficient, share best practices, invest in innovation and technology, and create synergies. Being part of an ITG allows businesses, especially small- and medium-sized enterprises (SMEs), to gain competitive advantages and access suppliers and eventually new segments and markets. At the end of the day for the SMEs, it is a significant reinforcement for their competitiveness, especially against the vehicle manufacturers and the big (trans-)national wholesalers. Examples of ITGs are ATR International, Nexus Automotive International, and Temot International (McKinsey & Company, 2017).
- OEM aftermarket concepts: OEMs have always tried to penetrate the aftermarket with their own services and products procured mostly from the OES or third parties. This was done primarily with and through their network and dealerships, claiming monopoly of quality and know-how and often discrediting independent providers either in the past as unsafe, qualitative minor, but also through pure aftermarket activities as WDs with repair networks. Renault introduced the Motrio network with Motrio garages (in more than 20 countries with 2500 garages) (Motrio.Com). PSA, in the strategy "push to pass," reinforced its activities with the introduction of Eurorepar concept as a wholesale activity, coupled with its own private brand and repair networks (BRAND; eurorepar.com). This activity has been reinforced after PSA merged with Stellantis. Using this strategy, the VM Stellantis continued acquisitions in the independent aftermarket as PSA before (Autobutler, MisterAuto), and several smaller ones in China and other regions, and acquired WDs in the market (DPK in Brazil) and cooperation with one of the ITGs (Nexus).

3 History of the Automotive Aftermarket

The automotive aftermarket owes its existence to the "machine that changed the world"[6]—the automobiles. Strangely, the very first cars produced in the 1890s were the most ecological: steam or electric driven! Technologically, the cars we maintain and repair today were built by Karl Benz in 1886: the gasoline-powered automobiles. By 1905, gasoline cars won the race over the steam or electric cars. They had the advantage that they still have today: good autonomy without the need for recharging, but just adding fuel. *(Just a piquant detail: the first cars were filling fuel at the pharmacies*; Wikipedia, n.d.-b) was All they thought they needed were driving skills and right from the beginning continuous repair and maintenance. Not a lot of things have changed ever since in the cars' technological powertrain principles.

The automotive aftermarket can be traced back to when the first car was invented. The car owners were practically the "guinea pigs" of their time as the drivers were the mechanics. Bertha Benz, the wife of Carl Benz, inventor of the first automobile, on her legendary travel from Mannheim to Pforzheim in 1888, had to do a lot of repair and solve technical adversities. She even invented the brake linings. The car was the noticeably big new issue but still widely unknown, and obviously with multiple problems and many challenges when needing repair, which were happening repeatedly even on the way. The aftermarket was also initiated by the fact that the vehicle manufacturers were practically unable to service the new inventions but were only next to their factories or big cities (to date, proximity remains one of the biggest advantages of independent repairs that can be found even in the remotest of villages). Once you bought a car, it was almost impossible to make any changes or keep moving once something happened (Cox, 2022).

In the beginning, car manufacturers provided no warranties or service contracts. According to an article on Wheels.ca (2016), the earliest days of the automobile era were guided by the principle of "caveat emptor" (Latin for "let the buyer beware"); the buyer is responsible for checking what he buys before purchasing; no warranty and no responsibility for any defects were possible after this "as it is" sale. According to the legend, the founder of Oldsmobile Ransom E. Olds, in 1901, when he started to sell his curved dash "Oldsmobile," meant to be the answer to Ford's Model T, offered to the buyers the only guarantee that "once the money changed hands, the car was theirs and the money his." In the beginning, the car owners practically addressed their problems to the local mechanics, but of course to the blacksmiths or even the bicycle mechanics or people with other skills for repairing the cars. Some car owners also learned to perform basic maintenance tasks such as changing oil, tires, or spark plugs themselves. These activities gave rise, in an eccentric but necessary way, to the development of the do-it-yourself (DIY) segment of the automotive aftermarket, which grew with World War II, and then especially in the United States (Wheels.ca, 2016).

As is known from marketing basics and entrepreneurial principles, the market needs appear as a gap or problem; finding ways to fill the gap or solve the problem

[6] See more in "The machine that changed the world" by Womack et al. (1990).

(of course discovering new approaches) is among the ways that start new business activities. This led to the following: cars needed service, and once the critical mass emerged, the first car repair shops emerged. In 1918 in Chicago, Walter Jacobs came up with the brilliant idea to rent out a dozen Model T Ford cars and also hired some practical "mechanics" to service them. History was written. (This is also a splendid example of entrepreneurship: Mr. Jacobs was so successful that he sold his business to John Hertz, founder of Hertz Corporation, one of the leading companies in car rentals worldwide today) (Wright, 2020).

Given the political and economic fragmentation and World War I, Europeans followed their own way. The remarkable inventor Robert Bosch,[7] a pioneer in developing electrical and electronic components, such as spark plugs, generators, starters, and ignition systems, revolutionized the market, which was only emerging, by founding the first Bosch Service Stations across Europe. This was much required as it was indispensable to provide the spare parts and technical support for his products. In 1921 in Hamburg, Germany, the first service station was opened, but since he was also selling the parts, this led to the emerging of the first wholesaler and retailer of the automotive market. Without Robert Bosch, the European aftermarkets would not be the same both in selling parts and servicing.

World War II as expected had a huge impact on the sector. In the United States, in 1940, almost 90% of the families owned a car, which was incredible at that time. The number was still increasing even when World War II was happening. The effects of the war were catalytic: it led to the curtailment of all materials essential to the automotive industry, such as rubber, metals, and fuel. The authorities classified these materials for military use, which dramatically reduced the production of new passenger cars and consequently limited the availability of spare parts.

Vehicle manufacturers continued to make cars and trucks for the military and government, at the same time expanding their product palette to the production of army arsenals such as tanks, engines, cannons, trucks, and aircraft. Car companies practically started producing exclusively for the war, and spare parts and fuel were obviously very limited. The most important part was, however, tires as rubber was unavailable since the rubber-producing countries were under the "enemy's" control.[8] The automotive industry shifted toward maintaining and repairing existing

[7] "Robert Bosch opened his 'Workshop for Precision Mechanics and Electrical Engineering' in Stuttgart 1 year later, Bosch built the first low-voltage magneto ignition device for internal combustion engines. This device set in motion the meteoric rise of his company. As the world became motorized, the business grew rapidly. Power tools and ignitions were at the core of the company from the start. But Robert Bosch was also quick to respond to the evolution of diesel engines, which did not require a magneto ignition device. In 1927, the diesel injection pump went into series production at Bosch. Even before the turn of the century, Bosch was expanding business abroad, first to the UK in 1898 and then to other European countries. By 1913, the company had subsidiaries in the US, Asia, Africa, and Australia, and 88 percent of its sales were outside of Germany." Source: Robert Bosch | Robert Bosch Stiftung (bosch-stiftung.de)

[8] "Soon after Pearl Harbor, Japan seized control of the Dutch East Indies (Indonesia) and British Malaya (Malaysia), cutting off roughly 90–95% of America's natural rubber supply." Source: Making Automobiles Last During World War II | The National WWII Museum | New Orleans (nationalww2museum.org).

vehicles, which reinforced the aftermarket that played a crucial role in keeping private and military vehicles operational. As in life, adversities, in this case, scarcity of parts, lead to innovative solutions. This was the time when remanufacturing of used parts and adaptation of nonautomotive materials for automotive use was starting to take dimensions. The culture of do-it-yourself (DIY) repairs and modifications, which started during the initial days of the automobiles, thrived during the war and continued to reach new heights after the war. This penury made drivers, companies, and local mechanics repair and maintain the vehicles and keep them running during the war.

Toward this were contributed efforts and campaigns as well as saving resources and maintaining the vehicles beyond their lifespan. The experiences and innovations that emerged during the war had a lasting impact on the automotive aftermarket. The emphasis on maintenance, repair, and innovation helped establish practices and business models that would become standard in the industry after the end of the war. Additionally, the technological advancements and mechanical skills developed during the war contributed to the postwar boom in automotive technology and customization. It is a sad paradox but not an exaggeration to say that the automotive aftermarket of the second half of the twentieth century owes its enormous success to World War II.

After World War II, and till the 1990s, the whole Western world experienced an economic boom. This period of prosperity led to increased car ownership among the middle class, creating a larger market for automotive parts, accessories, and services. The growth of the suburbs, and expansion of the highway system in the United States, and the boom in major European countries, with the rebuilding of Germany and its heavy industry, increased the need for passenger cars and utility vehicles. This, in turn, stimulated the demand for aftermarket products and services as people relied more on their cars for commuting and traveling. Advancements in automotive technology led to more complex vehicles with improved performance, safety, and comfort features. This complexity required a more skilled aftermarket service industry to maintain and repair vehicles, and also led to a wider range of aftermarket parts and accessories to upgrade and customize vehicles.

The rise of automotive specialty stores and chains in the 1960s made aftermarket products more accessible to the consumers. Service chains, like quick-lube shops and brake repair franchises, proliferated, offering specialized and convenient services. At the same time during the 1960s and 1970s, something significant happened: the introduction of environmental and safety regulations. The United States established the National Highway Traffic Safety Administration (NHTSA) and introduced the Clean Air Act, which influenced and changed the design and construction of the vehicles and made the repair and maintenance more complex and difficult as it required better aftermarket products and services and specialized trained mechanics to service the car.

The last part of the twentieth century saw the automotive industry become increasingly globalized, with more foreign cars on the roads in every country, marked by the success of particularly Japanese, Korean, and European models. This diversification broadened the aftermarket, requiring a wider variety of parts and expertise to service these vehicles.

During this period, the do-it-yourself (DIY) movement continued to grow among car owners, who preferred to maintain and customize their vehicles themselves, especially in North America. The availability of parts, tools, and information allowed more individuals repair and modify their own vehicles.

The end of the 1980s signaled the introduction of computerized engine controls and diagnostics that started revolutionizing the aftermarket. Electronic components became integral to vehicle operation, leading to the development of specialized diagnostic tools and equipment in the aftermarket. The distribution of parts became the decisive factor in finding and obtaining the correct parts but also sourcing their alternatives worldwide. Whereas quality issues were of paramount importance, many suppliers began engaging in the aftermarket and expanded their offers, adding several purchased additional products under their brand to complement their offers. The transition from paper catalogs to electronic catalogs is a major step toward efficiency but facilitated the expansion at the supplier's scene.

By 1990, the automotive aftermarket had evolved into a sophisticated and diverse industry, characterized by a wide range of products and services designed to meet the needs of vehicle owners and the challenges of new vehicle technologies. The industry had become essential to the overall economy, providing critical support for the automotive sector through maintenance, repair, and customization services.

The era around 2000 is characterized by a booming aftermarket with global sourcing of products and continuous improvements in quality and provision of high-quality services and support. The cars are getting "computer on wheels" as successfully described in 2002 by the Bosch Automotive Aftermarket division. The vehicles continued incorporating cutting-edge technologies and became the latest decade with the realm of electric and hybrid powertrains, very complex to maintain and repair. Advanced Driver Assistance Systems (ADAS) and increasing connectivity add to that. On the other hand, connected vehicles generate vast amounts of data. This is indeed the field where all the challenges start because of their importance. Original equipment manufacturers (OEMs) often control access to critical repair information, software, and spare parts as a kind of Trojan Horse to facilitate their expansion to the aftermarket further, an accomplishment that has eluded them for a century. Access to data and information is the key challenge of the last decades as obtaining the necessary data and tools is fundamental to the repairers to safeguard their ability to repair and maintain the vehicles.

During the last 20 years, the aftermarket suffered global supply chain disruptions during the Lehman Brothers crisis (Bernanke, 2018), as well as COVID-19. Despite this, the aftermarket was less affected compared with other sectors and exhibited an outstanding resilience as a critical sector for the global mobility. Despite the COVID-19 pandemic, it has impacted the availability of spare parts with huge delays in sourcing parts (He et al., 2020; Word Health Organization, 2020).

A unique development in the history of aftermarkets is the development of the e-commerce platforms and the rapid digitalization of the markets after 2000. The rise of e-commerce platforms has disrupted traditional distribution channels and especially the aftermarket key players needed to adapt to online sales, digital marketing, and customer engagement. The changing consumer preferences influence aftermarket demand, and the rise in shared mobility services (car-sharing, ride-hailing) affected vehicle ownership patterns, impacting repair and maintenance needs.

In the last 30 years, the environmental standards and regulations have become stricter by impacting the development of the aftermarket by means of availability of new products and service components that never existed. The word "sustainability" never existed in the vocabulary of the aftermarket and started having an impact by 2022 at least at the level of increased awareness and search for solutions for a more sustainable automotive aftermarket supply chain.

The adoption of new technologies, the shift toward electronics, connectivity, and the increasing number of electric vehicles (EVs) and hybrids require new expertise in EV-specific repairs and maintenance. Skilled technicians are essential for quality repairs, and there has never been a greater focus on training, catalogs, e-learning, hotlines, and platforms for the mechanics. However, developed countries with unfavorable demographic structures face a serious shortage of young talent and trained professionals, opting to be among the biggest challenges in the near future.

4 Vehicles in Operation (VIO) and Its Characteristics

The profound correlation between the rise in vehicles in operation (VIO) globally and the growth of the AA industry underlines the importance of the automobile market's strength. Without a car sold, there is no aftermarket. It was not only the issue of mobility that fueled the growth, but also technological advancements, consumer preferences, and the huge and extremely effective marketing machinery that have further "turbocharged" the industry in the last century. The aftermarket has grown continuously in the last 100 years and is recovering very fast from every single calamity of whatever size. The example of the United States is very characteristic as it is the country with the highest increase in history right from the beginning: in 1900, in the United States, there were just 8000 passenger cars; in 1920, nine million; in 1950, 40 million; and in 1960, 60 million.

In the 1960s, 75% of the global VIOs were in North America—22% in Europe and just 3% in Asia. The rest of the world, Africa, LATAM, and Oceania with just one million played no role whatsoever.

Things changed in the 1970s due to the constant surge in car sales worldwide. North America grew to a VIO of 173 million cars, but this was only 52% of the global VIOs. In the 1970s, the big market winner was Europe, with a massive 121 million vehicles, or some 36.0% of global VIOs! Asia had 29 million vehicles (9% of global VIOs), and South America nine million (3% of global VIOs). Africa with two million (0.6% of global VIOs) and Oceania with just one million (0.3% of global VIOs) played a minor role.[9]

[9] See more in ACEA (2022) and OICA reports and data sources.

Today, sources like the OICA or the S&P Global Mobility indicate[10] a global fleet of 1.54 billion registered passenger and commercial vehicles in 2022. Asia is not the largest anymore, the undeniable number 1, with 39% of the VIOs, followed by Europe with 27% and North America only third with 23%. Among the vehicle segments, passenger cars are the most dominant, representing about three-quarters of the global VIOs with 1.13 billion, whereas about 450 million were commercial vehicles (OICA, n.d.-a).

The country with the highest number of vehicles worldwide is not the United States anymore, where the vehicle parc is ca. 290.0 million vehicles, but China, with a VIO rd. of 320 million at the end of 2023. China has overtaken the position the United States held for over 100 years (see Table 1). Some other major countries are Japan, with 77 million, India, with 45 million, Russia, with 54 million, and Germany, with 53 million (Total-World-vehicles-in-use-2020.pdf (oica.net)).

As noticed from Table 1, a way to measure the density of vehicles in circulation in a country is the motorization rate, that is, vehicles per inhabitant or vice versa. This is an important indicator as it provides an idea of the potential size of the aftermarket per country and region, globally and locally. Hence, even if China is leading the VIO worldwide, it is still not highly motorized. The United States and New Zealand have the highest motorization ratio, with almost 900 cars per 1000 persons. Italy and Poland follow, with 770 cars per 1000 persons. China and India, as the world's most populated countries, have a rate of 223 and 33 cars per 1000 persons, respectively. If India had the same rate as the United States, it would have more than 1.1 billion vehicles in operation compared with the 60 million they have today (ACEA, 2022; OICA, 2020a).

The qualitative composition of the VIO is a further decisive factor in the aftermarket, considering the car manufacturers' brand share in every market. Not surprisingly, every manufacturing country has its local heroes that traditionally dominate the market. This is the result of the internationalization of production and further globalization, especially during the 1980s and 1990s, but it goes back even to the beginning of the automobile industry. Ford, for example, has plants in various European countries for many decades, with the first plant being in the United Kingdom in 1911. General Motors started its plants in Germany in 1929 with the brand OPEL (Wikipedia, n.d.-c; Socco, 2017; Supply Chain Management Review, 2010).

In the 1960s, there was the very first step of globalization of the Japanese VMs, which started exporting beyond Asia to Europe and the United States. The real impetus for the Japanese companies came in the 1980s with their American "transplants." Honda started building its model CIVIC in Marysville, OH, in 1981. In Europe, it started manufacturing in the United Kingdom in 1992 (but stopped in 2021). In Latin America, Honda began with a motorcycle plan in 1976 and cars in 1997, stepping into the Mexico market with car production in 1995. Toyota, the biggest of all Japanese companies, established a manufacturing plant in the United Kingdom in 1992 and in France in 2001. But the company was already in Brazil

[10] See more at https://www.spglobal.com/mobility/en/.

Table 1 Vehicles in use and population (source: author, compiled from OICA and World Bank statistics)

Total World Vehicles in Use (in thousand units) and World Population				
REGIONS/COUNTRIES	2015	2022	Number of Vehicles per 1000 Inhabitants	Population in respective areas (2022)
EUROPEAN AREA	*393.160*	*432.694*	*518*	*839.785.807*
EU 27 countries+EFTA+UK	307.759	337.518	641	528.996.127
RUSSIA, TURKEY & OTHER EUROPE	85.400	95.176	309	310.789.680
AMERICAS	*410.561*	*452.977*	*443*	*1.159.032.154*
NAFTA	324.763	360.912	722	499.721.590
UNITED STATES OF AMERICA	264.194	289.037	860	333.287.560
CENTRAL & SOUTH AMERICA	85.799	92.066	176	659.310.564
EAST/SOUTH ASIA	*433.336*	*644.048*	*143*	*4.294.510.219*
CHINA	162.845	318.034	223	1.412.175.000
INDIA	28.860	45.687	33	1.417.173.170
AFRICA/MIDDLE EAST	*49.978*	*60.557*	*49*	*1.704.349.589*
All countries/regions	*1.287.034*	*1.590.276*	*290*	*7.950.946.800*

since 1958 and started production in Argentina in 1997 (Toyota, n.d.). Nissan started producing in the United Kingdom in 1986, in Mexico in 1966, and in Brazil in 2014 (Nissan, n.d.).

If the 1990s were first dominated by the Japanese firms, they later signaled the era of the Korean expansion and transplants. Hyundai opened a plant in Turkey in 1997, another in the Czech Republic in 2008, in the United States (Montgomery, Alabama) in 2005, and in Brazil in Piracicaba in 2012. Kia started manufacturing in Slovakia in 2006 and opened a plant in Pesquería, Mexico, in 2016 (Hyundai, n.d.; Kia, n.d.). It was a copy-paste strategy of the Japanese and what the Americans had done throughout the last 100 years of automobile production. Ford has been in Brazil since 1919 and began manufacturing in 1953, in Argentina in 1962, and in Mexico since 1925 (Ford, n.d.). General Motors (as Chevrolet) began manufacturing in Brazil in 1925, in Mexico in 1935, and in Argentina in 1954 (General Motors, n.d.).

The impact of globalization and the successes of the VMs regionally have shaped the respective aftermarkets accordingly. European cars have a limited market share in the United States (mainly in the luxury segment with MB and BMW, Volvo), where Asian and American cars dominate the market and consequently define the demand for spare parts. Except for companies like ZF, Bosch, Schaeffler, Driv, Denso, NGK, etc., which were able to follow the VMs globally, OES suppliers were not particularly successful in other regions beyond their own area of specialization due to the challenges in their offer and program coverage, local competences, and overall competitiveness. Some second-tier suppliers have a mere global footprint with significant weaknesses from region to region (ACEA, 2023, May 2; Supply Chain Management Review, 2010, see also https://media.ford.com/content/fordmedia/fna/us/en/news/2021/01/11/ford-advances-south-america-restructuring.html).

China has massively disrupted the automobile sector ever since it first opened the country to Western car companies in 1979. In 2019, China accounted for more than 30% of the global auto production, with the United States second, at less than 17% (OICA, 2020b). With the increased investments in automotives in China and the growth of vehicle production, there has been a large increase in component production in the country, supplying worldwide, not just locally. This had as a result an overwhelming penetration of the aftermarket parts sold worldwide either by means of proprietary brands or even as procurement of the big brand names. Practically all the parts suppliers from producers to consolidators, regardless of their tier, procure their products from China to an extent that it has created a high level of dependency in the global supply chain. Geopolitical crises and tensions signaled the next wave of a rearranging supply chain, but China is way "too big to fail" as a global supplier in the medium run.

The average age of the vehicles in operation is a particularly important determinant in measuring the value of the aftermarket. The average age of the vehicles in use varies widely across countries and regions, depending on factors such as economic development, environmental regulations, vehicle taxation, consumer preferences, and maintenance practices (ACEA, 2019). According to ACEA, the average age of passenger cars was rd. 11.5 years in the European Union in 2019, while the average age of light commercial vehicles was 11.6 years, medium and heavy commercial vehicles was 13 years, and buses was 11.7 years (ACEA, 2021). Eastern Europe and the Baltics seem to have the oldest car fleets above 16 years old (Lithuania, Estonia, and Romania). The newest cars (on average, 8 years old) are in Luxembourg and Austria (ACEA, 2021). The oldest light commercial fleets (vans) were in Spain and Italy, with vehicles around 13 years old, whereas again Luxembourg and Denmark had the youngest (around 7 years old). (ACEA, 2021). Earlier (2000), scrapping incentives were a usual way to modernize and rejuvenate the circulating fleet. In the last 15 years, the average age has grown even further as there were no scrapping incentives or measures and of course the cars have been much better and are simply good even at this high age. Also, Frost & Sullivan presented similar results as postulated in their report in 2023 (Global Automotive Aftermarket Outlook 2023, p. 42).

5 The Commercial Vehicle Aftermarket

The commercial vehicle aftermarket is a very special category. The products and services address different types of vehicles, such as trucks, buses, vans, and trailers, that have nothing to do with the passenger cars as such. Medium-heavy commercial vehicles, buses, and their subsectors such as agricultural, earth-moving, etc., also differ from light commercial vehicles. Light commercial vehicles normally follow the powertrain of the passengers' cars with some modifications and constitute a "gray" commercial zone, concerning their repair and provision of parts done more by passenger car mechanics than the heavy-duty specialists. As proved during the COVID-19 pandemic, this market is extremely important for the overall mobility due to the professional characteristics and activities of these vehicles.

Also, in this sector, the market is practically divided into the parts form the OEM (original equipment manufacturer), which is made by the manufacturer or more often by a contracted supplier that manufactures it and has also the OEM brand on it. These are the original equipment suppliers (OES) that provide the parts and services to the vehicle manufacturers or their authorized dealers and the independent aftermarket (IAM) that are normally servicing all brands and all makes (UNECE, n.d.). Accordingly, the structure of the distribution follows the same pattern as passenger cars with the independent suppliers, warehouse distributors, specialized jobbers, and retailers, as well as the workshops.

Heavy-duty manufacturers produce these large and special vehicles and equipment for various industries, such as construction, mining, transportation, and agriculture. Unlike passenger cars, commercial vehicles are also important for the economy as they cater to the transportation of goods and people across various sectors and regions.

The commercial vehicle aftermarket is of paramount importance as it supports these functions. It requires high-quality parts that meet or exceed OEM standards, as well as reliable service and maintenance. Commercial vehicles operate under harsh conditions and heavy loads, so they need parts that can withstand wear and tear and ensure safety and performance (Frost & Sullivan, 2018a). The market is constantly evolving and innovating to meet the changing legislative demands and market challenges. Commercial vehicles are subject to environmental regulations, technological advancements, and customer expectations, so the aftermarket parts industry must adapt and improve accordingly.[11]

The market differs also from region to region, but it is more consolidated than the passenger cars market. Worldwide, you will find some of the same manufacturers' names everywhere despite the higher shares of some regional producers (Japan, Europe, the United States, China, Latin America). According to Statista, the leading truck manufacturer in 2022 based on revenue was Daimler Truck, with $55.8 ($60 in 2023[12]) billion (Daimler truck Daimler Truck Annual Report 2023). The company sells trucks under various brands, such as Freightliner in the United States, Mercedes-Benz, and Fuso in Asia (Daimler, n.d.). Volvo is the second largest truck manufacturer, with about US$51.7 billion in revenue. Also, Volvo owns different brands like Mack, Volvo, and Renault Trucks (Volvo Group, n.d.) and has a strategic partnership with Navistar (Navistar, n.d.; Financial Reports and Presentations | Volvo Group). The third largest is Volkswagen, named Traton, with about US$48 billion (TRATON | 2023 Annual Report) and brands like Volkswagen, Scania, and MAN SE (Traton, n.d.), and is also in a strategic alliance with Navistar (Navistar, n.d.). Ford, GM, Paccar, Navistar, with Mack, Freightliner, and Volvo, dominate the North American market. An emerging power in the world of heavy duty is China, with Dongfeng Motor and FAW, with about 50% share in China's heavy truck industry (Global Times, 2020). However, Dongfeng Motor and FAW are not yet among the top 10 global truck manufacturers by revenue (Statista, 2021a).

The global heavy-duty vehicle production in 2022 was estimated at around 16.6 million units, with Asia and Oceania being the largest producer, with about 9.4 million units (OICA, n.d.-b). The second largest producer was North America, with about 2.5 million units (OICA, n.d.-a), followed by Europe, with about 2.3 million units (OICA, n.d.-b).[13] The features of heavy commercial vehicles and buses vary by region, depending on factors such as the size and age of the vehicle fleet, regulatory environment, customer preferences, competitive landscape, and structure of the economy (e.g., OICA, n.d.-a).

According to Frost & Sullivan (Global Medium and Heavy-Duty Commercial Vehicle Aftermarket Outlook, 2023), there are some 50.0 million heavy commercial vehicles globally. The size of the medium- and heavy-duty commercial vehicles CV aftermarket is about $125 billion. The biggest part according to all estimates is IAM (accounting for 75% of the global market share according to Frost & Sullivan, 2023).

Calculating the size of the commercial vehicle aftermarket is not an easy task. One possibility is to measure the money spent per truck for repair and maintenance.

[11] See more in, e.g., https://www.continental-industry.com/de/industries/commercial-vehicles.

[12] All $ indicated in the chapter are US $.

[13] See more information at https://www.trwaftermarket.com/en/, https://www.spglobal.com/en/, https://www.frost.com and https://www.continental-industry.com/en.

Table 2 Medium and heavy commercial vehicles globally in 2022 (source: Author, TEMOT International Autoparts GmbH, approximations on the basis of Global Medium and Heavy-Duty Commercial Vehicle Aftermarket Outlook, 2023)

Region	Trucks (millions)	Spend per truck ($)	Total market revenues (billion $)
Europe	10.0	2.400	24.0
North America	10.0	2.600	26.0
South America	4.0	2.000	8.0
Asia Pacific	25.0	2.600	65.0
Africa	0.6	1.600	1.0
World	50.0		124.0

It sounds plausible but this varies greatly per region, condition, mileage driven, fleet composition, average age, etc. Accordingly, it has been calculated that in the Middle East, spending per truck reaches $3.000, whereas in Europe it is less than $2.500. If we take an average spend per truck of $2.500 with a number of vehicles in operation of 50 million, the value of the world aftermarket for medium-heavy-duty commercial vehicles should be rd. $124.0 billion. This is on the manufacturer level and provides an idea of how big the aftermarket is (Table 2).

Asia is the biggest and fastest growing and most diverse area in the world. The main drivers of the aftermarket demand are the rapid expansion of the vehicle fleet (especially in China and India), the low penetration of OES (except in Western and Central Europe) due to price sensitivity and lack of awareness, and the increasing demand for value-added services (such as mobile maintenance and fleet management) (Frost & Sullivan, 2023). The main challenges are the fragmented and unorganized nature of the market, lack of standardization and regulation, and impact of the black swans such as the pandemic (Frost & Sullivan, 2023).

The most mature market in the world is the European aftermarket, with a total revenue of ca. $24 billion in 2023 (own estimates). The European aftermarket unlike the rest is led by OES, which accounts for ca. 60–65% of the market share depending on the country and regions. The main drivers of the European CV aftermarket demand are the increasing average age of the vehicle fleet, stringent emission standards (Euro VI), and growing adoption of telematics and digital solutions. The main challenges are the cyclical market following closely the broader economic situation, which is in case of a slowdown always a major challenge, shortage of skilled technicians, and increasing competition from low-cost imports (Frost & Sullivan, 2018b).

According to ACEA, in 2021 the European trucks were on average 14.2 years old. Greece had, with 22.7 years, on average the oldest truck fleet, while the newest were found in Austria (6.6 years) and Denmark (7.5 years) (Report - Vehicles in use, Europe 2023 - ACEA - European Automobile Manufacturers' Association, 2023a).

The same report quotes for buses (a very particular category of commercial vehicles with different specifications and standards especially on the EU roads) an average age of 13.0 years, with Romanian buses leading the way as the oldest above 20 years of age (ACEA, n.d.), while the newest buses were in Austria and

Luxembourg, with vehicles around 5 years old (ACEA, 2023a; in 2021, out of the 27 EU countries, only 8 had an average age of less than 10 years old).

The North American aftermarket is also one of the largest and most mature in the world, with a total revenue of $26.0 billion in 2022 (own estimates). The aftermarket is more balanced between OES and IAM, with the IAM segment having about 60% of the market share (Frost & Sullivan estimate less rd. 50% in 2018). The main drivers of the aftermarket demand are the increasing average age of the vehicle fleet (above 12 years in 2022), high mileage driven in the American continent, and prevailing emission standards. The customer loyalty to brands and quality remains high, but has dropped in the last 5 years due to issues in availability and constraints posed by the pandemic. The main challenges were the economic slowdown due to the pandemic, which recovered gradually until 2023, labor shortage of drivers and technicians, and rising costs of parts and labor due to inflationary pressures.

The South American market is very volatile and small but rapidly growing, with a total revenue of $8.0 billion in 2022. The aftermarket is dominated by IAM, which vas varying from country to country but should be above 75%. There is a significant difference between the countries, with Brazil, Argentina, and Chile being the most important markets. Peru, Colombia, and Ecuador grew fast; Bolivia still has the oldest fleet. The main drivers of the aftermarket demand are the aging and deteriorating vehicle fleet (14 years in 2022), high dependency on road transport (due to poor infrastructure), and low availability of new vehicles (due to import restrictions). The Latin American mining industry is noticeably big and the respective market very important. The main challenges are the economic and political instability, high inflation and currency devaluation, and low-quality and counterfeit parts.

The smallest and most underdeveloped market is the African aftermarket, with a total revenue of about a billion dollars in 2022. South Africa, with the mining industry, is the biggest market. The aftermarket is characterized by the lack of any OE structure and the dominance of IAM, which accounts for 95% of the market share. This low penetration of OES, growing population and urbanization, increasing demand for mobility and transport, mining and forestry industry, remarkably high average age, and mileage driven are some of the drivers of the aftermarket demand and growth. The main challenges are the poor road and transport infrastructure, lack of regulation and enforcement, and prevalence of gray markets and counterfeit parts (World Bank Group, n.d.; Grand View Research, 2021; Maximise Market Research, 2021; GlobeNewswire, 2021; McKinsey & Company, 2017[14]).

[14] See more at https://www.spglobal.com/mobility/en/products/commercial-vehicle-aftermarket-parts-report.html, https://www.researchandmarkets.com/reports/5448386/global-mediumheavy-commercial-vehicle and https://store.frost.com/global-medium-heavy-commercial-vehicle-aftermarket-outlook-2021.html; more company-specific categories, see more in trwaftermarket.com, iamauger.com, meritor.com, spglobal.com, starmotorspares.com, retrieved multiple times.

6 The Automotive Aftermarkets Is a Huge Market

6.1 Some Considerations About the Revenues

Calculating the size of the automotive aftermarket is a very perplexing issue regarding the final values; estimations are used that are mostly based on several assumptions and speculative approaches as there is no central system or reliable data warehouse to calculate precise values. The published figures provide a good idea of the size of this overly complex sector.

According to Grand View Research, Inc., the total market size was estimated to be S427.51 billion in 2022. Frost & Sullivan[15] arrives at a similar estimation of $437.5 billion, estimating a compound annual growth rate (CAGR) of 6.6% between 2021 and 2027, including the influence of COVID-19. Fortune Business Insights (1) projected the sector to grow at 4.0% from 2023 to 2030.

Revenue includes only parts and accessories and excludes services unless otherwise noted; all revenues are measured at the manufacturer level and expressed in US$.

Another approach is to include labor for the services rendered, which is practically doubling the value of the market. Accordingly, McKinsey (www.mckinsey.com/industries/automotive-and-assembly/our-insights/ready-for-inspection-the-automotive-aftermarket-in-2030) spoke about an automotive aftermarket worth €800 billion in 2022 that could grow at a CAGR of 3% annually till 2030 to reach the astronomical amount of EUR €1.2 trillion (US$ 1.27 trillion).

Since 2020, the biggest replacement parts and accessories markets worldwide have been shifting. The biggest growth comes from Asia, basically from China, with a CAGR of 6.6% for the total aftermarket. It is projected to grow from approximately $400 billion in 2020 to $600 billion in 2027 (see Table 3).[16]

AASA and AUTOCARE agree that the aftermarket parts at the consumer level will reach the $400 billion figure in 2024 in the United States based on the total sales of both parts and services. This includes the sales of replacement parts,

Table 3 Replacement parts and accessories of global revenues (in billion $) (source: Author, TEMOT International Autoparts GmbH, on the basis of Frost and Sullivan, Frost and Sullivan, The Global Automotive Outlook 2023)

Area/year	2020	2023	2027
Europe	100	115	140
Asia	80	107	170
Latin America	32	39	47
North America	95	105	123
Rest	93	104	120
Total	400	470	600

[15] Frost and Sullivan, The global Automotive Outlook 2023.

[16] www.statista.com, www.grandviewresearch.com, www.fortunebusinessinsights.com, www.automotiveaftermarket.org.

accessories, chemicals, tools, equipment, and service and repair.[17] Following this estimation, we agree that the automotive aftermarket at consumer prices, including all, should exceed $1,7 trillion.

All the data service providers calculate the market size by using various data sources, such as vehicle registrations, miles driven, average age of vehicles, consumer spending, and industry surveys. If we take markets in the United States, China, Japan, Germany, India, France, the United Kingdom, Brazil, Canada, and Italy, we get a share of over 70% of the global market. These markets are predicted to retain their leading positions due to factors like high or rapidly increasing vehicle ownership, increasing disposable income, a surge in demand for vehicle maintenance and repair services, and the growing use of advanced technologies. China is poised to be the biggest aftermarket, and India should grow much faster in the next 15 years with its enormous potential.

More than 55% of the total value of the exported parts and accessories stems from Germany, China, the United States, Mexico, and Japan. The most exported auto parts categories are engines, transmissions, chassis components, electrical parts, and tires. Robert Bosch of Germany, followed by Denso of Japan, Continental of Germany, Magna of Canada, and ZF Friedrichshafen of Germany are the biggest companies worldwide.

The most sold auto parts segments are powertrain, chassis, interior, exterior, and electronics; the biggest product categories worldwide are tires, body parts, brake parts, batteries, filters, transmission, steering and suspension, electric parts, exhaust components, spark plugs, ignition, engine cooling air conditioners, lubricants, chemicals, accessories, and others. The most demanding auto parts by consumers in 2020 were brake pads, batteries, spark plugs, air filters, and oil filters. These parts are essential for the maintenance and performance of vehicles and need to be replaced regularly.

7 Factors Influencing the Growth of the Automotive Aftermarket

Throughout the history of the automotive aftermarket, at different times, several factors contributed to the growth of the industry. Some are universal and some change in significance with time, whereas new factors are expected to appear in the coming years (ADAS, services, monetization of connectivity, telematics, and other SaaS possibilities). Analytically some of the most noteworthy factors are as follows:

1. *The growth of the vehicle in operation* (VIO) and subsequently the increased traffic. Higher usage of vehicles leads to "wear and tear,"[18] necessitating main-

[17] See more in Auto Care Association (2023). 2024 Factbook. Auto Care Association.

[18] "Wear and tear" refers to the gradual degradation of vehicle components due to normal use and aging over time. Typical product categories are tires, endure friction, road conditions, and weather exposure. Regular wear leads to tread loss, reduced traction, and eventual replacement. Brake

tenance and replacement parts. With more vehicles on the road, the higher the demand for aftermarket parts and services.

2. *The VIO is getting older and* obviously older vehicles require more maintenance and replacement parts, strengthening the aftermarket sector. Related to these are other factors like.
3. The *mileage* (km) driven. Vehicles that have gathered more mileage typically experience more wear and tear, requiring more frequent maintenance and part replacements and vice versa. Mileage is also a common key performance parameter (KPI) used to schedule regular maintenance services such as oil changes, brake inspections, and tire exchange or replacement. Some service intervals require more specialized services like transmission fluid changes or timing belt replacements that are possibly more expensive. Vehicle owners attentive to mileage may be more proactive in seeking preventive maintenance, nurturing business for the aftermarket sector. Other consumers might respond to mileage-related wear or could also seek fuel efficiency and look for aftermarket products and services aimed at optimizing mileage and fuel efficiency.
4. *Resale value* and preservation: It is habitual that car owners aiming to preserve or enhance their vehicles' resale value might be inclined toward mileage-related maintenance.
5. *Technological advancements*: Modern vehicles come equipped with advanced technologies, requiring specialized parts and maintenance. Telematics and other advanced systems and apps that assist drivers in monitoring mileage and maintenance needs can indirectly promote aftermarket services.
6. *Consumer preferences*: DIY customers prefer to perform vehicle maintenance tasks themselves. Do-it-for-me (DIFM) consumers looking for convenience choose professional services for vehicle maintenance and repairs.
7. *Regulations and standards:* Change often and require new parts or services, influencing aftermarket demand.
8. *Access to data and information*: OEMs are not very open in sharing the car data and information, arguing on safety, proprietorship of data, tooling and design protection, etc. They provide them through their network, but the access might be extremely costly, difficult, and unsystematic. This limits not only the right to repair and the freedom of the consumer to choose where to maintain his vehicle but also often the possibility of repairing a car correctly with OES or other equivalent quality parts. Realizing the repair and maintenance at the OEM networks that have different structures, procedures, procuring, pricing, and availability of parts might be compared with the independent channels unfavorable to the consumer.
9. *Globalization*: Access to a global market has allowed for a more extensive range of products and competition, improving choice and pricing. When the powerful Chinese supplier industry was opened to the main world markets, that is, the United States and Europe, it penetrated them dramatically mainly with extremely competitive pricing. This influenced obviously all participants, and

---

components, filters, belts and hoses, suspension parts (shocks, struts, and control arms), fluids and lubricants, battery, and wiper blades.

WDs had access outside their local markets. OES had to rethink its manufacturing and procuring strategies for the aftermarket, moving a substantial number of factories to low-cost countries to remain competitive at least with their substitutable products. (Brand awareness, logistic advantages, service, information, training, support, tools, etc., added value to the products and counterbalanced the competitive offer of just a product. This maintained a pricing level and added to the overall size of the aftermarket.)

10. *Economic dynamics*: Economic growth or recession and recovery often correlates with increased vehicle usage and aftermarket services. Even if this is obvious, the automotive aftermarket has been extremely resilient to economic crisis, black swans, etc. The market in the main region, Europe and the United States, suffered only twice in the last 30 years, the time of the Lehman Brothers crisis (2008–10) and COVID (2020–21). In both cases, the crisis impacted mainly the OEMs and OESs due to their role in the production of new cars as their sales dropped significantly, but rebounded then immediately after reaching and exceedingly sooner than expected the previous levels.
    - *Purchasing power* plays a paramount role in consumer behavior and also across various aspects of the automotive aftermarket, from maintenance habits to technology adoption and sustainability preferences. It helps shape the market demand, consumer expectations, and competitive landscape of the automotive aftermarket industry. Consumers with more disposable income are likely to opt for regular maintenance, premium services, and quality replacement parts; with less might defer maintenance, opt for cheaper services, or choose lower-cost aftermarket parts. North–Central–Western European countries had always another pricing level with extraordinarily strong "brand" preference and different understanding concerning the products' quality and offered high level of services. Labor has been expensive, and discounting either it or the products was never really a choice. In Southern or Eastern European countries, the purchasing power has been always lower and consumers are very sensitive to that and always checking alternatives and bargaining every possible discount. Lax fiscal systems enabled often a different approach in which for generations the consumers had developed a different expectation once it was coming to their disposal income for repairs and maintenance. *Higher purchasing power* allows consumers to invest in upscale products, vehicle upgrades, customizations, and performance-enhancing parts and accessories. In contrast, consumers might prioritize essential repairs and maintenance over nonessential upgrades or customizations. Tuning is like a hidden champion in the *improvements* in vehicle appearance and performance: Deloitte (3) found that a significant part of the of US consumers (35%) are ready to invest more to improve their own cars with products like tires, alloys, batteries, brakes, accessories, tuning to boost their vehicles' performance, safety, and comfort. (Four-wheel alloys for a medium class can cost from $400 to $10.000!)
    - Although *the choice of service providers* depends on different reasons, it is usual that consumers with higher purchasing power and younger upscale cars use the authorized dealers or "premium" service centers and choose

upscale solutions. Lower purchasing power means that consumers look for equivalent quality but solutions from independent garages or DIY repair and maintenance that are more pricing sensitive.

11. Higher purchasing power enables the adoption of modern technologies, or new innovative products, and other smart solutions in vehicle maintenance repair many times, whereas of course lower purchasing power might limit the accessibility only to very necessary just to keep the car going. Driving up to the point of no return is very usual during times of crisis and recession when people have no means to repair and replace the vehicle. Consumers indeed will replace the vehicles more frequently if they have higher income, impacting the demand for aftermarket parts and services that are practically very low during the warranty period. The opposite might lead to longer vehicle ownership and increasing dependency on aftermarket products and services for vehicle longevity.
12. The availability of specialized repair and maintenance services, such as electric vehicle services, also boosts the industry.
13. There is a growing interest in sustainable and eco-friendly automotive parts, contributing to new market segments and influencing the size of the aftermarket as they are more expensive and scarcer. The electric vehicles (EVs) demonstrate the situation clearly: whereas it is still debatable how sustainable or eco-friendly choice is the electric vehicle (EV), especially if the power used is not from renewable sources, in some countries with high purchasing power, consumers exhibit the tendency to pay a premium price for sustainable aftermarket solutions, products, and services (i.e., Scandinavia). This is more obvious in societies that have already adopted respective legislation, created the infrastructure, and their consumers developed a higher sustainability attitude. Inversely, many regions with low purchasing power and cost considerations of the whole issue are very theoretical and sustainability preferences in product and service choices are nice to have but unattainable due to economic development, infrastructure, and geographical conditions. Offer of green parts[19] will definitely influence the cost of maintenance; however, currently, this is a rather theoretical approach as most of the key players discuss but do not act.
14. A factor that massively influenced at least the price level of parts worldwide was the growth, since the beginning of the twenty-first century, of e-commerce, the online platform that facilitates the selling of aftermarket parts and services, simplifying the delivery of aftermarket components, streamlining supply chains, and improving customer interaction (satisfaction or frustration). Online sales of automotive parts and accessories have seen significant growth, making the purchasing process easier for consumers. Grand View Research (n.d.-a) reported that unlike the rest of the sectors that will also grow, the global e-commerce automotive aftermarket was about \$58.32 billion in 2020 and will grow at an anticipated CAGR of 16.5% till 2028.

[19] "Green parts represent recycled, remanufactured or otherwise circular parts, e.g. „Green parts, which are made up of undamaged and reusable parts of end-of-life and written-off vehicles, are an area within the automotive aftermarket that has grown over the last few years." Source: https://aftermarketonline.net/green-is-good/ retrieved 1.12.2023.

We could possibly quantify the influence of each factor individually, but all of them collectively boosted and shaped the automotive aftermarket to grow so much and play the particularly important role that it has today.

8 The Regional Automotive Aftermarket Markets

8.1 North America

The North American car parts industry, although closely tied with Europe, has its own rich history. It refers mostly to the United States as a country continent, which led the way since the early 1900s, suffered but eventually came out strongly after World War II. During the 1950s, given the huge increase in car sales and customization, making cars a big part of the American dream and way of life, many people used also during the war and started fixing and upgrading their cars; this helped the car parts industry to grow faster, providing affordable and reliable parts to American consumers. A very particular characteristic of the American market is also the DIY sector, which was reinforced during these critical booming years.

Interestingly, during this period the car manufacturers faced a huge car demand and concentrated more on making and selling new cars, which made it very hard to take advantage of the very profitable downstream activity of repair and maintenance of the sold cars. It was a question of course of infrastructure, and it was impossible for them to be everywhere, especially in the remote parts of this vast country with own repair stations or dealerships.

In 2020, the United States' aftermarket sector was a notable business sector with revenues of approximately $383 billion. This contributed an estimated 2.3% to the GDP and provided employment to about 4.5 million people across various echelons from manufacturing to distribution and repair. From the Canadian context, the industry is also very important, with an estimated value of C$ 19.4 billion, employing over 400,000 people.[20]

The structure of the US aftermarket follows several layers we know, each characterized by distinct distribution levels and channels. On the top are the vehicle manufacturers like Ford, GM, Stellantis, Toyota, Nissan, Honda, and Hyundai that serve the market as original equipment manufacturers (OEMs). For these reasons, they have a rather secondary market presence with a limited market share. According to our estimates, this should not be more than 25% of the total aftermarket, varying obviously from product to product.

The OES manufacturers and suppliers in North America play an important role and produce their aftermarket parts and accessories themselves globally (less in the previous 20 years in the United States) or source them worldwide (mainly China) and complement their program in the product categories they compete in. They specialize in various areas such as mechanical parts, electronic components, body

[20] See more about the impact at https://www.aiacanada.com/wp-content/uploads/2019/10/AIA-Economic-Impact-Study-Final.pdf.

parts, or accessories. The United States has a big tradition of remanufacturing certain vehicle parts such as engines and transmissions, making it an important contribution to the US supply chain.

The warehouse distributors (WDs) play a very important role in the United States by procuring parts from VMs and various supplier tiers, or alternatively sourcing themselves from lower cost countries or those with technological advantages, that is, from Japan, South Korea, Taiwan, China, Mexico, Germany, etc. The American WDs are characterized by a wide national structure and distribute parts to everybody from jobbers and retailers to professional and individual consumers. They have grown so strong that they pursue the development of their own proprietary brand products (private brands) that possess in the meantime a dominant position in their portfolio, exceedingly often 80% of specific product category sales.

The WDs, with their extensive network, play a critical role in offering value-added services like technical support and training to their consumers, storing vast quantities of products from manufacturers, and supplying them to retailers and repair shops. AutoZone, Advance Auto Parts, and NAPA Auto Parts are examples (Grand View Research, 2020).

Similar to the WDs in the United States and Canada is the expansion of the retailers that sell aftermarket parts and accessories directly to consumers through their own outlets or franchises. They also operate through various channels, including physical stores and online platforms. Examples are O'Reilly Auto Parts, Pep Boys, Amazon, and eBay (Grand View Research, 2020).

Retailers in the United States have national, regional, and local structures. Some of the national structures are among the biggest worldwide, and in their procurement and proprietary brands are also wholesale because they also sell through their retail shops to professional installers and jobbers and not only to the consumers.

The two categories—big wholesalers and big retailers—are also called big boxes, a term used by stores like Walmart and Home Depot, with a large physical warehouse footprint. They compete with smaller WDs and retailers, offering lower prices, wide coverage, and convenience services.

Small- and medium-sized regional wholesalers are grouped into national networks with remarkable discipline and central coordination as the only solution to withstand the competitive pressure of the "big boxes." A particularly good example of the national groups is AAPA (Alliance), with members from all over the United States. It also has an international presence, with members in Canada, Mexico, and Latin America, and runs its own franchise systems for jobbers and repairers like AutoValue and Bumper to Bumper.

Jobbers is particularly an American term applied widely in the aftermarket literature. These companies bridge the gap between WDs and installers, typically managing smaller retail entities catering to installers and the DIY consumer demographic. They are often a member of the national groups (AAPA, Federated, Pronto, etc.), reinforcing their power and amortizing costs.

Repair shops and service providers: Service providers include workshops and garages that install and replace parts, perform maintenance, and repair vehicles.

They interact directly with consumers and procure parts from WDs, jobbers, or other distributors. Examples are Midas, Meineke, and Jiffy Lube (Grand View Research, 2020, (McKinsey et al., 2023).

e-commerce or -tailers are big in the United States, and sell parts and services through the Internet from business-to-business (B2B) and business-to-consumer (B2C) (we would dare say B2everywhere) (Deloitte, 2019). Example is Amazon, with strong activities also in the aftermarket, Internet webpages, huge range of products, aggressive conditions, and numerous shipping centers. They provide comprehensive offers, strong pricing, outstanding Internet presence and efficient distribution of products or services, as well as deep customer data analytics, capturing a good part of the market (Deloitte, 2019). Other examples are RockAuto, PartsGeek, and AutoAnything (Grand View Research, 2020; Hayes, 2021).

Third parties with immediate interests in the US aftermarket are insurance companies and national accounts, which influence the choice of parts and service providers for vehicle owners and streamline the consumers to their networks. Examples are State Farm, Geico, and Progressive (Grand View Research, 2020).

8.2 Latin America (LATAM)

In Latin America (LATAM), the automotive aftermarket extends from Mexico in the north to the southernmost part of South America. This part of the world has a high dynamism with a lot of potential and growth but also significant challenges.

These countries suffer from hyperinflation, and, although they have a great potential for repair and maintenance, the lack of availability of parts due to the quantitative import restrictions is keeping the aftermarket at a moderate growth level. Given the success of the Asian vehicle manufacturers, suppliers of these applications enjoy a high level of recognition and grow over-proportionally. European suppliers with OES pedigree occupy the premium level of spare parts not being further competitive either because of pricing attractiveness or lack of comprehensive coverage. The structure of the market follows the same pattern as in the other regions. Wholesalers are growing and covering the countries nationwide, and offering comprehensive coverage, high product availability, and in segments by following the rather old legacy of good, better best. Private brands have started penetrating the market. In Brazil, big retailers (like Fortbras) have emerged, with more than 13,000 own or franchised shops close to the final consumers. Due to the size, it has been a very attractive market also for the entry of companies from abroad, mainly American companies (e.g., AUTOZONE) (Auto Care Association, 2023; corporate reports).

Mexico is the biggest Central American market and has an increasing VIO of more than eight million cars. Nissan is the undisputable market leader, but, in general, the Asian car manufacturers are doing particularly well, making Mexico a good market for Asian parts and their respective brands. The imports of secondhand cars from the United States of mainly American brands like Chevrolet and Ford have created another segment, whereas the European car manufacturers are represented by Volkswagen (its iconic "Sedan" Beetle is still very visible everywhere though not produced anymore). Due to its position, the country has a number of vehicle manufacturer production plants but also OES. Through this and the favorable cost structure and proximity to the United States, Mexico is a significant part of the distribution of spare parts in the American continent. It is estimated that this position will be stronger in the next 10 years due to the geopolitical pressures on the Asian supply chain and mainly the Chinese.

Brazil and Argentina lie along the "Atlantic rim" and have a strong presence in the automotive industry. These are rather European powertrain-dominated countries with very particular characteristics. More than 45 million vehicles are on the Brazilian roads, making it the largest market in the region. Grand View Research estimates that the Brazilian automotive aftermarket was about $11.45 billion in 2020 and should reach $17.00 billion by 2028, growing at a CAGR of 4.9% from 2021 to 2028. Despite some setbacks due to COVID-19, the market has achieved and exceeded this growth in 2022 and 2023. Beyond allegations of corruption and illicit practices, Brazil is among the most inefficient aftermarkets due to the fiscal federal system. There are 26 federative units and 1 federal district in Brazil. They all have their own fiscal systems with own tax laws, revenue collection mechanisms, and budget management. The systems vary significantly from one state to another and function as if there are different countries, imposing duties upon entry from one state to the other. In this way, all participants bear enormous costs, cannot really procure correctly and optimize stocks, and need specific policies and infrastructure in every single state. It is worth mentioning that due to its size Brazil is also very strong in commercial vehicles and has the biggest LATAM wholesalers, like Pacaembu (Pacaembu Autopeças - Página inicial (pabu.com.br)).

Argentina, with more than 14 million passenger cars and a good commercial vehicles fleet, faced a severe economic crisis in 2001, leading to defaulting on more than $80 billion of debt. Since then, the country has been facing a continuous crisis with hyperinflation and continuous devaluations despite the pegging of paso to the US dollar. The demand for parts is very high, but the quotas limit the import of goods in an attempt to limit the public debt, which is out of control. Illicit practices and corruption create an explosive mixture in a country where European cars are dominating the market, followed by Japanese and Korean cars.

The most advanced economy and probably developed society in the region is Chile. As it is overlooking the Pacific rim, the automotive aftermarket is undergoing modernization and the efforts toward sustainability and pollution control, especially in Santiago, are influencing the market. The average vehicle age in Chile is about 11 years, which is the lowest in Latin America. There are about six million cars in Chile, and unlike the Atlantic countries of the continent, in Chile (as well as in the countries of the Pacific Ocean), the Asian brands are leading. The Asian manufacturers hold over 70% of Chile's vehicle in operation (VIO), which includes passenger cars and light trucks (excluding medium- and heavy-duty trucks). WDs without a solid offer of Asian applications have a limited role and competitiveness.

Interestingly, the most important American manufacturer Chevrolet maintains a strong presence by not using the US-specific powertrains but is part of global development projects (like the General Motors Global Emerging Markets Project, GEM; https://gmauthority.com/blog/gm/gm-platforms/gem/); others are tailored to regional needs and may use different platforms or powertrain. As usual, this leads to the creation of individual markets for parts only for LATAM. It has the most developed WDs in the aftermarket, that is, (Chile | Grupo Refax) possesses a good infrastructure and due to its geographical uniqueness a dynamic heavy-duty sector with good WDs for sectors like CAREN (CAREN | REPUESTOS Y ACCESORIOS MERCEDES VOLVO SCANIA FORD), etc.

Among the rest of the countries, Colombia's aftermarket, with approximately seven million units to serve, is characterized by a low purchasing power that pushes the market to affordable solutions. The country is, however, growing and has a relatively economic stability after decades of political upheaval.

Together, these countries bring in over 90% of all car parts and accessories in the area. Each of these countries has its unique market features, such as the number and types of vehicles, market size, how products are distributed, and competition.

Peru, Ecuador, Bolivia, Uruguay, and Venezuela are emerging markets with a relatively low percentage of motorization and significant growth potential. As their economies develop, the demand for vehicles and related services will rise. All are facing socioeconomic and political challenges. The common characteristic is that they are all so unique that they include varying numbers and types of vehicles, market sizes, distribution channels, and competitive landscapes. Each one of these countries is very individual as it is the whole of Latin America.

Latin America's sales of new vehicles are not growing very fast, reflecting the economic circumstances. It is slow growth in an area with a high potential, but as everywhere, the cars are getting older and the demand for car parts and services has increased. Out of the 110 million vehicles in Latin America, it was calculated that more than 50% are above 12 years old.

The costly economic situation is increasing as well as the price sensitivity of the consumer, with a growing demand for affordable solutions in the IAM. Inflation levels influence the consumer buying behavior significantly in local markets and limit its already low purchasing power. Online shopping habits are spreading gradually in the automotive aftermarket, particularly visible in the popularity of marketplaces selling parts of passenger cars. One major player is Mercado Libre, which has been growing every year in spare parts sales over-proportionally. As in Latin America, there is a significant lack of data availability for good part identification, and the e-commerce platforms have invested in cataloging and standardizing, which supports their market penetration massively.

8.3 Asia Pacific (APAC)

The Asia-Pacific (APAC) regions display the world's most vibrant and fast-expanding automotive aftermarket sector. It is estimated that the Asia-Pacific automotive aftermarket will reach approximately $190 billion by 2027, with a CAGR of over 7.1% from 2022 to 2028.

Data from the International Organization of Motor Vehicle Manufacturers (OICA) determined China as the top nation in 2020, with 281.5 million vehicles (latest TEMOT International Autoparts GmbH estimations show the 2023 VIO exceeding 320 million). The average age of vehicles is still about 7 years, which means that the Chinese aftermarket is yet to grow as it has not yet reached its sweet spot (the sweet spot represents the most favorable situation in which all market factors are optimally balanced and aligned, resulting in the best possible outcome, creating the nest presuppositions for that); Japan follows, with 78 million, India, with 70 million, and Australia, with 20 million units. Almost everywhere, older vehicles are not scrapped but are maintained and getting older and older. This adds to the demand for spare parts due to increased maintenance needs for a special segment of very old cars, the owners of which are normally people with very low purchasing power and price-sensitive, looking for the cheapest solution just to keep the cars running. However, overall, the increase in the VIO fed the demand for spare parts across the countries, and the engagement of the OESs' and IAM players increased professionalism and the offer of better parts and services. Some consolidation among parts distributors especially in the Oceania has been following the European and American trends. Japan and Korea are mature markets; the Australian market structure heavily resembles that of Western Europe; the Philippines and Vietnam, with 100 million people, have each only six million cars; Thailand has 50% pickups; Indonesia has one Toyota model that is above 50% of the market; and Singapore phases out the cars that are in use for 10 years of age. This is a fascinating region, and each country offers a unique market scenario, influenced by vehicle ownership rates, market maturity, consumer trends, and regulatory influences. There is no APAC aftermarket, but 15 unique aftermarkets with incredible potential.

Japan, with the reputation for having high-quality vehicle exports, has dominated the aftermarket as far as the traded parts are concerned. Leading OEMs in the region comprise Toyota, Honda, Hyundai, and Nissan, among others. The success of the Koreans with Hyundai /Kia has increased the interest in Korean applications and the expansion of the respective specialists. This follows practically the same pattern as once the Japanese companies like Meiji Sangyo started trading outside Japan to cover the needs of the market. In the aftermarket parts front, giants like Denso, NGK, Bosch, and Continental lead the charge. Warehouse distributors bridge the gap between manufacturers and the decisive point of sale, ensuring product availability. Despite its export of vehicle success, in the last few years China has penetrated massively (as throughout the world) with its aftermarket parts. Many of the parts sold in the Asia Pacific have a Chinese origin, even if they are from a famous brand, followed by the Japanese, Korean, Indian, European, and other countries. Japanese parts suppliers have dominated the market for ages given the success of companies like Toyota, Honda, Nissan, Suzuki, and Mitsubishi. (Forbes, 2023; McKinsey & Company, 2017[21]).

[21] See more in OICA. (n.d.-b). Vehicles in use. https://www.oica.net/category/vehicles-in-use/; https://www.grandviewresearch.com/industry-analysis/automotive-aftermarket; https://www.mor-

8.4 China

The Chinese automotive aftermarket (CAA) is the biggest and fastest evolving sector worldwide, offering multifaceted market opportunities. Deloitte predicts that by 2025 the maintenance market capacity is going to grow to 1.8 trillion yuan ($250 billion), mainly attributed to the sales of new cars and the projected increase in average vehicle age from 5.3 years in 2020 to 7.0 years by 2030.

Historically, the industry, at its beginning during the late 1990s, prospered alongside a boom in car sales. All key players are present in the Chinese market from original equipment manufacturers (OEMs) to independent aftermarket (IAM) service providers and e-commerce platforms. The e-commerce platforms are the biggest worldwide and very characteristic in China, reaching an incredible value of 500 billion yuan (ca. $70, billion), nearly 30% of the total aftermarket revenue.

To date, OEMs have exerted a power position, especially in the segments concerning service and maintenance of new and premium vehicles, characterized by offerings such as high-quality products and services, authorized warranties, and customer loyalty programs. This has been the result of the widely implemented 4S model, a retail concept that was introduced by Volkswagen in the late 1990s to integrate the sales and service network of its vehicles in China.

4S stands for sales, service, spare parts, and surveys (i.e., the customer feedback/input). The 4S model aimed to provide a one-stop shop for customers, offering quality products and services, as well as building customer loyalty and satisfaction. The 4S model became the dominant retail format for most automotive brands in China as it allowed them to control the distribution and aftersales service of their vehicles, as well as collect customer data and feedback. However, the 4S model with time was confronted with many challenges, such as high operating costs, low profitability, intense competition, changing customer preferences, and digital disruption. As it has lost a significant grip in the market, some automotive brands have started exploring alternative or complementary retail models to the 4S model, such as online platforms, direct sales, experience centers, or new retail concepts. All the models have addressed particular cultural and sociological dimensions of China, offering through these models a very high level of digitalization, more convenience, flexibility, personalization, and innovation to customers, as well as above all reduce costs and increase efficiency for dealers (Hayes, 2021).

E-commerce is a big issue in China that functions like a laboratory and is characterized by huge entities such as Alibaba (Vehicle Parts & Accessories (alibaba.com), JD.com JD.com, Inc., and TUHU 途虎养车 (tuhu.cn)). They have become a crucial power in the aftermarket offering O2O solutions for car owners. They have revolutionized the Chinese aftermarket by offering a mix of competitive pricing, a wide choice of products and availability, fast deliveries, and easy access to related

dorintelligence.com/industry-reports/global-automotive-aftermarket-distributors-market; https://www.marketresearch.com/Mordor-Intelligence-LLP-v4018/Asia-Pacific-Automotive-Aftermarket; Statista. (n.d.-b). Average age of passenger cars in Central and Eastern Europe in 2022. https://www.statista.com/statistics/1272928/cee-average-age-of-passenger-cars/.

information that was difficult to obtain before. As a further development following these, they offered repair and maintenance at their connected service repair workshops network.

TUHU, for example, emerged as a significant player, enhancing market connectivity by linking car owners with a network of service providers, thereby enabling a streamlined procurement of a variety of services and products ranging from tires to inspections. Alibaba is the largest e-commerce platform in China, with a dominant position in various sectors, such as retail, cloud computing, and digital entertainment, and operates also in the aftermarket car services market. Through its subsidiary Tmall Auto (an online marketplace that sells new and used cars, as well as aftermarket products and services, such as tires, batteries, oil changes, and inspections), it has partnered with various service providers, such as 4S shops, independent repair shops, and roadside assistance companies, to offer O2O solutions for car owners. Alibaba has an advantage over TUHU in terms of its massive user base, strong brand recognition, and financial resources. However, Alibaba also faces challenges in the aftermarket car services market, such as regulatory scrutiny, antitrust investigations, and data security issues.

Some other players like JD.com operate in the aftermarket car services market through their subsidiary JD Auto. JD Auto is an online platform that sells new and used cars, as well as aftermarket products and services, such as tires, batteries, oil changes, and inspections. JD Auto also partners with various service providers, such as 4S shops, independent repair shops, and roadside assistance companies, to offer O2O solutions for car owners. JD has an advantage over the rest in terms of its efficient logistics network, reliable customer service, and quality assurance. However, JD also faces challenges in the aftermarket car services market, such as high operating costs, low profit margins, and intense competition.

Cango (see CANGO-leading automotive transaction service platform (cangoonline.com)) is a leading online platform that provides auto financing and aftermarket services in China. Cango targets low-income car buyers who have limited access to traditional financing channels. Cango offers auto loans, auto insurance, and auto leasing services to its customers, as well as aftermarket products and services, such as tires, batteries, oil changes, and inspections. Cango has as advantage its niche market segment, strong risk management, and data analysis capabilities.

It is important to have a look at the O2O model for the aftermarket. OTO stands for online-to-offline, which means that customers can use online platforms to access offline services, such as booking appointments, choosing products, and paying for services. It uses various techniques to integrate online marketing with offline experiences, such as in-store pickup, online reservations, and QR codes[4]. The O2O model is exceedingly popular in China, especially in the automotive aftermarket, which includes services such as repair, maintenance, and accessories. There is no clear consensus on who invented the O2O model, but some sources credit the Chinese e-commerce giant Alibaba as the pioneer of this concept. Alibaba launched its O2O platform called Koubei in 2015, which connects online consumers with local offline services such as restaurants, cinemas, and hotels (e.g., see Lee et al., 2022).

The O2O model provides convenience and efficiency for customers, who can compare prices, read reviews, and select services online, and then enjoy high-quality and customized offline service. It helps service providers to attract more customers, increase their brand awareness, and optimize their operations by using big data and artificial intelligence to analyze customer behavior and preferences. It creates a more integrated and diversified platform that combines product offerings with service quality and allows for cross-selling and upselling opportunities.

The O2O is becoming more sophisticated and comprehensive as service providers integrate their resources and enrich their service content to meet the evolving needs of customers who depend on mobile Internet. It is focusing on the new retail concept, which emphasizes the integration of modern logistics with online service and offline experience to create a superior customer journey. It is adapting to the technology changes in the automotive industry, with the transformation from conventional to electric or hybrids, fully connected and autonomous. This will radically change the mix of cars being serviced and the demand for aftermarket services (Chang et al., 2020; Iotasol, 2021; Yan et al., 2022).

The Chinese aftermarket is a unique case in the modern automotive business. It is an unmatched case with an explosive growth. It is not without its own challenges and nuances, especially when contrasted with markets in the United States and Europe. China's aftermarket is marked by its immense size, explosive growth course, and complex market structure involving a multiplicity of competing entities across diverse segments. The policy and regulatory framework are still in an evolutionary phase, fighting with issues such as quality inconsistency, information asymmetry, and the proliferation of counterfeit products. At the same time as the sale of new cars is almost 50% electrical ones, there is the emergence of a new market and new participants (vehicle manufacturers, suppliers, garages, and consumers expected that the Chinese ecosystem will be at the epicenter of developments in the coming decade, and if geopolitics does not change the world's commercial pattern of trade, it will massively influence the global aftermarket).[22]

In conclusion, China's automotive aftermarket is the most dynamic aftermarket ecosystem, forecasted for substantial growth and marked by the proliferation of various business models and participants, from traditional OEMs to emergent e-commerce platforms.

8.5 India

The history of the Indian automotive aftermarket started during the old days of automobile manufacturing in the country. The first car was imported to India in 1897 by Jamshedji Tata. The first car factory was established in Mumbai in 1928 by General Motors. The first local car was launched in 1942 by Hindustan Motors. The first major policy intervention for the automotive sector was the Automobile Policy of 1956, which aimed to promote the local industry and substitute the imports. This

[22] See more in, e.g., Russo et al. (2023) and YCP Solidiance (2022).

policy led to the establishment of public sector enterprises, such as Ashok Leyland, Bharat Heavy Electricals Limited (BHEL), Hindustan Machine Tools (HMT), and Scooters India Limited (SIL). The policy also encouraged joint ventures between Indian and foreign companies. The liberalization of the Indian economy started in 1991 and was an important turning point for the automotive sector and its aftermarket. The government reduced import duties, removed licensing restrictions, and allowed foreign direct investment (FDI) in the sector. This led to an influx of global vehicle manufacturers and component suppliers into the Indian market, creating more competition and opportunities for innovation.

India is extremely low motorized and growing fast as an automotive market. It has enormous potential for aftermarket growth. Many analysts see India as the next China but with a much higher potential as it is still halfway behind its great rival (in population). The aftermarket is characterized by a high share of two-wheelers, low vehicle penetration, high vehicle age, minimal maintenance awareness, high price sensitivity, and low brand loyalty. According to Motor India Online, two-wheelers are the top contributors to the total aftermarket at 46%, followed by passenger cars at 34%, commercial vehicles at 15%, and three-wheelers at 5%. The aftermarket with about 50 million VIOs reached revenues of about $15 billion probably at the end of 2023. According to a report by McKinsey, the total aftermarket revenue in India is expected to grow by a CAGR of 12%, reaching US$ 17 billion by 2025. It is estimated that the increasing number of vehicles, growing average age of vehicles, growing demand for used cars, and rise in modern technologies and eventually new business models for such a complex market will increase the market greatly. Maruti Suzuki, with about 43%, has the highest brand share in the market, followed by Hyundai, Tata, and Mahindra & Mahindra, each with market shares of above 12%.

The Indian automotive aftermarket will be one day one of the largest markets in the world, but it still lags behind some of its peers in terms of size, quality, and efficiency. It faces challenges such as low customer awareness, poor infrastructure, counterfeit products, and regulatory uncertainties. To overcome these challenges and tap into the potential of the Indian automotive aftermarket, they invest in digitalization and innovation, strengthen their distribution and service networks, and collaborate with each other to create value for the customers and the industry. There are big investments in India's infrastructure, particularly its roads and national highways, and it is expected that people will commute more and drive more kilometers, which will raise the need for maintenance. Also, the new pollution standards and safety rules introduced in 2023 could result in prolonged ownership of existing vehicles, eventually fueling the aftermarket demand for vehicle maintenance.

The Indian automotive aftermarket comprises different segments, such as parts, accessories, lubricants, tires, batteries, and services. The OEMs (Tata, Mahindra, etc.) have a very high market share in the aftermarket as they have developed a dense network and are very competitive against the independent players.

The IAM parts are supplied by various players, such as component manufacturers, importers, distributors, wholesalers, and retailers, to independent workshops and retailers. However, the Indian independent aftermarket is very fragmented and regionally exceptionally long. More than 80% of the total aftermarket sales are to unorganized players. The rest (growing) 20% are the organized players such as the vehicle manufacturers, component manufacturers, branded distributors, large retailers, e-commerce platforms, and some service chains. The organized players are gradually increasing their presence and penetration in the market by offering better quality, warranty, service, and convenience to the customers. In the last few years, many WDs have started growing nationally and invested in their expansion inside and outside India.

8.6 Africa

The African automotive aftermarket demonstrates the most complex collection of different aftermarkets that make it impossible to draw some common characteristics. It has a rich history connected with the continent's colonial past. The market beyond the same segments as everywhere (passenger vehicles and commercial vehicles) is really different geographically (south: South Africa; north: Morocco, Algeria, and Egypt; Sub-Sahara: Nigeria, Ghana, Kenya, and the Rest of Africa)[1].

The Mediterranean countries like Tunisia, Algeria, and Morocco are quite developed because of the colonialization times and the strong French affinity, the South African market also for the same reasons. Sub-Saharan Western, Central, and Eastern African countries follow their complicated political and economic upheavals, whereas demographic developments and huge growth of population in Nigeria and Kongo Angora make these markets probably the most promising aftermarket of 2040–50.

There are approximately 60 million VIOs, and the market size is around $35 billion, but nobody really knows. Whereas worldwide there are more than 200 vehicles per 1000 persons, in Africa there are only 40. This shows that Africa has a vast potential. Current vehicle ownership might be very low, but the middle class is starting to grow and has an appetite for individual mobility.

The market is driven by several factors, such as the growth of an emerging middle class, increasing demand for safe and reliable transport, overaging of the vehicle fleet, and nonexistence of electric vehicles. Lack of investments, skills shortages, infrastructure demands, unstable energy supply, low intra-African trade, and prevalence of fake parts and corruption are the big challenges for quick solutions. On top of most of the cases, the political situation is very unstable and does help build any operative framework. The continent also faces a difficult environmental challenge, but for that there is no light at the end of the tunnel.

Beyond the Mediterranean African countries and South Africa, the countries with the highest potential in the Sub-Saharan African automotive aftermarket are Nigeria and Kenya. Nigeria has 12 million vehicles on the road, while Kenya has only two million, most of which are more than 15 years old. Both countries rely on imports to keep their fleets moving, with China and Dubai being the main supplier of parts. Kenya also serves as a regional hub for the aftermarket as it exports parts to neighboring countries such as Uganda and Rwanda.

The history of the African automotive aftermarket dates to the colonial era when European powers introduced motor vehicles to their colonies. Since then, the market has evolved through various phases, such as import substitution, liberalization, regional integration, and localization.

The key players in the African automotive aftermarket are the importers that play the role of the warehouse distributors; they import and distribute parts to retailers and workshops. Repair and maintenance concepts, which provide services to the drivers, are rare to be found and only in South African and North African francophone countries.

As in every market, the products sold in the African automotive aftermarket are those that are essential for the functioning of a vehicle, such as engine components, transmission parts, brake systems, suspension systems, steering systems, electrical systems, cooling systems, fuel systems, exhaust systems, and filters, tires, batteries, belts, hoses, spark plugs, wiper blades, bulbs, fuses, and lubricants. Accessories (such as audio systems, navigation systems, security systems, air conditioners, seat covers, floor mats, and roof racks) are highly required as well as most of the cars are in a desolate situation.

The distribution channels in the African automotive aftermarket can be divided into two types—official and unofficial—and often illegal ones are the majority. Official channels are rare and follow legal and regulatory frameworks, such as authorized dealerships, franchised workshops, importers, independent wholesalers, retailers, and online platforms.

Unofficial channels are those that operate outside or below the formal system, such as street vendors, roadside mechanics, and gray market traders. The role of fake parts in the African automotive aftermarket is significant and detrimental. Fake parts are those that are counterfeit or substandard and are sold at lower prices than genuine or quality parts. They pose a threat to the safety of drivers and passengers as they can cause accidents or breakdowns due to deficient performance or durability. They also damage the reputation and profitability of legitimate players in the market as they erode customer trust and loyalty. They also undermine the development of local manufacturing capabilities as they discourage investment and innovation.

The corruption level in the African automotive aftermarket is high and widespread. It can take various forms in the market, such as bribery, extortion, fraud, embezzlement, nepotism, and collusion, which affect all stakeholders in the market, such as government officials, customs officers, tax authorities, police officers, judges, warehouse distributors, repairers, retailers, and consumers. It creates inefficiencies and inequalities in the market, such as delays, costs, risks, barriers, distortions, and losses.

8.7 The Middle East

The Middle Eastern aftermarket is a rather broader dynamic and diverse area that covers Turkey, Lebanon, Israel, Jordan, Syria, Iran, Iraq, Kuwait, Bahrain, Qatar, Saudi Arabia, the United Arab Emirates, Oman, and Yemen. Sometimes people are also from Afghanistan and Pakistan, and reach out to Maghreb countries, excluding Turkey. Each country has its own characteristics and challenges that affect the growth and development of the aftermarket. The Middle East has a very intriguing past of car history, starting back to the early days of oil exploration and development. Cars are seen as a symbol of status, wealth, and power in many countries, and owners tend to customize and enhance their vehicles with various accessories and modifications. The region also has a strong tradition of motorsports, such as rally racing, drifting, drag racing, and off-roading, which create demand for performance parts and tuning services.

The key players in the market are the same as in other regions including OEMs, OES, importers, warehouse distributors (WDs), retailers, independent workshops, franchised workshops, and specialized workshops, such as bodywork, paintwork, tuning, or restoration. 24.

According to Frost & Sullivan data, the Middle East aftermarket was around $17.5 billion in 2019 and is expected to grow at a compound annual growth rate (CAGR) of 5.8% to reach $23.8 billion by 2025. The same source estimates that the passenger car segment accounted for 77% of the market value in 2019, while the commercial vehicle segment accounted for 23%. The largest product categories in terms of revenue were tires (24%), batteries (12%), lubricants (11%), brake parts (10%), and filters (9%).

There are some specific types of repair and maintenance concepts in the Middle East aftermarket. Some of the most common ones are as follows:

1. Quick service centers (QSCs): These are the outlets that offer fast and convenient services such as oil change, tire rotation, and battery replacement, usually without an appointment. Some examples of QSCs are Speedy Lube, Express Service Centre, and Quick Fit Auto Center.
2. Multi-brand service centers (MBSCs): These are outlets that offer comprehensive services for various makes and models of vehicles, usually with a warranty and quality assurance. Some examples of MBSCs are Bosch Car Service, Autopro Service Center, and Midas Service Center.
3. Specialty service centers (SSCs): These are outlets that offer specialized services for specific types of vehicles or needs, such as luxury cars, sports cars, classic cars, and electric cars. Some examples of SSCs are Al Tayer Motors Service Centre.

Turkey had the highest number of registered vehicles in the Middle East in 2019, with 23.1 million units, followed by Iran, with 19.7 million units, and Saudi Arabia, with 12.4 million units.

Turkey is in the meantime a very strong producer and exporter of automotive spare parts and has penetrated almost all over the world with economical products of good quality. Geopolitical problems have undermined or reinforced trading activities for the region to a broader base often worldwide. Iran or Russia, which are subject to economic sanctions, are both served widely by UAW or Turkey that have not undersigned the respective imposed sanctions. UAE is an international base of exporters that specializes not only in the rest of the Arab countries but also almost all over Africa.

8.8 Europe

The European automotive aftermarket is internationally a vast sector, and probably the most developed one globally. It is responsible for the repair, maintenance, and enhancement of a very colorful VIO differing very much from country to country. It features key industry players like Bosch, Schaeffler, Continental, Valeo, Mann, and Mahle from the OES side, and Stellantis, MB, BMW, Land Rover, Volvo, and VW from the OEM side. These players are driving innovation and growth globally, with the market value exceeding US$ 166 billion in 2021 and is projected to grow at a 5% CAGR through 2028. The OES and distribution echelons separately get consolidated, firstly, on a national level and then also internationally. Big WDs have acquired a number of peers and become bigger. There are some five WDs with revenues above $1.0 billion and activities in many European countries, two belonging to the American listed in the Stock exchange organizations LKQ and GPC, the rest from Poland, France, Germany, and Switzerland. However, there are still about 400,000 privately owned mostly family companies in the distribution of spare parts that are bringing the needed products to the service providers and the repairers and the professional or private drivers procuring them from the OES and other different sources of equivalent quality.

The European automotive aftermarket is a really multifaceted landscape with diverse national models. The locomotive is definitely German, which emerged after World War II mainly through its mighty VMs and rapid motorization reaching the highest worldwide. With a VIO of 53 million vehicles, it is the leading motorized nation in Europe. The legacy of manufacturing excellence is proven by the fact that German automobile manufacturers (VW, MB, BMW, etc.) produced over 15 million vehicles globally in 2021. The country's strong vehicle manufacturing foundation is based on a solid, innovative, and steady supply of original equipment by the respective suppliers (OES) such as Bosch, ZF Friedrichshafen, Mann, Mahle, Schaeffler, and Continental that contributed massively to the VMs' success and expanded with them all over the world being with a handful of Japanese, Korean, French, Italian, and American ones the only real global OES. As a matter of fact, more than 30% of the world's biggest automotive suppliers are of German origin. This success makes Germany one of the top three export countries around the world. Almost $300 billion was the export value of the German OEM and OES in 2021.

The German automotive aftermarket is a mature market and is increasing marginally. By 2025, it will reach about $25 billion. The market is getting more consolidated, and big WDs from the United States (LKQ, GPC) have made significant acquisitions in the distribution chain in the last 10 years. Also, local matadors (W&M, Winkler, Europarts) have grown in the same way and expanded to other countries. E-commerce is somehow less than in other regions. According to Statista in 2021, there were some 36,570 repair shops in Germany covering all the sectors of OEMs IAM, passenger cars, and commercial vehicles. Over the past two decades, the number has gradually declined from 45,800 in 2001 (source: Car workshop numbers in Germany 2001–2021 | Statista).

France, Italy, and the United Kingdom are the three countries after Germany as far as the size of the market and the importance is concerned. They come from a strong car manufacturing basis with strong OEMs and OES industry.

France has a very big tradition, especially with the success of the French OEMs in several countries. The French OEMs are strong but also versatile as they have established alternative multibrand garage chains like Motrio (Renault) and Eurorepar (PSA, now Stellantis) that coexist with independent garages. The French OES constitutes a vivid sector, and companies like Valeo and Faurecia have grown to be real global players.

The market is very well structured, characterized by a huge number of garage concepts and WDs as groups or national players. The WD Autodis has been expanding in the last 10 years through acquisitions to the adjacent countries, Belgium, Spain, and Italy being the top five biggest WDs in Europe.

Italy has had a very rich aftermarket history ever since Turin became the automotive capital of Italy in the 1880s. The market is still very fragmented especially at the level of distribution with strong geographical differences (north–south) and practices. Fiat has been the pillar of the market and used to have almost 60% market share of the circulating VIOs, which made the respective service network very homogeneous for its cars. Ever since, however, the globalization of Fiat and the merger of PSA with Stellantis, the local shares have changed radically and dropped below 35% of the total (around 40 million vehicles). As a result, this led the previously strong OES to lose significant space, with the exception of some very specialized ones (BREMBO). There are still a vast number of repairers (some say about 120,000 or all types) that mostly reflect the structural challenges of the Italian economy.

The United Kingdom is a case of its own. The aftermarket has been among the first European ones to get into the consolidation process and used as the springboard for the entry of the American companies to Europe (LKQ, GPC, UniSelect). The loss in the 1980s of the own OEMs together and the growth of other brands like Ford, BMW, Nissan, Toyota, Jaguar, GM/Opel, PSA, and Honda through their own local plants have created a complex and very price-competitive automotive aftermarket.

Among the other European countries, Poland has had a noticeable success as it is where most of the OES started building their plans for a more affordable costs structure in the 1990s. At the same, the VIO increased dramatically from eight million cars (of mostly Eastern European origin) in 2008 to more than 35 million cars in 2022. This surge is mainly attributed to the massive import of used cars from different European countries, mainly Germany, and that continuous for more than 15 years (although also the sale of new cars has been very vivid). The commercial policy of the suppliers toward the Polish WDs and the huge increase in the VIO drove the growth of robust WD structures that soon expanded all over Europe and became among the biggest ones. Examples are InterCars with activities in 20 European countries (see Inter Cars, Intercars.eu) and MotoProfil (see Moto-Profil | Car parts).

Overall warehouse distributors (WDs) and above all the small and medium enterprises (SMEs) are especially important in the European aftermarket for being the backbone of the economy and playing an important social role. SMEs, providing repair and maintenance services, may stand alone or affiliate with trading groups, offering both specialized and general services. The international trading groups (ITGs) ATR, GAUI, Global One, ADI, NEXUS, and TEMOT International shelter WDs with national and international activities and other national or regional structures under their umbrella. For example, TEMOT International Autoparts GmbH unites mainly SMEs and family-owned WDs to improve their competitiveness as these create economies of scale, enhanced purchasing power, and marketing support while allowing SMEs to remain agile and locally focused.

In the distribution, the expansion in e-commerce has revolutionized how customers interact with the aftermarket, providing a convenient platform for comparing and purchasing automotive parts. The biggest companies in Europe are Amazon and eBay, but every single country has its local champions.

In Europe, technology is playing a crucial role and has transformed the aftermarket in the last 30 years. The innovations of the OEM/OES gave rise to a different VIO, drove the expansion of electronics, and separated the fleet into modern diesel and gasoline vehicles. In the last 10 years, it has been all about connectivity and its digital services, telematics, vehicle diagnostics, etc., which are indicative of an industry transforming toward the digital age.

The future of the European aftermarket is characterized by trends that can be identified already now. The shift from diesel and gasoline engines to electrified powertrains disrupts the traditional aftermarket revenue streams. The repairers should reskill and upskill for EV diagnostics and software-related issues. The gap in skills and labor shortage are a major issue upfront and are already noticeable. It is a demographic issue as a huge number of technicians are retiring, and attracting new talent poses challenges as the sector is not enjoying any particular attractiveness or awareness.

As Europe has entered a time of connectivity and all connected vehicles generate vast data, the access of the aftermarket players to these data to be able to connect and repair is essential. At this level, a lot of discussions are taking place in the decision-making centers, that is, the European Union, to introduce the right legislation and safeguard the right of the consumer to repair.

The sustainability imperative is also expected to be decisive for the further development of the sector and the role of legislation and OES with green parts initiatives, that is, remanufacturing of parts and the eco-friendly practices toward the WDs and the repairers are of critical importance.

In conclusion, among others, the European automotive aftermarket thrives on diversity, innovation, and resilience. The aftermarket must navigate in the coming years in a landscape where the OEM competition is fierce, mainly through their direct access to car data and information and an environment, full of opportunities, formed by digitalization, electrification, and new mobility services.

The market players can only succeed if they adapt their strategies and operations to meet the shifting needs and expectations of the car owners. The automotive aftermarket is balancing growth and transformation with innovation and new approaches concentrated around the customer needs and sustainability concerns. Thus, a new narrative is being constructed for the industry at the crossroads of time and change.[23]

Bibliography

AAG. (2021). France | Alliance Automotive Group. Retrieved from https://allianceautomotivegroup.eu/our-network/france/

ACEA. (2019, July 1). *The automobile industry pocket guide 2019–2020*. https://www.acea.auto/publication/automobile-industry-pocket-guide-2019-2020/

ACEA. (2021, January 20). Report - Vehicles in use, Europe 2021. https://www.acea.auto/publication/report-vehicles-in-use-europe-january-2021/

ACEA. (2022, January 19). *Vehicles in use, Europe 2022*. https://www.acea.auto/publication/report-vehicles-in-use-europe-2022/

ACEA. (2023, September 22). *Employment trends in the EU automotive sector*. https://www.acea.auto/figure/employment-trends-in-eu-automotive-sector/

ACEA. (2023a, May 1). *World commercial vehicle production*. https://www.acea.auto/figure/world-commercial-vehicle-production/

ACEA. (2023b, May 2). *Motorisation rates in the EU, by country and vehicle type*. https://www.acea.auto/figure/motorisation-rates-in-the-eu-by-country-and-vehicle-type/

ACEA. (2023c, September 22). *Employment trends in the EU automotive sector*. https://www.acea.auto/figure/employment-trends-in-eu-automotive-sector/

Analysis and Forecast 2027. Retrieved from https://www.maximizemarketresearch.com/market-report/automotive-aftermarket-market/13057/

ANFIA. (2020). Il mercato dell'aftermarket automobilistico italiano nel contesto europeo e mondiale. Retrieved from https://www.anfia.it/it/aftermarket

At the Crossroads: The European Aftermarket in 2030. https://www.bcg.com/crossroads-european-aftermarket-2030.

Auto Care Association. (2023). *2024 Factbook*. Auto Care Association.

Automotive Aftermarket Industry Size, Growth, Outlook 2028. https://www.fortunebusinessinsights.com/automotive-after-market-102613.

Automotive Aftermarket Industry Size & Growth Report, 2030, Grand View Research, https://www.grandviewresearch.com/industry-analysis/aftermarket-automotive-parts-market.

Automotive Aftermarket Industry Trends, Size, Share, Analysis. https://www.alliedmarketresearch.com/automotive-aftermarket-market-A17049.

Automotive Aftermarket Market, Size, Growth 2022 to 2027. https://www.marketdataforecast.com/market-reports/global-automotive-aftermarket.

Baines, T., Ziaee Bigdeli, A., Bustinza, O. F., Shi, V. G., Baldwin, J., & Ridgway, K. (2017). Servitization: Revisiting the state-of- the-art and research priorities. *International Journal of Operations & Production Management, 37*(2), 256–278. https://doi.org/10.1108/IJOPM-06-2015-0312

[23] See more in "At the Crossroads: The European Aftermarket in 2030." This is a comprehensive study, conducted by BCG in collaboration with the European Association of Automotive Suppliers and Wolk After Sales Experts.

Ballou, R. H. (2004). *Business logistics/supply management: planning, organizing, and controlling the supply chain* (5th ed.). Prentice Hall.

BCAR Auto Parts. (n.d.). *How is the automotive aftermarket structure?* https://bcarparts.com/en/how-is-the-automotive-aftermarket-structure/

Bellis, M. (2019, July 6). *A history of the automobile: The evolution of the car dates all the way back to the 1600s*. ThoughtCo. https://www.thoughtco.com/who-invented-the-car-4059932

Bernanke, B. S. (2018). The real effects of disrupted credit: Evidence from the global financial crisis. *Brookings Papers on Economic Activity, 2018*(2), 251–342.

Black, A., Makundi, B., & McLennan, T. (2017). *Africa's automotive industry: Potential and challenges. Working Paper Series N° 282*. African Development Bank. https://www.afdb.org/fileadmin/uploads/afdb/Documents/Publications/WPS_No_282_Africa%E2%80%99s_Automotive_Industry_Potential_and_Challenges.pdf

Breitschwerdt, D., Cornet, A., Kempf, S., Michor, L., & Schmidt, M. (2017, June). *The changing aftermarket game—and how automotive suppliers can benefit from arising opportunities*. McKinsey & Company. https://www.mckinsey.com/~/media/McKinsey/Industries/Automotive%20and%20Assembly/Our%20Insights/The%20changing%20aftermarket%20game%20and%20how%20automotive%20suppliers%20can%20benefit%20from%20arising%20opportunities/The-changing-aftermarket-game.ashx.

Casper, R., & Sundin, E. (2021). Electrification in the automotive industry: Effects in remanufacturing. *Journal of Remanufacturing, 11*, 121–136. https://doi.org/10.1007/s13243-020-00094-8

Chang, J.-R., Chen, M.-Y., Chen, L.-S., & Chien, W.-T. (2020). Recognizing important factors of influencing trust in O2O models: An example of OpenTable. *Soft Computing, 24*, 7907–7923. https://doi.org/10.1007/s00500-019-04019-x

Chen, F., & Terken, J. (2023). *Automotive interaction design: From theory to practice*. Springer. https://doi.org/10.1007/978-981-19-3448-3

Clark, C. (1940). *The conditions of economic progress*. Macmillan.

Coffey, D. (2023, March 9). *It's time to develop better automotive value chains in Africa*. OECD Development matters. https://oecd-development-matters.org/2023/03/09/its-time-to-develop-better-automotive-value-chains-in-africa/#:~:text=With%20a%20large%20and%20growing,of%20203%20per%201%20000

Competitive Analysis and Forecast to 2026. https://www.marketresearch.com/MarketLine-v3883/Automotive-Aftermarket-France-Summary-Competitive-30853722/. Retrieved multiple times.

Cornet, A., Deubener, H., Möller, T., Schaufuss, P., & Tschiesner, A. (2019, January 8). *A long-term vision for the European automotive industry*. McKinsey & Company. https://www.mckinsey.com/industries/automotive-and-assembly/our-insights/a-long-term-vision-for-the-european-automotive-industry

Cox, L. (2022, March 25). *Who invented the car?* Live Science. https://www.livescience.com/37538-who-invented-the-car.html

Daimler. (n.d.). Daimler Trucks & Buses. https://www.trwaftermarket.com/en/commercial-vehicles/

Deloitte. (2022). Future of automotive sales and aftersales.

Deloitte & Touche LLP. (n.d.). The changing aftermarket game—and how automotive suppliers can benefit from arising opportunities. Retrieved from https://www2.deloitte.com/content/dam/Deloitte/de/Documents/consumer-industrial-products/CIP-Automotive-Aftermarket-Study.pdf

Durugbo, C. M. (2020). After-sales services and aftermarket support: A systematic review, theory, and future research directions. *International Journal of Production Research, 58*(6), 1857–1892. https://doi.org/10.1080/00207543.2019.1693655

European Environment Agency. (2016, March 15). Average age of road vehicles per country. https://www.eea.europa.eu/data-and-maps/daviz/average-age-of-road-vehicles-6#tab-chart_1

Fisher, A. (1935). *The clash of progress and security*. Macmillan.

Forbes. (2023, June 8). *The Global 2000*. https://www.forbes.com/global2000/#6b0c4a7d335d

Ford. (n.d.). *Ford in South America*. Corporate website.

Fortune Business Insights. (n.d.). Automotive aftermarket industry size, share & COVID-19 impact analysis. http://www.fortunebusinessinsights.com/automotive-after-market-102613

Fourastié, J. (1949). *Le Grand Espoir du XXe siècle. Progrès technique, progrès économique, progrès social.* Presses Universitaires de France.
Frost & Sullivan. (2018a). *Global automotive aftermarket outlook, 2018.* https://store.frost.com/global-automotive-aftermarket-outlook-2018.html
Frost & Sullivan. (2018b). *Strategic analysis of global automotive aftermarket, 2018.* Frost & Sullivan.
Frost & Sullivan. (n.d.) *OEMs and the Independent Aftermarket (IAM) Competing to Gain Commercial Vehicles (CV) Aftermarket Share.* https://www.frost.com/frost-perspectives/oems-and-independent-aftermarket/
General Motors. (n.d.). *GM in South America.* Corporate website.
Ghisellini, P., Cialani, C., & Ulgiati, S. (2016). A review on circular economy: The expected transition to a balanced interplay of environmental and economic systems. *Journal of Cleaner Production, 114*, 11–32. https://doi.org/10.1016/j.jclepro.2015.09.007
Global Times. (2020). *China's heavy truck sales hit record high in 2020.* https://www.spglobal.com/mobility/en/products/commercial-vehicle-aftermarket-parts-report.html
GlobeNewswire. (2021, October 04). *Global medium/heavy commercial vehicle aftermarket outlook report 2021 - Rapid expansion mobile-on-demand service models in the wake of the COVID-19 pandemic.* https://www.globenewswire.com/news-release/2021/10/04/2308080/0/en/Global-Medium-Heavy-Commercial-Vehicle-Aftermarket-Outlook-Report-2021-Rapid-Expansion-Mobile-on-demand-Service-Models-in-the-Wake-of-the-COVID-19-Pandemic.html
Gorton, D. (2021, December 8). 6 Countries that produce the most cars.. Investopedia. https://www.investopedia.com/articles/markets-economy/090616/6-countries-produce-most-cars.asp
Grand View Research. (2020). Automotive Aftermarket Size, Share & Trends Analysis Report by Replacement Part (Tire, Battery, Brake Parts, Filters, Body Parts, Lighting & Electronic Components, Wheels, Exhaust Components), By Service Channel (DIY, DIFM, OE), By Certification (Genuine Parts, Certified Parts, Uncertified Parts), By Region, And Segment Forecasts, 2020–2027.
Grand View Research. (2021). Automotive aftermarket industry size & growth report.
Grand View Research, Automotive Aftermarket Industry Size & Growth Report, 2030, https://www.grandviewresearch.com/industry-analysis/aftermarket-automotive-parts-market.
Grand View Research. (n.d.-a). *E-Commerce Automotive Aftermarket Size, Share & Trends Analysis Report by Replacement Parts (Lighting, Engine Parts, Braking Systems, Transmission & Steering), By End Use, By Region, And Segment Forecasts, 2021–2028.* https://www.grandviewresearch.com/industry-analysis/e-commerce-automotive-aftermarket-report
Grand View Research. (n.d.-b). *Automotive Aftermarket Industry Size, Share & Trends Analysis Report by Replacement Part (Tire, Battery, Brake parts, Filters, Others), By Distribution Channel, By Service Channel, By Certification, By Region, And Segment Forecasts, 2024–2030.* https://www.grandviewresearch.com/industry-analysis/aftermarket-automotive-parts-market
Grand View Research. (n.d.-c). *E-Commerce Automotive Aftermarket Size, Share & Trends Analysis Report by Replacement Parts (Lighting, Engine Parts, Braking Systems, Transmission & Steering), By End Use, By Region, And Segment Forecasts, 2021–2028.* https://www.grandviewresearch.com/industry-analysis/e-commerce-automotive-aftermarket-report
Hayes, A. (2021, October 8). *Online-to-Offline (O2O) Commerce Definition and Trends.* Investopedia. https://www.investopedia.com/terms/o/onlinetooffline-commerce.asp
He, S., Elo, M., Zhang, X., & Zhang, J. (2020). China gets triple hit by COVID-19. In *Covid-19 and international business* (pp. 307–320). Routledge.
Heid, B., Huth, C., Kempf, S., & Wu, G. (2018, August 27). *Ready for inspection: The automotive aftermarket in 2030.* McKinsey & Company. https://www.mckinsey.com/industries/automotive-and-assembly/our-insights/ready-for-inspection-the-automotive-aftermarket-in-2030
Hyundai. (n.d.). *Hyundai Motor Company Worldwide Network.* Corporate website.
IEA. (2020, June). *Global EV Outlook 2020.* https://www.iea.org/reports/global-ev-outlook-2020
IEA. (n.d.). CO2 emissions from fuel combustion 2020—Highlights. Retrieved from https://www.iea.org/reports/co2-emissions-from-fuel-combustion-2020-highlights

Iotasol. (2021, July 1). *O2O Business Model*. https://www.iotasol.com/blog/O2O-business-model.

Izzy, L. (2016, January 22). *From carriages to cars: The early Canadian automotive industry.* Wheels.ca. https://www.wheels.ca/news/from-carriages-to-cars-the-early-canadian-automotive-industry

Jiang, R. (2013). A new bathtub curve model with a finite support. *Reliability Engineering & System Safety, 119*, 44–51.

Kia Worldwide. (n.d.). Corporate website.

Kornafel, P. (2004). *Inventory management and purchasing: Tales and techniques from the automotive aftermarket*. 1st Books.

Lee, P. T. Y., Feiyu, E., & Chau, M. (2022). Defining online to offline (O2O): A systematic approach to defining an emerging business model. *Internet Research, 32*(5), 1453–1495. https://doi.org/10.1108/INTR-10-2020-0563

MarketLine. (2018, January). Automotive aftermarket in France. https://store.marketline.com/report/automotive-aftermarket-in-france/

MarketLine. (2021). Automotive aftermarket in France - Market summary.

Maximise Market Research. (2021). Automotive aftermarket market: Global industry.

McKinsey. (2023). A road map for the automotive industry in Europe. https://www.mckinsey.com/industries/automotive-and-assembly/our-insights/a-road-map-for-europes-automotive-industry

McKinsey & Company. (2017, July 11). *The changing aftermarket game—and how automotive suppliers can benefit from arising opportunities*. https://www.mckinsey.com/industries/automotive-and-assembly/our-insights/the-changing-aftermarket-game-and-how-automotive-suppliers-can-benefit-from-arising-opportunities

McKinsey Heid, B., Huth, C., Kempf, S., & Wu, G. (2018, August 27). *Ready for inspection: The automotive aftermarket in 2030.* McKinsey & Company. https://www.mckinsey.com/industries/automotive-and-assembly/our-insights/ready-for-inspection-the-automotive-aftermarket-in-2030

Mordor Intelligence. (n.d.). Africa automotive market size & share analysis - Growth trends & forecasts (2024–2029). https://www.mordorintelligence.com/industry-reports/sub-saharan-africa-automotive-market

Navistar. (n.d.). Navistar and TRATON GROUP reach agreement for TRATON to acquire Navistar for $44.50 per share in cash.

Nguyen, B., Simkin, L., Canhoto, A. I., & Dibb, S. (2019). The state-of-the-art of CRM: A literature review 2008–2018. *Journal of Strategic Marketing, 29*(1), 1–30. https://www.tandfonline.com/doi/full/10.1080/0965254X.2019.1672639

Nissan. (n.d.). *Nissan global*. Corporate web-site.

OICA. (2020a). *2020 Production Statistics*. https://www.oica.net/category/production-statistics/2020-statistics/

OICA. (2020b). Production statistics. https://www.oica.net/category/production-statistics/2020-statistics/

OICA. (n.d.-a). *Motorization rate 2020—Worldwide.* https://www.oica.net/category/vehicles-in-use/

OICA. (n.d.-b). *Vehicles in use*. https://www.oica.net/category/vehicles-in-use/

Opportunities and challenges facing the automotive aftermarket today. https://fixico.com/blog/opportunities-and-challenges-facing-the-automotive-aftermarket-today/.

Reportlinker.com. (n.d.). Africa automotive market size & share analysis - Growh trends & forecasts (2023–2028). See https://www.reportlinker.com/p06469573/?utm_source=GNW

Rodrigue, J-P. (2020). *Automobile production, selected countries, 1950–2022*. The Geography of Transport Systems. https://transportgeography.org/contents/chapter1/the-setting-of-global-transportation-systems/automobile-production-world/

Rolan Berger GmbH. (2018). *The shifting balance of power - New customer structure in the automotive aftermarket calls for new distribution methods*. Roland Berger GmbH.

Russo, B., Zhang, T., & Fan, B. (2023, February 28). *The impact of technology shifts on China's automotive aftermarket*. https://automobility.io/2023/02/the-impact-of-technology-shifts-on-chinas-automotive-aftermarket/

Schiller, T., Kummer, P., Berdichevskiy, A., Weidenbach, M., & Sadoun, J. (2020). *The future of automotive sales and aftersales*. Deloitte. https://www2.deloitte.com/de/de/pages/consumer-industrial-products/articles/future-of-automotive-sales-and-aftersales.html

SMMT. (n.d.). *UK automotive aftermarket reports*. https://www.smmt.co.uk/reports/uk-aftermarket/

Socco, V. S. (2017, October 17). *Globalization and its impact on the automotive industry*. Visor. https://visor.ph/industry/globalization-and-its-impact-on-the-industry/

Statista. (2021a). Leading truck manufacturers worldwide based on revenue in 2022 (in billion U.S. dollars).

Statista. (2021b). Number of heavy commercial vehicles in use worldwide from 2006 to 2022 (in millions). Retrieved from 1.

Statista. (n.d.-a). *Automotive industry. Worldwide - statistics & facts*. https://www.statista.com/topics/1487/automotive-industry/#topicOverview

Statista. (n.d.-b). *Average age of passenger cars in Central and Eastern Europe in 2022*. https://www.statista.com/statistics/1272928/cee-average-age-of-passenger-cars/

Statista. (n.d.-c). *Total sales of the global automotive aftermarket from 2016 to 2021*. https://www.statista.com/statistics/581758/size-of-global-automotive-parts-aftermarket/

Statista. (n.d.-d). *Worldwide revenue of selected truck and bus manufacturers in FY 2022*. https://www.statista.com/statistics/270293/worldwide-leading-truck-manufacturers-based-on-production-figures/#:~:text=Germany%2Dbased%20Daimler%20Truck%20continued,as%20industrial%20and%20marine%20engines

Supply Chain Management Review. (2010, December 14). *Globalization making major impact on automotive industry*. https://www.scmr.com/article/globalization_making_major_impact_on_automotive_industry

The Car Aftermarket Europe: New data from 34 countries. https://wolk-aftersales.com/the-car-aftermarket-europe-new-data-from-34-countries/The-Car-Aftermarket-in-Europe-2022-c143460450.

Top 10 automotive aftermarket companies of 2020 - Bizvibe Blog. https://blog.bizvibe.com/blog/top-automotive-aftermarket-companies.

Toyota. (n.d.). *Toyota global*. Corporate web-site.

Traton. (n.d.). *Our brands*. Corporate web-site.

UNECE. (n.d.). Periodical technical inspections (PTI). See more in https://wiki.unece.org/display/trans/Periodical+Technical+Inspections+%28PTI%29+34th+session

Volvo Group. (n.d.). *Our brands*. Corporate web-site.

Wikipedia. (n.d.-a). *Automotive aftermarket*. https://en.wikipedia.org/wiki/Automotive_aftermarket

Wikipedia. (n.d.-b). *History of the automobile*. https://en.wikipedia.org/wiki/History_of_the_automobile

Wikipedia. (n.d.-c). *List of countries by vehicles per capita*. https://en.wikipedia.org/wiki/List_of_countries_by_vehicles_per_capita

Womack, J. P., Jones, D. T., & Roos, D. (1990). *The machine that changed the world*. Harper Perennial.

World Bank Group. (n.d.). The motor vehicle parts industry in low- and middle-income countries: A new dataset of firms and trade flows. Retrieved from http://documents1.worldbank.org/curated/en/487011588503753890/pdf/The-Motor-Vehicle-Parts-Industry-in-Low-and-Middle-Income-Countries-A-New-Dataset-of-Firms-and-Trade-Flows.pdf

World Health Organization. (2020). *Impact of COVID-19 on people's livelihoods, their health and our food systems, 13.10.2020*. World Health Organization. https://www.who.int/news/item/13-10-2020-impact-of-covid-19-on-people's-livelihoods-their-health-and-our-food-systems

Wright, I. (2020, May 24). *The evolution of modern car technology*. CarBuzz. https://carbuzz.com/features/evolution-of-car-technology-and-features

Yan, H., Wang, S., Yang, Y., Guo, B., He, T., & Zhang, D. (2022). O^2-SiteRec: Store site recommendation under the O2O model via multi-graph attention networks. In *2022 IEEE*

38th international conference on data engineering (pp. 525–538). https://doi.org/10.1109/ICDE53745.2022.00044

YCP Solidiance. (2022, July). *E-commerce automotive aftermarket trends in China*. https://ycp-solidiance.com/article/china-ecommerce-automotive-aftermarket

ZDK. (2021). Zahlen & Fakten 2020. https://www.kfzgewerbe.de/verband/zahlen-fakten?tx_kfzapiconnector_apidownload%5Baction%5D=downloadIFile&tx_kfzapiconnector_apidownload%5Bcontroller%5D=ApiDownload&tx_kfzapiconnector_apidownload%5BfileId%5D=17375&tx_k

Online Sources

https://ihsmarkit.com/research-analysis/average-age-of-cars-and-light-trucks-in-us-rises-again-in-2021.html.

https://www.aiacanada.com/wp-content/uploads/2019/10/AIA-Economic-Impact-Study-Final.pdf.

https://www.continental-industry.com/de/industries/commercial-vehicles.

https://www.fortunebusinessinsights.com/automotive-aftermarket-102613.

https://www.grandviewresearch.com/industry-analysis/automotive-aftermarket.

https://www.marketresearch.com/Mordor-Intelligence-LLP-v4018/Asia-Pacific-Automotive-Aftermarket.

https://www.marketresearch.com/Mordor-Intelligence-LLP-v4018/Asia-Pacific-Automotive-Aftermarket-Growth-13966481/.

https://www.mordorintelligence.com/industry-reports/global-automotive-aftermarket-distributors-market.

https://www.statista.com/statistics/275443/average-age-of-passenger-cars-in-selected-countries/.

https://www2.deloitte.com/us/en/pages/manufacturing/articles/automotive-aftermarket-trends.html.

Fotios Katsardis is an experienced professional with extensive international experience in the automotive aftermarket. He holds a Bachelor of Arts (BA) degree in Economics, a Master of Arts (MA) in European Economic Studies, and an MBA in International Management, and was a Robert Schuman Scholar by the European Parliament. His working experience gained in different countries relates to multinational enterprises, strategic alliances, purchasing groups, SMEs, and family businesses. He has worked in different international positions for a renowned global Original Equipment Supplier and for the past 19 years as the CEO and President of a global automotive aftermarket alliance that operates across 110 countries. Fotios combines industry expertise, academic qualifications, and a passion for the automotive aftermarket. He is an acknowledged speaker and panellist of the sector. He is also the Chairman of the initiative Talents4AA, initiated with his sector peers, aiming at raising awareness of the automotive aftermarket and attracting new talents to the sector.

Understanding the Automotive Aftermarket

Hasmeet Kaur

Abstract

This chapter provides an overview of the value chain, key stakeholders, and geographical differences in the automotive aftermarket industry and offers insights into the industry dynamic that can be analyzed to gain an understanding of this business. The chapter breaks down the different layers of the value chain to help business/management students better understand how different stakeholders interact in this industry and what roles each of them are playing. It also highlights the key trends shaping the industry and in turn outlines the impact they have on the strategic positioning of market players.

1 Understanding the Automotive Aftermarket as a Market

The aftermarket industry for vehicles is complex, fragmented across multiple links in the value chain. Our primary focus here is the aftermarket for passenger cars and light-duty vehicles, with a secondary focus on commercial vehicles. Nevertheless, much of what we say here applies equally to both segments—in particular, the fact that the aftermarket for both segments operates according to very specific rules that are not typical of other automotive segments or indeed other industries.

Most of the money that owners spend on their vehicles is actually spent after the initial purchase. Outgoings include costs related to fuel, various types of insurance, repairs, maintenance, tuning, etc. The moment the vehicle leaves the dealership and hits the road, it becomes a part of what we consider the "addressable market" for the automotive aftermarket industry.

H. Kaur (✉)
Roland Berger, Munich, Germany
e-mail: hasmeet.kaur@rolandberger.com

M. Elo, F. Katsardis (eds.), *Automotive Aftermarket*, Management for Professionals, https://doi.org/10.1007/978-3-031-62419-3_3

For the various players operating in this market, this level of complexity can be highly challenging. There are more than 1.5 billion cars on the road today, of over 100 different makes and brands. The vast majority of vehicles are aged between 0 and 30 years. Each of these different vehicles and distinct segments has unique requirements and preferences. Consequently, repair shops must stock more than 100 aftermarket-specific components, and for each of these components, multiple brands and quality levels are typically available. The challenge of providing customers with a comprehensive product offering is almost insurmountable for typical companies in the aftermarket (Roland Berger Aftermarket Component Model).

2 A Complex Market Structure

Compared to the series vehicle production market and the components that are installed in the vehicle during the manufacturing process, the value chain of the vehicle aftermarket is highly fragmented. The aftermarket comprises two primary channels. The first is referred to as the *OE channel*, in other words the channel supported by the authorized original equipment manufacturer (OEM) network, which consists of dealerships and workshops such as those run by Volkswagen, BMW, Toyota, and Hyundai. The second channel is the independent aftermarket or *IAM channel*, which is supported by a large group of component manufacturers, including both OE suppliers and the like, whose components are installed in the vehicle when it is manufactured, and generic manufacturers, whose components are not used in series production but are instead sold via the aftermarket.

This two-channel structure (OE and IAM) is found right around the world, across all major geographies. But how the aftermarket is structured beyond this differs by region. Significant variation also occurs in the level of consolidation in the aftermarket, by which we mean the number of players in each market that control the value chain. In the United States, for example, five retailers dominate the automotive aftermarket, together controlling around 55% of distribution. In Europe, on the other hand, the IAM channel dominates, accounting for roughly 60% of the aftermarket. Consolidation is well advanced in countries such as Germany, France, and the United Kingdom, and is beginning to take hold at a pan-European level, where the top 10 IAM groups make up around one-third of this channel.

Smaller markets around the globe each have their own peculiarities, too. Market dynamics vary widely depending on the region. Thus, in South America the aftermarket is dominated by 10 wholesale chains, which together make up three-quarters of the market. The region is characterized by many local repair shops, but these are slowly losing ground to the bigger chains. In the Middle East and Africa (MEA), many used cars are imported from North America, the European Union, and Asia, with sales of such vehicles accounting for a significant portion of overall growth in the vehicle parc. Partly as a result of this focus on imports, the vehicles on the road vary greatly in terms of model, brand, age, and even if they are right- or left-hand drive, making it extremely difficult for players in the aftermarket to apply a single strategy across the entire region.

From a business/management perspective, the best way to identify a favorable position in the aftermarket is to look at which types of players exert the greatest control or influence over channel strategy, customer brand preference and pricing structure, and are consequently best able to optimize their own profit margins. The power structure of the aftermarket gradually shifts over time; it also differs depending on geography and whether we are looking at passenger vehicles or commercial vehicles. For passenger vehicles, wholesale distributors in Europe are rapidly gaining influence at the expense of IAM workshops. The wholesale distributors are a layer between the component manufacturers and the garages/workshops that install the parts in the vehicles. The purpose of the wholesale distributor is to buy different parts from multiple suppliers and offer a full spectrum of products for workshops. In the United States, influence lies firmly with pure retailers, while in China control is split between OEM and IAM workshops, and to a lesser extent wholesale distributors (see Fig. 1). For commercial vehicles, car parc managers have the most influence over purchase decisions in Europe and are also gaining power in the United States. However, in both the United States and China, most influence still lies with wholesale distributors.

It is worth pointing out that across regions and segments, vehicle owners/drivers often rely heavily on the recommendations of workshop mechanics when deciding which vehicle or model to buy. For passenger vehicles, the end customer is the one who will ultimately drive the vehicle and foot the bill, and the judgment and advice of a mechanic with whom they have a personal relationship can be a critical factor in their purchase decision. Moreover, the importance of workshop mechanics and their personal interface with end customers is likely to increase in the future as vehicle complexity skyrockets and the ability of customers to carry out their own repairs diminishes. Having said that, with evolving vehicle ownership models (e.g., car leasing, car sharing) the decision-maker is no longer the car driver. Instead, it is a professional fleet manager.

Like patterns of influence, the levels of consolidation in the aftermarket also change over time. Compared to some of the most mature industries in the world, the global automotive aftermarket is still in the early stages of consolidation, although the picture varies greatly depending on the market, region, and precise type of player. In North America, for instance, aftermarket retailers are highly consolidated, with the top three players controlling more than 50% of the market. This contrasts with Europe, where the top three wholesale distributors still command less than a 15% share of the market—although this, too, is a picture that is changing rapidly. For its part, China has a consolidated digital landscape, with selected players integrating the entire value chain, but these players still command a smaller market share than in Europe.

Automotive component manufacturers remain rather unconsolidated overall. However, when we look at specific component segments, the picture is sometimes quite different: For example, in the premium OE space, two or three very large players control 50–80% of the market, and the rest of the market is made up of generic manufacturers and private labels.

In the increasingly important area of e-commerce, four or five large players currently enjoy a major market share. However, this market is still in its infancy, and as

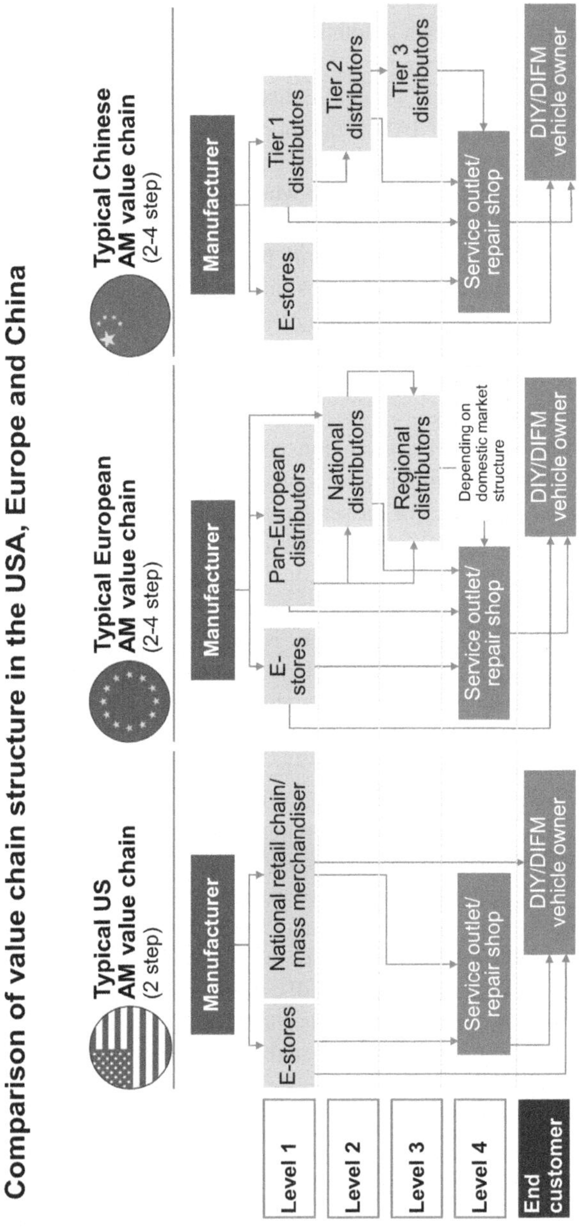

Fig. 1 Comparison of value chain structures (Source: Roland Berger, used with permission)

it grows, the share held by this handful of players is likely to shrink. We foresee the emergence of more integrated platform solutions, involving collaborations between multiple industry stakeholders, replacing today's standalone solutions.

Overall, the demand for parts will decline in the future owing to technology trends such as electrification and the fact that electric vehicles (EVs) typically have fewer moving parts than traditional internal combustion engine (ICE) vehicles. Increasing penetration of advanced driver assistance systems (ADAS) will also reduce the volume of parts sold because fewer vehicles will be involved in road accidents. Technological advances such as these mean that price pressure will decrease over time and the workshop landscape will consolidate, putting further pressure on wholesale distributors to consolidate, too.

3 Reflecting Different Contexts

This section illustrates how different contexts demonstrate structural variations, different business environments, and distinct market-specific characteristics.

3.1 European Automotive Aftermarket

The European aftermarket is characterized by a large number of value chain participants. This is mainly due to the existence of both national and regional distributors in Europe, adding two extra levels to the value chain between the manufacturer and the end customer (see Fig. 2). As a result, profit pools are more distributed in Europe than in the United States, say.

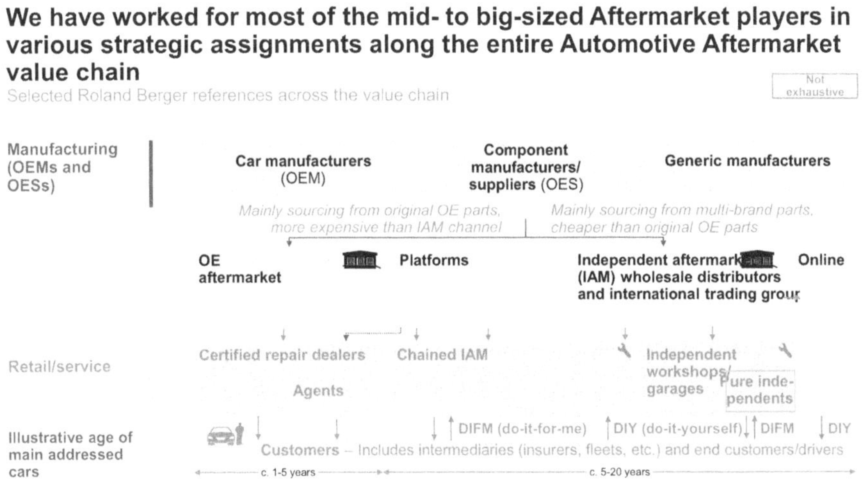

Fig. 2 European automotive aftermarket value chain (Source: Roland Berger, used with permission)

This unique structure is likely to eventually give way over time as national and regional distributors are gradually acquired by pan-European distributors such as LKQ Corp and Alliance Automotive Group (AAG). Consolidation will lead to fewer market participants overall, with a more consolidated supplier base (large wholesale distributors prefer to work with fewer suppliers) and potentially higher efficiency in the value chain. Simplification of the distribution structure and its requirements will also lead to a reduction in operating costs.

We can identify six major types of players or "archetypes" across the distribution channel in the European aftermarket value chain today, besides the OEMs themselves. These are (1) parts suppliers, including both original equipment suppliers (OESs) and generic manufacturers; (2) IAM wholesale distributors; (3) international trading groups (ITGs); (4) e-commerce players; (5) IAM initiatives by OEMs; and (6) emerging players, such as insurance companies or logistics companies. The path lines connecting the different players can be complex, as shown in Fig. 3.

Within these six archetypes, there are many different players and a wide variety of business models. The first archetype, parts suppliers, sells components to distributors or garages. They are mainly branded suppliers in Europe, although "white label" specialists are the preferred option in the United States and are also emerging in Europe. Next, we have IAM wholesale distributors, who buy vehicle parts from suppliers and resell them to garages and e-commerce players. These players offer a one-stop shop and also often have logistics capabilities. The third archetype, ITGs, are international groups that represent a group of wholesale distributors—typically numbering between 50 and 60 in total—around the world. They offer bundled purchasing volumes to OESs, who gain access to global markets by cooperating with them.

The fourth archetype, e-commerce players, can be online pure players or generalists. They source parts from wholesale distributors and sell them on to garages or individuals. Within this group, online intermediaries are also emerging, linking garages and individuals. Fifth are IAM initiatives by OEMs—mostly garage concepts or e-commerce platforms that allow OEMs to capture additional value along the life cycle of the vehicle. Finally, emerging players include insurance companies that wish to reduce the cost of repairs and fleet operators that want to cut the total cost of ownership (TCO) by optimizing their sourcing of vehicle parts.

3.2 China's Automotive Aftermarket

China has an unofficial system of ranking cities, with four large tier 1 metropolitan cities (Beijing, Shanghai, Guangzhou, and Shenzhen), followed by multiple tier 2 and tier 3 cities. The country's automotive aftermarket to a large extent mirrors this multitier structure, addressing different parts of the structure in a fragmented fashion. As a result, roles are often duplicated in the value chain.

A number of factors hinder consolidation of the aftermarket, from the high gross margin enjoyed by OESs and the lack of maturity of the consumer market to the complex Chinese certification system. As a result, no single retailer or wholesaler is able to bargain along the entire value chain. Distribution rules and price sensitivity

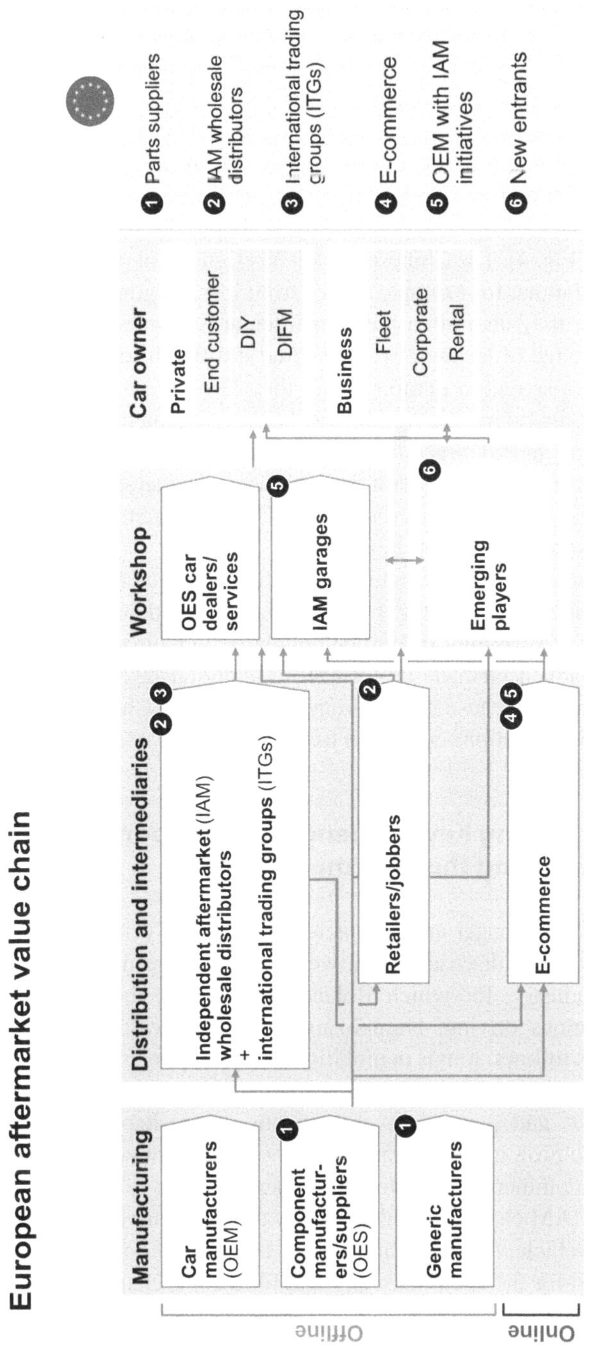

Fig. 3 An illustration of the European aftermarket value chain (Source: Roland Berger, used with permission)

also vary considerably between urban and rural areas. Any player hoping to consolidate the market will need to master strategic product positioning and effectively establish proximity and credibility with local workshop networks in order to manage the high level of complexity arising from regional and quality-related diversity. Collaboration will be key for a sustainable disruption model.

The Chinese aftermarket is highly fragmented and includes six key archetypes: (1) parts manufacturers (OESs); (2) tier 1 distributors in tier 1 cities; (3) tier 2 and tier 3 distributors present in smaller cities; (4) 4S retailer and repair shops that belong to the OEMs; (5) IAM retailer and repair shops; and (6) digital aggregators and platforms (see Fig. 4). The Chinese market has significant pain points across the value chain. Distributors, for example, suffer from low margins due to the multiple distribution layers; they also often have insufficient resources to drive expansion and limited knowledge of technology. OESs find it difficult to control the channel due to price and cross-region dumping, and they also face the challenge of fake parts. Aggregators, including digital players, have the problem that customers can effectively skip them and go direct to the workshops or parts manufacturers. Players in the service network must cope with the high cost of generating leads, their own lack of CRM skills, nonstandardized operations, management and marketing, and general difficulties expanding their business.

The European market is more mature than China in terms of overall consolidation and professionalization of the IAM. Also, the brand loyalty to OES parts among workshops is very high in Europe compared to China. However, China is significantly more advanced than Europe in terms of digital aggregator and platform-based business models because of the strong preference of Chinese customers and businesses for digital solutions compared to Europe.

4 Understanding Key Challenges and Opportunities: What Is Driving the Aftermarket?

The aftermarket for passenger and commercial vehicles is driven by multiple factors. Elsewhere, we have discussed what we see as the four major trends affecting the automotive industry, for which Roland Berger uses the acronym MADE: Mobility, Autonomous driving, Digitalization, and Electrification. In brief, our observations are as follows: trends in mobility are developing globally; autonomous technology has firmly arrived on the scene; digital features are impacting on the automotive market; and powertrain electrification is reshaping markets. These developments have profound implications for both automotive manufacturers and the passenger and commercial vehicle aftermarkets.

The size of the IAM channel is influenced by three key drivers: the vehicle parc, annual spend per vehicle, and the channel share of the IAM. From a channel standpoint, what is good for the IAM is usually bad for the OE channel, and vice versa; as one channel expands, the other shrinks proportionately.

Let us look at the first driver of the IAM, the *vehicle parc*. This expands in line with sales of new cars and imports of used cars to local markets. It shrinks in line

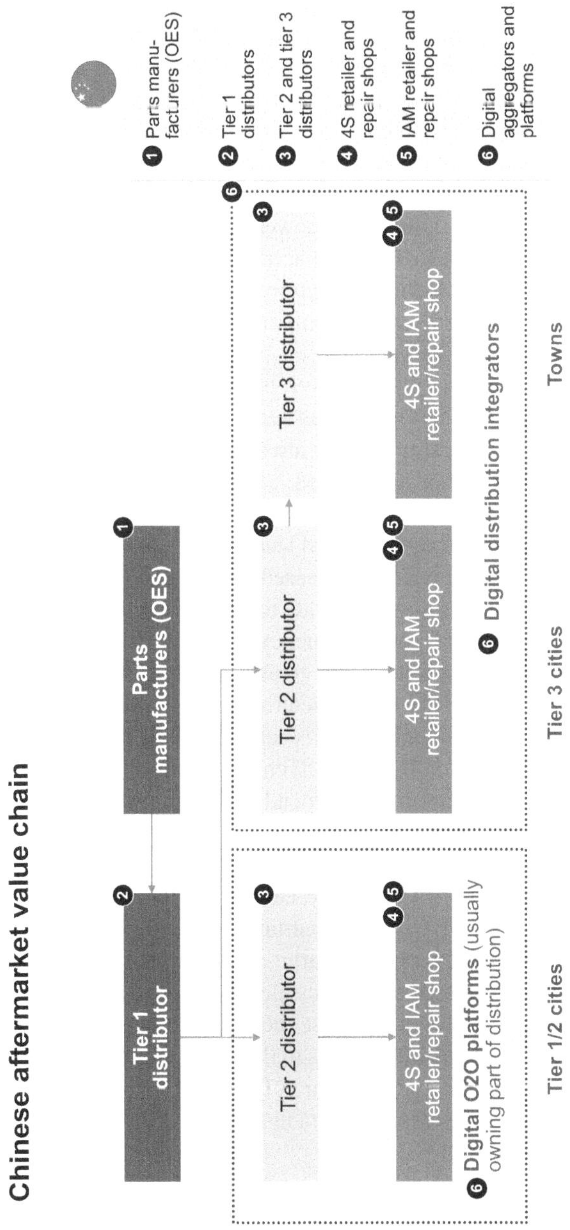

Fig. 4 An illustration of the Chinese aftermarket value chain (Source: Roland Berger, used with permission)

with increases in the scrappage rate, usually determined by the number and extent of high-impact accidents and local legislation on car parc renewal. The vehicle parc is particularly affected by two of the MADE trends. New mobility models potentially have a negative impact on new car sales as utilization—the number of kilometers driven—is significantly higher per vehicle for shared mobility than for private cars that are not shared. But at the same time, the rise of autonomous driving and increasing penetration of ADAS leads to reductions in the number of accidents and thus pushes down the scrappage rate.

Annual spend per vehicle is driven by three further factors. First, vehicle architecture, that is, whether the car is an ICE, hybrid, or battery electric vehicle (BEV). ICE and hybrid vehicles have traditional powertrain architecture and so feature a larger share of parts that may need to be replaced or repaired. By comparison, BEVs have 30% fewer such parts. Vehicle architecture also determines what technology is used in the vehicle in terms of electrical components, electronics, etc., and thus has an impact on the addressable market.

Second, annual spend per vehicle is driven by the price of parts, which depends on factors such as the complexity of the parts being replaced, whether or not the part in question is an innovative technology or already commoditized, and whether or not the part is remanufactured or refurbished.

Third, annual spend per vehicle depends on the replacement rate, which is in turn driven by the number of accidents, wear and tear on parts, and the prescribed maintenance frequency of the components. Increased penetration of ADAS lowers crash rates and so brings down the replacement rate for components, while wear and tear on parts and prescribed maintenance frequency depend largely on the number of miles driven: The more the car is used, the greater the need to replace components.

Lastly, the *channel share of the IAM* is determined by both the age of the vehicle parc and the amount of vehicle content that is addressable by the IAM. After 3 years on the road, cars typically switch from the OE channel to the IAM channel (although the precise length of time varies greatly depending on geography and the warranty periods offered by OEMs). In other words, the older the vehicle parc, the higher the channel share of the IAM.

Not only does the content of the vehicle need to be addressable by the IAM, it also needs to be accessible. OEMs control certain captive parts of the vehicle, such as body parts and software solutions, to a large extent. All other parts of the vehicle are accessible to the IAM. As the electrics/electronics (E/E) content of vehicles increases, OEMs are trying hard to increase their captive share of the vehicle, and hence the aftermarket, by controlling access to the data required to diagnose vehicle problems accurately. Here, the channel share of OE versus IAM will ultimately be decided by legislation on how open this data must be.

4.1 Industry Trends

We identify 10 main trends affecting the industry, each made up of a number of subtrends (see Fig. 5). The first four come under the general heading of *new customer*

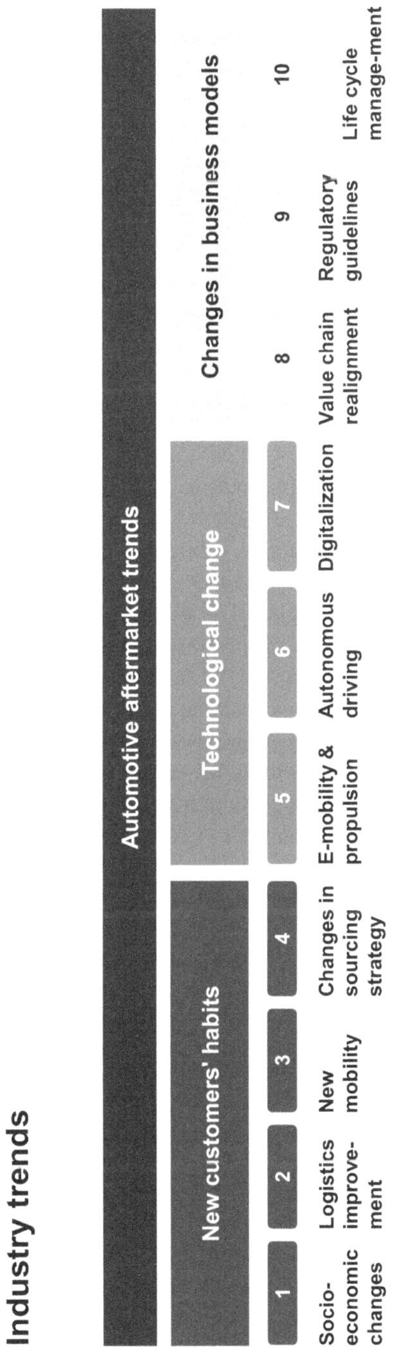

Fig. 5 Industry trends in the automotive aftermarket (Source: Roland Berger, used with permission)

habits and include socio-economic change (urbanization, growth of the car parc, increasing vehicle age, the circular economy, inflation, decrease in manual gear boxes), logistics improvements (new logistics, fleet operation and utilization), new mobility (shared mobility, loss of significance of ownership, micro-mobility), and changes in sourcing strategies (modularization, 3D printing, unification of the platform).

New customer habits will make the aftermarket more competitive and reduce the amount of room for differentiation. Socioeconomic changes and new mobility concepts are redefining the ownership of vehicles as a result of increased urbanization and the emergence of ride hailing and ride sharing, which will ultimately have a negative effect on the total car parc. At the same time, the average mileage and age of vehicles in circulation will increase, leading to more frequent replacements. New mobility concepts will result in greater standardization of products, thus reducing the complexity of parts and increasing commoditization.

Next, we have three trends related to *technological change,* namely e-mobility and propulsion (new energy vehicles or "NEVs," the NEV supply chain, complexity of repairs, decrease of diesel), autonomous driving (ADAS, development of sensors and algorithms, increased safety), and digitalization (connectivity, big data, automatic repair diagnostics, digital features, mobile service).

Technological change will lead to the introduction of new types of vehicles and tools that will decrease overall market size, put pressure on margins and favor the OE channel. In terms of annual spend per vehicle, e-mobility means less complex powertrain and transmission products, resulting in lower maintenance costs. Autonomous driving will also reduce accident rates, which means fewer repairs per vehicle in the future, as we pointed out above. However, predictive maintenance and other digital concepts may increase replacement frequency and content per vehicle. Digitalization will also likely lead to products and services with higher margins.

Finally, we have three trends related to *changes in business models.* These are value chain realignment (nontraditional entrants, new or changing intermediaries, OEM aftersales initiatives, the role of e-commerce, buying groups, consolidation of the value chain, business model redevelopment), regulatory guidelines (emissions regulations, block exemption, access to data, type approval), and life cycle management (awareness of TCO, proliferation of quality).

No major impact is expected on changes in business models within the car parc, although increased regulation on vehicle emissions will lead to the exclusion of noncompliant vehicles. In terms of annual spend per vehicle, new developments in business models—especially in the area of services—are expected to increase annual spend per vehicle, bringing higher value to customers. But this may be balanced out by a drop in prices resulting from block exemption regulation. Ongoing changes in the value chain will also increase customer bargaining power and hence price pressure, with a negative impact on margins. Additionally, regulatory actions will foster increased competition in the aftermarket, contributing to a decrease in prices and margins, although new business models based on services will counter this downward trend to some extent. All in all, the evolution of business models in the automotive landscape will open up opportunities for both the OE and the IAM channel to gain or maintain channel share, although overall the IAM is expected to gain market share at the expense of OE due to increased transparency over TCO.

5 Conclusion

The automotive aftermarket plays a pivotal role in ensuring the longevity of different means of mobility across the globe long after vehicles are out of their warranty period. This promotes the affordability of vehicles by increasing their lifetime and usage as second- or third-hand used cars in different parts of the world. It also contributes to a more sustainable industry where cars can be used for much longer, reducing the climate impact from manufacturing. At the same time, the industry offers significant employment across all layers of the value chain. It also offers the luxury of choice to car drivers/owners who would like to customize or tailor their vehicles according to their needs at different price points. Lastly, the automotive aftermarket industry is home to multiple small and medium-sized businesses across the value chain, especially in wholesale distribution and at workshop/garage level. These businesses in turn make a major contribution to local economies.

References

Roland Berger Aftermarket Component Model, Roland Berger GmbH.
Roland Berger GmbH publications and documents.

Hasmeet Kaur, Global Managing Director and Partner at Roland Berger, operates from the Munich office. In this capacity, she is part of the consultancy's global management team, overseeing Innovation and Products. With a specialization in the automotive aftermarket industry, she has led projects for over 70 global automotive clients across 25 countries. Her extensive experience encompasses business growth, market entry strategies and business model turnaround initiatives. Additionally, her portfolio of clients has been quite holistic, covering the whole automotive value chain, including automotive OEMs, especially new EV OEMs, automotive suppliers, wholesale distributors and ITGs as well as e-commerce players.

An Expert Vignette from the American Perspective: John R. Washbish, President and CEO

Maria Elo and Fotios Katsardis

Abstract

The American automotive aftermarket has a particular role in the global development of the automotive aftermarket business. The North American Free Trade Area Agreement (NAFTA) countries represent a major business market globally. Specifically, the United States is also seen as an innovative context with special characteristics, for example, in management, distribution, marketing, and business organization. This chapter offers an expert view with a long-term industry perspective that contributes to understanding this business as well as its potential and challenges. A direct voice from this setting is illuminating and informative in highlighting the American aftermarket and its particularities and dynamics.

This book provides a range of perspectives on the automotive aftermarket. While doing so, it is important to incorporate practitioners' voices from leading experts that may offer a broad understanding as well as a deep insight into this business.

M. Elo (✉)
University of Southern Denmark (SDU), Odense, Denmark

Belt and Road Institute of International Business, School of Economics, Shanghai University, Shanghai, China

Abo Akademi University, Turku, Finland

Turku school of Economics, University of Turku, Turku, Finland

Institute Arbeit und Technik, Gelsenkirchen, Germany
e-mail: melo@sam.sdu.dk

F. Katsardis
Temot International Autoparts GmbH, Ratingen, Germany
e-mail: f.katsardis@temot.com

M. Elo, F. Katsardis (eds.), *Automotive Aftermarket*, Management for Professionals, https://doi.org/10.1007/978-3-031-62419-3_4

We are delighted to invite John R. Washbish, President and CEO of Aftermarket Auto Parts Alliance, to offer his personal experiences and views on the automotive aftermarket, its potential, and challenges. John R. Washbish is an industry expert with an extensive career in the automotive industry, particular knowledge of the automotive aftermarket from a top position, but also with personal insights on the American family business development and the American AA education. He is coined as "an icon of the automotive aftermarket" by Northwood University, USA,[1] and has received their Outstanding Business Leader Award in 2019.[2]

The vignette by John R. Washbish has an interview format that enlightens several aspects of the automotive aftermarket. Many of these aspects connect to discussions that are found also in the contributions of other authors in this book. We cordially thank John R. Washbish and give him the arena.

We start with a personal storyline:

Could you shortly describe:

1. How did you enter this business and excelled in your career?

   In 1956 my parents relocated from Miami, Florida, USA to Louisville, Kentucky. My father became the general manager of Louisville Auto Spring Co., Inc. (LASCO) This company was owned by my mother's family. LASCO operates as a specialist in wholesaling and installing steering and suspension parts for light and heavy vehicles. Leaf springs for heavy vehicles is one of their unique services. In 1960, my father opened an automotive aftermarket distribution business adjacent to Louisville Auto Spring Co., which he christened Moog Louisville Warehouse, Inc. MOOG operates as a warehouse distributor of a wide range of aftermarket auto and truck parts. Following in my father's footsteps, my brothers, David and Doug, own and operate LASCO and MOOG respectively. They have grown the businesses to multi-location organizations and leaders in the market. BTW, my youngest brother, Don, owns and operates Ferrels Car Care, which is a multi-bay aftermarket repair centre in Lexington, Kentucky. With a small family business and six kids (I am oldest), all the Washbish children became involved in the family business as soon as we could, literally, read, write and count! As a small family business I became involved in the business at a very early age.

2. What has made the automotive aftermarket in the United States attractive for you as a career path?

   There was never a question about a career in the automotive aftermarket but a couple events change my aftermarket career path. In 1971, the Wix Filter Corporation awarded me a scholarship to study Automotive Replacement Parts Management at Northwood Institute in Midland, MI. During the1973 spring trimester at Northwood, I was selected by Monroe Auto Equipment Company to work at their Sint-Truiden, Belgium manufacturing and distribution facility. The

[1] https://www.northwood.edu/obl/archives/john-r-washbish retrieved 24.10.2023.

[2] https://www.counterman.com/alliance-ceo-john-r-washbish-accepts-northwood-outstanding-business-leader-award/ retrieved 24.10.2023.

international aspect of the aftermarket was intriguing. Upon graduation from Northwood Institute with a Business degree in addition to my Aftermarket degree, there were many job opportunities in the aftermarket. My father and I agreed that it would be nice to work with a supplier channel partner for a couple years before returning home to the family business. The aftermarket had been attractive to me because it serves such a basic but generally unappreciated service; keeping the world moving, one mile at a time! I enjoy the aftermarket because of the close relationships that were initially established through my parents and blossomed through all my family members who currently work in the automotive and heavy-duty aftermarket. And, it has provided a fair wage with decent benefits and the opportunity to have a global perspective.

3. What were your dreams when you started?

   To get a couple years of North American travel, entry level experience and broad exposure to a wide variety of aftermarket suppliers and distributors before returning to the Kentucky family business. The plan was for the family business to dominate the automotive and heavy-duty parts distribution business in the "Kentuckana" marketplace. My brothers have accomplished this task!

4. What is your position today and your favorite achievements?

   Currently I am the President and CEO of the Aftermarket Auto Parts Alliance, Inc. We are an international aftermarket information technology and marketing support organization with shareholders in the United States, Canada, Mexico, Latin America and TEMOT in Europe. Recently, we formed a joint venture with Federated Auto Parts Inc., the Federated Alliance, LLC. This joint venture is affiliated with the Automotive Parts and Services Group, LLC. (APSG), which includes the Auto Value, Bumper to Bumper, Federated, Parts Plus and Pronto organizations. APSG will benefit from shared resources, especially in the information technology and national account sales arenas. My time at the Alliance has been rewarding. After 33 years on the supplier side, it has been fun to 'grow' the headcount here, especially in the information technology department. To provide a better view of my perspective, my resume is attached for your perusal.

   Here, we shift the focus to the industry-sectoral development:

   Could you shortly provide your views as an expert for the American AA?

5. How would you describe the AA as a sector in the United States (economically, socially, politically)?

   Economically it is the second largest business in our economy. Socially a vehicle is necessity and is the second largest expenditure after a home. We have over four (4) million people working in the automotive and/or heavy-duty aftermarket sector. Politically we could motivate the country but do not have a united front. Probably never will. Too many 'individual' objectives.

6. Could you share some key data in terms of the US AA market, its size, and segments?

   The market has over 280 m vehicles with an average age of over 12.5 years old! The good news for the independent aftermarket is this, by their own admission, car dealers (National Automobile Dealers Association) only have enough service bays to repair 25% of the current fleet. The dealer bay count was considerably reduced in the 2008–2009 timeframe, resulting from many dealership closings.

7. Who are those key players that lead AA in the American market? And are there particular companies or stakeholder groups of interest?

   Our shareholders represent some of the most successful aftermarket companies in the United States, Canada and the LATAM regions. Our shareholders operate over 170 distribution centres, 2100 wholesale/retail outlets, supporting over 4000 certified service centres. All the shareholders excel at customer service providing a unique supply of inventory tailored to their marketplace. We are governed by a Board of Directors that operates according to the fundamental by-laws established in 2000, when Auto-Value and Bumper to Bumper merger to form the Aftermarket Auto Parts Alliance, Inc. We can provide a list of or current Board of Directors if requested.

8. Could you illuminate shortly the key differences of OEM and AA business and the meaning of the American automotive manufacturing in influencing the AA landscape in the United States?

   Without question, autos and trucks have shaped America more than anything! OEM is more about the NEW CARS and they do repairs because they are forced to! AA is about parts, service and complete repair because it is our livelihood.

9. What kind of major changes in the car service and maintenance business do you perceive?

   Whatever the maintenance provider can do to enhance and improve the experience. They realize that a flawless service experience will sell more new cars! Car dealers offer to come to your location and pick up your vehicle and provide a loaner vehicle.

10. What have been the major critical events happening to the American AA if you look back? In terms of strategy, technology, innovation, markets, globalization, other?

    Without question, technology has been the major improvement in the North American aftermarket. Technology enables an incredibly broad selection of parts and services to be available, in urban markets, usually in 30 minutes or less! Technology also drives the data tools that manage the inventory at every step of the channel. And technology supports the various BTB and BTC efforts. Technology assists in innovation and advanced globalization. Many parts sold in North America are produced outside Canada and the USA. China and Mexico would represent the primary 'outside' of North America supply countries.

11. What factors drive the AA sector in your opinion? And are there particular American factors influencing the development (market, culture, regulation, other)?

    Currently, RIGHT TO REPAIR legislation in our Federal Congress is essential for our technicians to have complete access to repair data on all vehicles; automotive, heavy-duty, etc. The culture is favourable for aftermarket parts and service based on the data already presented. The politicized advance of the EV has furthered frightened the average American, which will cause them to depend on their current vehicle longer! By my estimate, the independent aftermarket represents 75% of the North American aftermarket. Recently while attending a MEMA OES/OEM Seminar in Troy, Michigan, NADA suggested they represented 31% of channel is very effective at supplying parts to car dealers!

12. How does the American AA differ from the AA in other contexts, like the EU or Asia?

    Major difference is the European independent aftermarket is only 25% of the aftermarket parts and service, while the OEM represents 75% of aftermarket volume. The vehicle parc in Asia is so young the aftermarket is just developing. It is not developing with many EU or North American characteristics. Ecommerce trade propels the majority of Asian aftermarket activity.

13. Could you briefly describe the relation of the US AA with the Canadian, Mexican, and other countries and its importance?

    Canada and the USA operate in identical markets, with the exception of currency and import taxes. Mexico operates as two markets. Northern Mexico, the area within 100 miles of the USA is populated with a high percentage of USA manufactured vehicles. Central to southern Mexico, has a high percentage of Asian vehicles. The proximity of the three countries makes us natural trading partners.

14. What are the key challenges for the future of the American AA as a sector?

    Consolidation at the supplier base is creating challenges for distributors. Many legacy family-owned North American suppliers have been sold to private equity owners. These PE owners are not investing in the future. The distributor base is also consolidating, which is difficult for suppliers to deal with!

15. Are there particular megatrends or concerns that influence AA in the United States?

    Speed and availability. We live in a country with instant gratification.

16. How do you imagine the American AA in the future, say the next 15 years, how does it look like, and who is participating?

    The North American aftermarket will be especially strong in 15 years. Best case EV projections are less than 10% of the total vehicle park in 2035. The ICE platform will represent over 300 million vehicles in the USA alone. Since there will be no focus on improving or changing the ICE platforms on the highway, the aftermarket for these vehicles alone will thrive for years. Additionally, the class 4 and 5 vehicle growth has been and will continue to be strong, providing a growing aftermarket opportunity.

17. What about the institutional side and regulation? And how do you see it impacting the future of the American AA?

    As mentioned, RIGHT TO REPAIR must get approved. We need support from our aftermarket partners around the world to be unified in this approach.

18. What kind of concerns do you have about the development of the sector and its challenges and opportunities in the United States?

    Private Equity activity is growing in our sector.

From the sectoral view, we are moving on to the family businesses and new startups:

19. Entrepreneurship and family businesses are said to be part of the American DNA, how does that demonstrate itself in the US aftermarket?

    Virtually all North American aftermarket suppliers and distributors began as family held operations. This 'fad' is going away as many family businesses in the aftermarket sector have been consolidated over the last 30 years. I mentioned the existence of private equity on the supply side of the market. 75% of the independent distribution side has been consolidated with four (4) publicly traded retailer organizations. 25% remains with family and small private equity investments.
20. Could you give some examples of American family businesses and startups?

    There has not been a significant North American family business startup in the last 20 years! At the same time, there have been a number of family businesses that continue to grow and expand; particularly in rural markets.
21. How can American family businesses and smaller AA companies compete today and in the future?

    They are experts in their backyard. They have a hometown advantage but must affiliate themselves with the Aftermarket Auto Parts Alliance, Inc. to ensure access to all the tools necessary to compete with the OES and big box retailer channels. The Alliance has all the tools necessary and, with our membership in APSG, allowed to transact business with the national repair chains looking for a national contract!
22. What about the geographies of the American AA? And what role will location and local business ecosystems have in the era where digitalization continuously increases?

    Whatever the case may be, punch line is the same, who has the broadest range of parts and services that can be delivered the fastest?
23. How do the startup hubs, incubators, and accelerators influence the growth of new AA actors?

    If these start-ups amount to anything, folks react.
24. What about the consolidation and global investments happening? And how do the American AA investors shape this business landscape?

    As mentioned, most private equity firms concern us.

    a. In the United States and internationally? *Concerns in both areas.*
25. What kind of investors are producing a sustainable impact for the AA?

    There have been a handful of investors that actually use the word SUSTAINABLE and they seemed focused on this platform. These companies are impressive. A couple of the best are remanufacturers.
26. What do you see as the major issues for business succession in the United States?

    Convincing young people to stay with one organization long enough to learn the culture of the organization and work to enhance it.
27. What would you like to give as advice to a young AA business owner in the United States?

    Focus on your backyard and all the opportunities in it.

28. Could you share your views on how education and training are organized in the United States for the AA?
    a. For employees: *Most organizations have extensive training programs with extensive training libraries. Suppliers are also expert trainers in regard to the aspects of their products and installation techniques. The Alliance participates in the Group Training Academy, which is an incredible training resource tool for each level of the distribution and service chain.*
    b. For business owners and entrepreneurs: *The University of the Aftermarket, which is affiliated with Northwood University in Midland, Michigan is considered the Harvard of the automotive and truck parts aftermarkets. Degreed courses are available and taught by a combination of university professors and aftermarket business professionals. Recently I lectured at a University of the Aftermarket 101 Introduction to the Aftermarket Class. The class consisted of entry level professionals at various distributors and suppliers in the aftermarket community.*
29. What would you like to give as advice to a young AA manager in the United States?
    Get involved. There are four million people in the North American aftermarket. It is a huge market with significant opportunities for hard working, well educated, willing to relocate individuals! Go where they need you! Do what they need you to do!
30. What would you like to give as advice to a young AA student?
    It is a solid industry for a vibrant career and you are home on the holidays!

Then, we move to the final comments:

31. Is there something else you would like to add?
    I could go on forever but have much to do. Good luck.

Thank you very much for the valuable perspectives John R. Washbish!

Maria Elo full professor of international business and entrepreneurship at the University of Southern Denmark, professor at the BRIIB Institute of International Business at Shanghai University, Senior Research Fellow, University of Turku, Adjunct professor at Åbo Akademi University, and Research Fellow at Institut für Arbeit und Technik, works in the area of international business, -entrepreneurship, and -migration with topics such as sustainability, business ecosystems, policy frameworks, networks, internationalization, mobility forms, human stickiness, migrant entrepreneurship, transnational and family businesses, diaspora investment, and remittances. She has published books and articles, for example, in Journal of World Business, Journal of International Management, Journal of International Business Policy, Industrial Marketing Management, Regional Studies, Journal of International Entrepreneurship, and International Journal of Entrepreneurship and Small Business. Maria is one of the founders of Talents4AA and a Talents4AA ambassador leading the collaboration with the education sector.

Fotios Katsardis is an experienced professional with extensive international experience in the automotive aftermarket. He holds a Bachelor of Arts (BA) degree in Economics, a Master of Arts (MA) in European Economic Studies, and an MBA in International Management, and was a Robert Schuman Scholar by the European Parliament. His working experience gained in different countries relates to multinational enterprises, strategic alliances, purchasing groups, SMEs, and family businesses. He has worked in different international positions for a renowned global Original Equipment Supplier and for the past 19 years as the CEO and President of a global automotive aftermarket alliance that operates across 110 countries. Fotios combines industry expertise, academic qualifications, and a passion for the automotive aftermarket. He is an acknowledged speaker and panellist of the sector. He is also the Chairman of the Talents4AA, initiated with his sector peers, aiming at raising awareness of the automotive aftermarket and attracting new talents to the sector.

"The Biggest Hidden Champion": A Vignette from Michael Söding, Industry Veteran from Schaeffler Group

Maria Elo and Fotios Katsardis

Abstract

The automotive aftermarket has gained surprisingly little interest in management and academic research literature. Yet, it is a global business with a multifaceted range of business actors, many of those representing also highly global organizations. Particularly those leading suppliers operate in numerous markets across the world while innovating new and better solutions. Innovative products and services often stem from these suppliers and aftermarket businesses that are underexamined, if not hidden in terms of awareness and managerial-scholarly attention. This chapter brings up an expert insight that portrays automotive aftermarket features, dynamics, and future trends and contributes by illuminating also strategic aspects of the global aftermarket supply chain.

M. Elo (✉)
University of Southern Denmark (SDU), Odense, Denmark

Belt and Road Institute of International Business, School of Economics, Shanghai University, Shanghai, China

Abo Akademi University, Turku, Finland

Turku school of Economics, University of Turku, Turku, Finland

Institute Arbeit und Technik, Gelsenkirchen, Germany
e-mail: melo@sam.sdu.dk

F. Katsardis
Temot International Autoparts GmbH, Ratingen, Germany
e-mail: f.katsardis@temot.com

M. Elo, F. Katsardis (eds.), *Automotive Aftermarket*, Management for Professionals, https://doi.org/10.1007/978-3-031-62419-3_5

This book attempts to offer multiple perspectives and voices from the automotive aftermarket, including different positions, types of firms, and regions. Hence, we have invited leading experts to illuminate their views, share their personal experiences, and reflect the automotive aftermarket as a sector with their expertise. In addition to family businesses and groups that dominate many areas of the AA, there are also notable multinational enterprises that represent global players supplying parts, technology, and innovation across the world, in the AA business and beyond. Schaeffler Group is one of the leading AA actors, one of the largest family-owned companies in the world, and among the most innovative companies in Germany.

Hence, we are pleased to invite Michael Söding, former CEO of Automotive Aftermarket, member of the Board of Managing Directors at Schaeffler Group, and an industry veteran as he coins himself. Michael Söding has a long expertise in the sector covering diverse elements from European to global business and to sustainability concerns. It is important to notice that business contexts and processes matter, and the views of this industry are often limited to particular aspects. Yet, beyond these limitations there are numerous perspectives that deserve attention. For example, the role of the AA industry is multifaceted regarding sustainability by addressing the life cycle and the respective processes, as Michael Söding notes: "*Through our innovative repair solutions, we extend the life cycle of vehicles. Thanks to our system expertise, supporting services are performed quickly and efficiently to increase vehicle mobility to the maximum extent.*"[1] Such statements show the relevance of AA, its particular roles and solutions, and underline the respective expertise inherent in this industry. Schaeffler Group and its Automotive Aftermarket division represent an example of a multinational enterprise operating on many levels. For example, it develops holistic systems and solutions serving the aftermarket and builds on an understanding of the entire vehicle system for the repair business.

The vignette by Michael Söding follows an interview format that addresses several aspects of the automotive aftermarket, both opportunities and challenges, but also personal views and talent issues. He titled his vignette "The biggest hidden champion," reflecting the automotive aftermarket as a sector and extant awareness of it. His experiential perspective from the multinational enterprise side offers valuable insights. As the readers may notice, several aspects explained by Michael Söding may also relate to discussions that appear in other book contributions as there are numerous overlapping dynamics not limited to a specific company or a position.

We cordially thank Michael Söding and give him the arena.

We start with a personal storyline:

1. Could you briefly describe your career in the AA? And how did you enter the business?

   Back in 1988 I graduated from Technical University Darmstadt as industrial engineer, majoring in automotive engineering. I then started my career at

[1] https://www.schaeffler-sustainability-report.com/2020/to-our-stakeholders/statements-of-the-board-of-managing-directors.html retrieved 24.10.2023.

Pirelli in the OE technical environment. Soon after I moved over into product management, taking responsibility for the tuning segment. My second job already focused on spare part business at the battery producer Hoppecke. With stages at Knorr Bremse and Exide I finally ended up at LuK Aftermarket Service, which merged in Schaeffler Automotive Aftermarket at a later stage. In my 20 years at Schaeffler I grew from Head of Global Sales & Marketing in 2002 to President Automotive Market in 2009 and finally CEO Automotive Aftermarket and Member of the Executive Board in 2018.

All in all, more than three decades in an industry of mobility services, spare parts and automotive aftermarket (AA).

2. What pulled you to this business and kept you there?

   It basically started as a coincidence, but soon after I realized AA offers a significant amount of autonomy in an OE centric environment. Not too many people understand what you are doing, not too many interfere, and so you find a decent level of independence as long as you deliver to the targets. On top I always loved to connect with decision makers from multiple nationalities, which is an inherent given in the AA.

3. What kind of a skill package did you have when you started? Education, languages?

   As an industrial engineer I could combine technical and business skills. Such more generalist education always helped me to think and act across different disciplines. As an example, I learned to discuss with inhouse product developers about necessary features in their technical language while the negotiation with customers was obviously commercially driven.

   My English was rather improvable, to say the least. And only English as a foreign language turned out to be a severe showstopper namely in Eastern European markets right after the iron curtain was torn down.

   What negotiation partners appreciate most you don't learn at schools or universities: Decisiveness and Integrity! Neither you learn how to manage teams and survive in large corporations: Leadership and Inclusion! To me life-long learning is a mantra. It ain't over till it's over.

4. What were your dreams when you started?

   To have a stake, to have a say, to develop my team and to bring people to their max.

5. Did the sector contribute to your personal development, if yes, how?

   Half of my life relates to that industry. While I tried to influence it's development it influenced me vitally. I could travel the whole world, connect and interact with an incredible number of inspiring and entrepreneurial people. They taught me a lot. Next, the more I could contribute to company results the more entrepreneurial I could become myself. Can you imagine what it feels like, when the top management decides to upgrade your organization to a division with full external reporting and nominates you as a CEO?

   But the biggest influence was created by the awards that individual customers and International Trading Groups eventually threw in your direction to express their satisfaction with your company's performance. Such awards are an incredible stimulus for high performance organizations as well as for me personally.

6. How would you describe AA as a sector?

   The biggest Hidden Champion on Earth! Honestly. What is obvious to see is the very high degree of fragmentation both horizontally and vertically. Even the relatively big players need many peers to associate and partner up with. And everyone is in need of service providers for versatile sector specific solutions and trading groups to bundle demand and negotiation efforts. All sums up to an incredibly performant and successful Ecosystem with partners across the globe.
7. What have been the major critical events happening to AA if you look back?

   AA itself is quite robust when it comes to changing market dynamics. Important to comprehend is, AA isn't a stand-alone industry, but highly embedded into the larger context of the automotive industry and the whole sector of mobility. When car makers or their suppliers either boom or suffer, it effects the AA big scale in terms of product availability, investment funding or marketing expenses. When governments launch scrapping schemes to boost new car sales, AA will lose significant market potential. Remember "Cash for Clunkers" back in 2009? The end of the cold war and new trade policies in the early 90's probably marked the biggest single growth driver with huge new market opportunities from Eastern Europe all the way to China.

   The biggest poison to AA though is erratic decisions by policy makers, that change the business model sometimes almost overnight. Trade barriers such as import taxes or country specific safety and technical norms and most recently Brexit are shining examples for this.

   The constant shift in technology as well as the enhancement of quality levels show strong impacts on producers and service providers. Electronic injection killed carburetor technology, while producers of exhaust systems lost more than 50% of their spare part business due to the introduction of stainless steel.
8. What drives this sector in your opinion?

   First and foremost, the business is determined by the demand side. By trying to push markets way above you may lose money three times: first, because you need to reduce prices or increase terms to boost sales, second, you need to give extended payment terms and third, you don't sell in the next period due to overstock but sit on idle capacities. That's why logistics competences and demand forecasts are such a strong asset to our sector. You plan and produce against a virtual demand with a lead time of up to 9 months, but then deliver within 24 hours along the whole value chain.

   Secondly, we are confronted with a significant variety of technology and huge number of stock keeping units, but very low volumes on part level. All at the time. What makes it even more demanding from a production perspective is the time offset between parts that serve latest OE demand and technology while AA pars may care for twenty-year-old applications.

   And thirdly the way to market. It differs significantly across different countries, regions and individual sales strategies even in times of growing globalization and business standards. There are countries with an impressive OES dominance, while other markets are mainly served by independent Aftermarket operators through three steps, two steps or even one step distribution models. And E-commerce still must proof its relevance for that industry.

9. Who are those key players that lead AA? And are there any particular companies or stakeholder groups?

   Let me first reflect about the inherent value chain of AA. The ultimate goal is to grant safe, secure, clean and affordable mobility for more than 1 billion cars and trucks at any point in time around the globe. In such value chain you need participants, which understand the nature and remedy of technical maintenance and service, you need those, who produce parts and develop technical instructions, you need large and small size distribution to organize warehousing and logistics across continents as well as at the last mile and you need workshops and technicians who finally fix the car. All these are crucial elements for any proper service, but undoubtedly you need to connect them into a process that is only successful, when all elements or chain links function well. To bring such process orientation into action you finally need facilitators, who connect all those bits and pieces million times a day. Examples for such facilitators are national and international trading groups, which connect thousands of regional warehouse distributors and local retailers with large suppliers and producers of parts as well as technical and digital service providers, that create trading standards, offer dealer management systems and standardize raw data, business processes and E-tailing solutions.

   Against this backdrop I would simply deny the dominance of single individuals. The critical success factor is a vibrant and swift ecosystem, that connects all stakeholders, small or large, within minutes. A workshop mechanic usually doesn't know beforehand, what type of service and spare parts a car requires, when it pulls in in the morning. What he knows is, the owner wants to pick up the car latest in the afternoon "fixed first visit". So, everything from failure diagnosis and parts ordering to parts supply, repair process and electronic resets happens within hours.

10. What are the key challenges for the future of AA?

    AA is a strong margin provider and offers rather stable revenue streams. That always called for strong competition from inside and outside the industry and will so going forward. Car manufacturers try to increase the market share of the OES Channel, E-Commerce giants squeeze in distribution networks, Intermediates try to control the transaction process and absorb profits, and all lobby on regulatory level to hamper market access and limit the number of competitors. A never-ending battlefield! And of course, each new technology puts stress on all players to learn, adapt and contribute to latest requirements.

11. Do you perceive some critical events to come, if yes, are those of external (like the COVID pandemic) or internal nature (e.g., corporate processes)?

    Like any given industry AA is leveraged by a vast number of influences and unprecedented crises almost daily. The good news here, mobility is a constitutional right, almost everywhere. The Covid pandemic has proven its relevance and the aftermath of lower car production supports our industry strongly. From my perspective the major strain will come from a technology shift away from combustion engines to xEV. The faster production sites are shut down the more demanding the manufacturing of spare parts for older applications will be. But all in all, I'm certainly optimistic for our industry.

12. Are there particular megatrends or concerns that influence AA?

    The undeniable need for a swift decarbonization of mobility obviously effects the running car park significantly. As synthetic fuels are unlikely to mitigate the CO2 emission on a larger scale, governments may subsidies xEV with negative repercussion on car park age and service demand. The imperative to reduce CO2 also effects the production of spare parts. How industrial remanufacturing can leverage emissions, waste and pollution is highly discussed in our industry these days.

    In the larger context of sustainability, it's mandatory for our sector to focus on gender equality and education. Initiatives such as "Talents4AAM" and "Next Automotive Aftermarket Generation" are shining examples for our will to raise awareness and attractiveness of our industry.

13. How do you imagine AA in the future, say the next 15 years, how does it look like, and who is participating?

    I like those projections that speak to a far time horizon. No one will remember me or my prediction anymore. But yokes apart, the future aftermarket will still need parts, will still need physical service and will still need logistics to connect both sides. Due to rising quality standards and predictive maintenance the volume is likely to shrink. The predictability itself will influence margin levels negatively, because the end user finds enough time to negotiate the necessary service and repair. Who ones the data will definitely win market share and profitability.

14. What about the institutional side and regulation? And how do you see it impacting the future of AA?

    The impact of regulation is fundamental and rising and so is the effort of associations and lobbyists to advocate and influence institutions and law makers. A prominent example of today is the battle around "Access to Data (AtD)". While it's mission critical for the Independent Automotive Aftermarket to get access to in car data, car producers counter with flimsy arguments like hacking, data protection and cyber-attacks. To make it even worse it's uncertain, where lobbying would make more sense. On national level, on EU-level, at UN standardization level? On top, different regions and economic zones regulate differently which multiplies complexity and bureaucracy. But to make your lobbying efforts valuable your story of value creation, affected jobs and professionalism needs to be overly convincing.

    And please allow one side remark. As the automotive industry with all its layers is somewhat German centric, also many associations, initiatives and single actors are German driven. That causes a certain suspicion about the "German Club". Let me therefor encourage and invite participants from all other countries to engage and lead such activities going forward. As it is mandatory for our industry to stay alert and advocate ourselves better than others do, we need all hands on deck.

15. How can smaller AA companies compete in the future?

    Size doesn't matter. Speed does, performance does. If you overdeliver to customer expectations, there is a place for your business. Joining peer groups and pooling competences and resources stays mission critical. I pretty much like the quote of Harald Kostial: "Who wants something, finds ways, who doesn't, finds reasons".
16. What about family businesses and geographies of the business? And what role will these elements have?

    Every market, every business environment is different, therefor my answer can't be general. Family businesses typically show higher resilience and longer strategic horizons but depend on competences and resilience of few. It sounds like a contradiction in terms but those who love to be autonomous and independent need ever so often to rely on others, delegate more and put themselves into dependence for the sake of faster and robust solutions.
17. What about consolidation and global investments happening? And how will this development change the landscape?

    The business cycle of cheap money has attracted many investors and pushed market consolidation at low risk. In future, companies probably will prioritize investments in technology, IT, digitalization, e-commerce and sustainability, leaving few resources for mergers and acquisitions. The immanent risk though is rising numbers of insolvencies and company crashes.

    I realized at various Capital Market Days, where my company exclusively explained strategy and performance to analysts and investors, that they per se *have very little insights into our industry. It's obvious our industry needs more visibility at their level.*
18. What kind of concerns do you have about the development of the sector and its challenges and opportunities?

    A Hidden Champion is - hidden! Not the best preconditions to attract talents for our industry. As many countries see more people retiring from than entering occupation it seems vital for our industry to boost awareness and attractiveness both on company and industry level. When potential candidates opt for the apparently better job opportunity, what arguments do we have to sell our case? Are career paths transparent, do they have access to industry specific education, even academies? For sure the AA ecosystem offers attractive jobs and careers, but it is mandatory to invest into visibility and specific education. And women are highly welcome!
19. What would you like to give as advice to a young AA manager? And what to a young AA business owner?

    Regardless, whether you are a young business owner, entrepreneur or manager in the AA, my advises would be simply the same. The character of the business and the character of your function must fit to the character of your personality and competences. The story of your life should be as consistent as possible. In thirty years in the industry, I came across very different types of business models and functional jobs. I saw people failing and I saw people succeeding and growing. But I didn't find the superior business model or way to market that I could recommend.

Besides those fundamental decisions what to do and where to make a difference I recommend to build up teams, to empower people and to delegate decision power as early as possible. The faster and the better you act as a leader rather than a mere manager or owner, the more your organization will trust you and contribute to the company targets. You can concentrate on strategy, stake holders and growth drivers.

But no lasting success comes fast. The more you believe in yourself, the more resilient you are, the bigger your success will be. With optimism, discipline and self-consciousness you are best prepared!

Thank you very much Michael Söding for the contribution and perspectives on the AA!

Maria Elo full professor of international business and entrepreneurship at the University of Southern Denmark, professor at the BRIIB Institute of International Business at Shanghai University, Senior Research Fellow, University of Turku, Adjunct professor at Åbo Akademi University, and Research Fellow at Institut für Arbeit und Technik, works in the area of international business, -entrepreneurship, and -migration with topics such as sustainability, business ecosystems, policy frameworks, networks, internationalization, mobility forms, human stickiness, migrant entrepreneurship, transnational and family businesses, diaspora investment, and remittances. She has published books and articles, for example, in Journal of World Business, Journal of International Management, Journal of International Business Policy, Industrial Marketing Management, Regional Studies, Journal of International Entrepreneurship, and International Journal of Entrepreneurship and Small Business. Maria is one of the founders of Talents4AA and a Talents4AA ambassador leading the collaboration with the education sector.

Fotios Katsardis is an experienced professional with extensive international experience in the automotive aftermarket. He holds a Bachelor of Arts (BA) degree in Economics, a Master of Arts (MA) in European Economic Studies, and an MBA in International Management, and was a Robert Schuman Scholar by the European Parliament. His working experience gained in different countries relates to multinational enterprises, strategic alliances, purchasing groups, SMEs, and family businesses. He has worked in different international positions for a renowned global Original Equipment Supplier and for the past 19 years as the CEO and President of a global automotive aftermarket alliance that operates across 110 countries. Fotios combines industry expertise, academic qualifications, and a passion for the automotive aftermarket. He is an acknowledged speaker and panellist of the sector. He is also the Chairman of the Talents4AA, initiated with his sector peers, aiming at raising awareness of the automotive aftermarket and attracting new talents to the sector.

Right to Repair in the Automotive Industry

Álvaro de la Cruz Tomás

Abstract

The automotive aftermarket keeps 1.5 billion vehicles globally on the road while contributing $1.8 trillion to the global economy. Throughout the car's life, multi-brand repair shops perform a huge majority of the repairs needed. This vibrant industry and the consumer choice that it creates are being threatened by a lack of access to wirelessly transmitted vehicle repair and maintenance data and information.

Without the convenient and affordable choice of independent parts and repair, consumers will have limited access to affordable vehicle services and repair. These restrictions can have negative effects on local economies and the well-being and safety of millions who rely on vehicle transportation daily.

In the chapter, we will go through some of the most significant legislative and regulatory dossiers being adopted at the European Union and United Nations level, affecting the automotive aftermarket and consumers' right to repair their vehicles at a workshop of their choice.

Á. de la Cruz Tomás (✉)
FIGIEFA, Brussels, Belgium
e-mail: Alvaro.delacruz@figiefa.eu

M. Elo, F. Katsardis (eds.), *Automotive Aftermarket*, Management for Professionals, https://doi.org/10.1007/978-3-031-62419-3_6

RIGHT TO REPAIR IN THE AUTOMOTIVE INDUSTRY

(Source: Shutterstock, used with permission)

1 Introduction

Globally, the automotive aftermarket keeps 1.5 billion vehicles on the road while contributing $1.8 trillion to the global economy. Throughout the car's life, multi-brand repair shops perform a huge majority of the repairs needed. This vibrant industry and the consumer choice that it creates are being threatened by a lack of access to wirelessly transmitted vehicle repair and maintenance data and information.

Without the convenient and affordable choice of independent parts and repair, consumers will have limited access to affordable vehicle services and repair. These restrictions can have negative effects on local economies and the well-being and safety of millions who rely on vehicle transportation daily.

Within the Global "Right to Repair" campaign, FIGIEFA has been working tirelessly to secure the businesses of thousands of European aftermarket players, including independent parts distributors, independent workshops, and a wide variety of mobility service providers.

In this chapter, we will go through some of the most significant legislative and regulatory dossiers being adopted at the European Union and United Nations level, affecting the automotive aftermarket and consumers' right to repair their vehicles at a workshop of their choice. In all these pieces of legislation, FIGIEFA has contributed with evidence gathering, research and position papers, advocating directly to different levels of policymakers, writing articles and open editorials, and joining forces with other stakeholders to deliver successful campaigns.

2 Dossiers and Fields of Action

1. **Motor Vehicle Block Exemption Regulation (MVBER).** This is *the* pilar legislation for our industry, granting the playing field for independent operators.

On April 17, 2023, the European Commission adopted the revised and updated MVBER and its Supplementary Guidelines ('SGL'), which were set to expire on May 31, 2023. The new MVBER regime is applicable until May 31, 2028.

In the accompanying press release, the European Commission clarified that this limited prolongation will allow timely adaptations to possible market changes, such as those resulting from vehicle digitalization, electrification, and new mobility patterns that are expected to consolidate in the coming years.

An overarching aim of the new MVBER regime is to ensure that independent aftermarket operators, including workshops and other independent market players within the automotive supply chain, can have access to the relevant inputs that are necessary for vehicle repair and maintenance.

The key provisions in the regulation as such remain unchanged. They can be summarized as follows:

- A vehicle manufacturer may not prevent its authorized repairer from selling OEM-branded spare parts to an independent repairer *(this rule aims at ensuring access of independent repairers to OEM captive parts in particular).*
- A vehicle manufacturer may not prevent its original suppliers of parts, repair tools or diagnostic or other equipment from selling the same also to independent wholesalers or repairers *(this rule helps independent wholesalers access parts from manufacturers who also supply the vehicle manufacturer, except in limited circumstances related to tooling costs, IP or know-how).*
- A vehicle manufacturer may not prevent its component suppliers from placing their own trademark or logo on the parts supplied *(dual branding).*

Thanks to the work done by FIGIEFA and other aftermarket stakeholders, some important updates were made in the new MVBER, with great impact on the industry.

Main Changes in the Revised Text.

In essence, compared to the previous MVBER regime, the revised SGL entail a number of changes aimed at

a) Providing a stronger legal framework to avert market foreclosure risks for independent operators.
b) Enhancing clarity of the rules and alignment to technological progress and market realities, in key areas of specific interest for FIGIEFA.
c) Nevertheless, they have also failed to address some significant gaps.

When assessing the competitive impact of vertical agreements on the motor vehicle aftermarkets, the European Commission aims at preserving competition both between the members of authorized networks and between those networks and independent operators. To this end, the revised the MVBER Sector-specific Guidelines (SGL) establish the principle whereby the selective distribution agreements entered into between a vehicle manufacturer and its authorized repairers and/or parts distributors may fall foul of Article 101(1) if the vehicle manufacturer prevents *"access for independent operators to essential inputs"* that are necessary for repair and maintenance.

This concept of "essential inputs" includes all items available to vehicle manufacturers the withholding of which may have an *"appreciable impact on the ability of independent operators to carry out their tasks and exercise a competitive constraint on the market."*

In addition to technical information, this concept explicitly covers also tools, training, and vehicle-generated data (para. 62). Moreover, as the list of essential inputs in paragraph 62 is merely illustrative and nonexhaustive, also vehicle manufacturers' captive parts could fall within its scope.

Importantly, the revised SGL include additional and useful examples of what may qualify as technical information, namely information required to work on advanced driver-assistance systems (ADAS) and battery management systems, as well as activation codes that are needed to install certain replacement parts.

2. **Access to Data.** Connected cars allow innovative remote functions that enable multiple new use cases and business models. To ensure fair competition and being able to keep offering digital services in the independent aftermarket, FIGIEFA is working on both, the horizontal legislation (**Data Act**, to be approved by June this year) and a **Sector-Specific legislation on access to in-vehicle data** (commission proposal to be published by Q3).

The **Data Act** will provide a horizontal regulatory framework and cover almost all categories of connected products, with the objective of ensuring fair access to and use of machine-generated data for users, tackling barriers to access for both consumers and businesses.

The automotive aftermarket and mobility services sector represents an example of an industry being affected by the current lack of "access to data". Fairly regulated access to vehicle data could support the digital transformation of an industry with over 4.5 million jobs, most of which are in SMEs. Currently only one player (i.e., the vehicle manufacturers) uses its privileged position to prevent other stakeholders from effectively accessing data. This has hampered independent service providers in the aftermarket and mobility services sector from the ability to provide competitive digital services to users, applying their own innovation potential.

Since the publication of the Data Act proposal by the European Commission, FIGIEFA has strongly supported its objectives.

The Data Act represents a step forward in the effort to promote a genuine Internal Single Market for data based on fair and competitive rules. However, as a horizontal piece of legislation, it is not, in itself, sufficiently tailored to respond to the specifics of the automotive aftermarket. This is why a future sector-specific legislation for the automotive sector is still needed, as clearly affirmed by various commission studies and the ITRE committee of the European Parliament.

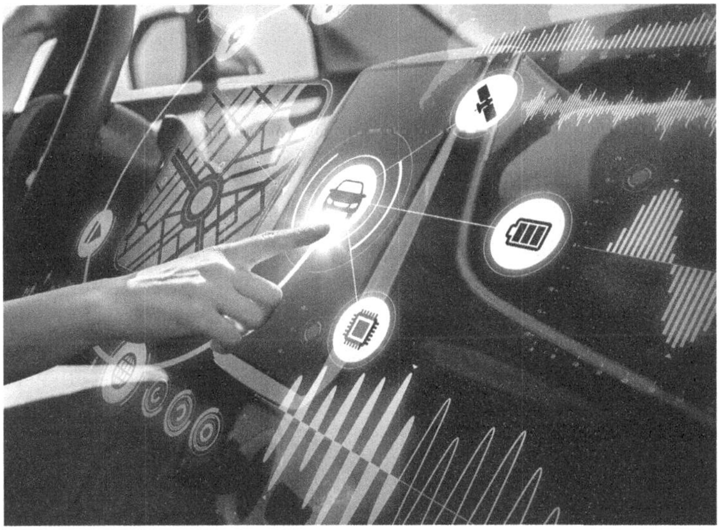

(Source: Shutterstock, used with permission)

3 What Does the Sector Need?

The European Commission has documented through several studies and, most recently, through a Public Consultation and a Targeted Stakeholder consultation, that there are systemic barriers (intrinsic to the vehicle design—and many other structural and behavioral impediments) to fair access of Independent market operators to vehicle-generated data, functions, and resources. Such systemic gatekeepers and other barriers cannot be remedied through the horizontal Data Act.

The Data Act itself emphasizes this very point: "... new rules are needed to ensure that existing vehicle type-approval legislation is fit for the digital age and promotes the development of clean, connected and automated vehicles. Building on the Data Act as a framework for the access and use of data, these rules will address sector-specific challenges, including access to vehicle functions and resources."

Without the definition of stand-alone sector-specific legal and technical rules covering data, functions, and resources, it will not be possible for independent service providers to develop their own competing digital services in advance. Without such rules, there will simply be no competing service providers to whom users could consensually transfer their data, as intended by the Data Act.

The commission forecasts that the vehicle-generated data market could potentially reach €400 billion by 2030. Preventing fair competition in Europe, delaying innovation, and loading more costs on European consumers and businesses for vehicle-generated data services seems an unlikely recipe for Europe leading in that global market.

The commission originally scheduled this legislation for adoption in 2021, and the commission's proposal for the Data Act itself restated the need for sector-specific legislation to address these issues.

The access to in-vehicle data, functions, and resources is not only crucial but absolutely imperative for automotive suppliers in Europe to survive and thrive in an ever-evolving industry. Without fair access, the entire automotive ecosystem, including suppliers and the aftermarket sector, cannot remain competitive versus tech companies that are already dominating the infotainment systems of vehicles, nor can we continue to innovate the components and services that meet new digital demands. Therefore, a sectoral and legally binding regulation on access to in-vehicle data is urgent to guarantee users' freedom of choice, fair competition but also to enable both the deployment of zero-emission mobility and the development of a genuine European business ecosystem that is independent, efficient, and competitive.

The European Commission confirmed they are working on a legislative proposal in this regard that would be ready in the second half of 2023.

3. **Repairs Clause Under the Design Protection**. The Repair Clause is an indispensable precondition to achieve an accessible and affordable right to repair at the service of the circular economy. Consumers should always be able to choose between competing suppliers of spare parts to repair their products. After +30 years of legislative attempts, the European Commission has brought back this issue to the table and FIGIEFA is fighting to secure a real right to repair for European consumers, also when it comes to **visible car parts**.

(Source: Shutterstock, used with permission)

On November 28, 2022, the European Commission published proposals for the revision of the EU Design Directive (COM(2022) 667) and the EU Design Regulation (COM(2022) 666). FIGIEFA appreciates the proposals, especially the introduction of an EU-wide Repair Clause in the Design Directive (Art. 19) and the confirmation of a permanent Repair Clause in the Design Regulation (Art. 20a).

Both legislative proposals ensure on one side full protection of manufacturers' design rights over their products (such as a vehicle) and avoid on the other side monopolies in aftersales markets on visible spare parts (such as vehicle body panels, headlights, and windscreens) by excluding those for the purpose of repair and replacement from such protection. Today such Repair Clauses exist only in some EU Member States. We strongly support the intention of the European Commission to deploy this approach throughout the entire European Union. Harmonized rules for design protection and exemptions are a major step forward to simplify the EU regulatory framework, and a well-rounded Repair Clause would ensure competition and consumer's choice for visible spare parts, submit their prices to competition, promote innovation, and make the right to repair becoming a reality for all European consumers1.

FIGIEFA believes, however, that the intention behind this inclusion could materialize more efficiently with some targeted improvements. Therefore, we call on the European Parliament and the Council to pay attention to the following issues:

- An efficient Repair Clause needs to cover both new and existing designs to fully benefit the consumers. The European Commission itself recognized that there is "no broad economic justification" for maintaining design protection on visible spare parts. Therefore, the European Commission's proposal of a compulsory 10-year transition period for Member States to implement a Repair Clause in national law is not justified. This will prevent the EU-wide Repair Clause in the Design Directive from having any effect on all existing products for another 10 years, denying consumers the benefit of a right to an affordable and accessible vehicle repair. FIGIEFA urgently call on the European Parliament and the Council to agree on a shorter and flexible transition period, leaving Member States the choice of applying the Repair Clause to all designs in advance. A maximum transition period of up to 3 years would provide sufficient time for Member States to transpose the Directive into national law, while still being able to apply the Repair Clause to existing designs at an earlier date.
- It is also important to ensure clarity and legal certainty in the wording of the Repair Clause so that it applies uniformly and unequivocally to the benefit of all European consumers and businesses. The undue restriction of the Repair Clause to "form-dependent component parts of complex products only" (Recital 35), "upon whose appearance the design of the component part is dependent" (Art. 19(1))4, and the unclear and redundant consumers information requirements5, should be removed.

4. **Cybersecurity.** With the rise of connected and automated driving, on one side, and the general increase of new cyberthreats, on the other side, legislators worldwide felt the need to introduce regulation for addressing the issue of cybersecu-

rity in the automotive sector. Whilst fully supporting the protection against cybersecurity threats, FIGIEFA upholds that legislation should, however, not lead to granting vehicle manufacturers an arbitrary control of the cybersecurity implementation.

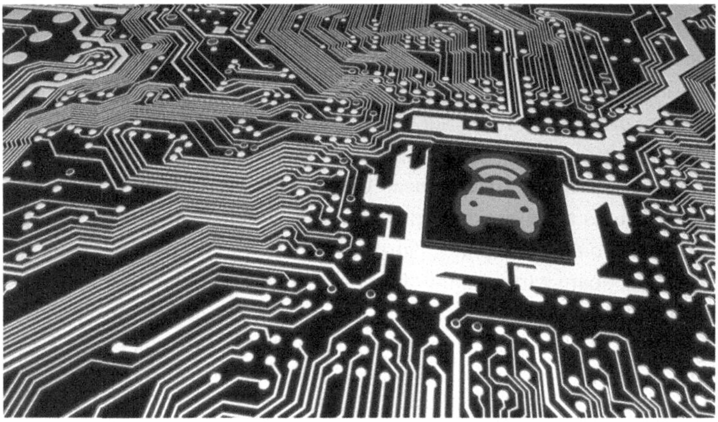

(Source: Shutterstock, used with permission)

UNECE, a body of the United Nations dealing with mobility issues (among other topics), finalized in June 2020 two pieces of legislation on the matter, Regulation No. 155 on "Cybersecurity" and Regulation No. 156 for "Software Updates." These two regulations were transposed into the European Union's legislation in mid-2021 and have been applicable since July 2022 for newly type-approved vehicles and will be applied from 2024 for the existing vehicle park.

With Regulation No. 155, UNECE has established an initial inventory of potential cyberthreats and corresponding mitigation measures. These mitigation measures, however, are not concrete implementation measures, but give the freedom for vehicle manufacturers to implement their own proprietary security controls. They are now allowed to set their own benchmark (i.e., "what is adequate security?") and implement their own proprietary cybersecurity measures as part of vehicle type approval. Thereby, each vehicle manufacturer will create its own cybersecurity management system to set up organizational processes and implement security/software update-related measures for each vehicle type. As a result, vehicle manufacturers can implement access control mechanisms and practices to address cybersecurity concerns.

The proprietary cybersecurity strategy of the vehicle manufacturers could make it impossible to use spare parts from independent sources as they could be rejected by the vehicle in the name of "security." This exclusion could have a profound and negative impact on the entire portfolio of spare parts identified as "cybersecurity relevant" (e.g., any part with electronic components).

First examples are access restrictions to the OBD port via proprietary vehicle manufacturers' security certificates, proprietary vehicle manufacturers' codes (QR codes or software) needed for the activation of spare parts (often with vehicle manufacturers' own diagnostic tools), or the general prevention of remote communication with the vehicle and its data. All these restrictions could now be imposed widely under the legal requirements of cybersecurity protection.

This would prevent independent, multi-brand businesses to conduct a wide range of repair and maintenance services and drive further consumers into the vehicle manufacturers' contracted networks.

FIGIEFA fully supports measures to protect connected vehicles against cybersecurity threats. However, the process of including UNECE Cybersecurity Regulations into the European Union's legislation should not lead to granting vehicle manufacturers an arbitrary control of the cybersecurity implementation. The European Union must take the necessary measures to avoid that the entire automotive aftermarket (and related digital and mobility value chains) is disrupted. This is why FIGIEFA, together with other aftermarket organizations gathered in "AFCAR" (Alliance for the Freedom of Car Repairs), is currently in the process of informing European Union's officials and Member States representatives about this issue. The objective is to ensure that the transposition of the UNECE Regulations into the European Union's legal framework is accompanied by robust implementation clauses to ensure that all stakeholders continue to have the ability to operate, in a nondiscriminatory and competitive manner, whilst addressing cybersecurity. Without such measures, the aftermarket would be at risk. In more detail, FIGIEFA is calling upon decision-makers to ensure, that is,

- Cybersecurity compatibility and interoperability for replacement parts.
- Cybersecurity compatibility and interoperability for multi-brand diagnostic tools.
- Setting up a European Union-wide certification scheme by extending the current SERMI scheme to cybersecurity (including also an approval and authorization scheme for diagnostic tools and any other operator involved in providing mobility services).
- Amending the provisions on the OBD port to define rules for the period of proprietary issuing of security certificates by vehicle manufacturers until the European Union-wide certification scheme is setup.

In parallel, FIGIEFA is organizing technical meetings with spare parts experts to work on concrete implementation requirements for cybersecurity and is arranging informational webinars to prepare FIGIEFA members for cybersecurity. Last but not least, FIGIEFA commissioned an independent cybersecurity study, with the aim to show that it is perfectly possible to have the highest level of cybersecurity protection, whilst at the same time allowing independent communication with the vehicle, its data, and resources.

5. **Sustainability.**

(Source: Shutterstock, used with permission)

Environmental, social, and governance (ESG) issues are increasingly affecting the automotive aftermarket as well. Against this background, the two sector associations—CLEPA, the European Association of Automotive Suppliers, and FIGIEFA, the European Association of the Independent Parts Distribution—initiated the Forum on Automotive Aftermarket Sustainability (FAAS). It is a collaborative endeavor aiming to provide a platform where all participants in the automotive aftermarket can exchange ideas, cultivate, and adopt sustainable practices.

The FAAS aims to establish itself as the foremost sustainability initiative within the aftermarket sector. To accomplish this, the FAAS established the following shared objectives:

- Engage and unite all relevant stakeholders in the automotive aftermarket sector, leveraging our collective influence to foster a more sustainable aftermarket.
- Propel transformation within the automotive aftermarket supply chain by offering sustainable solutions and contributing to global initiatives such as the energy transition, low-carbon production, pollution reduction, and the circular economy.
- Bridge knowledge gaps within the sector and equip companies with the necessary tools to actively participate in sustainability efforts.
- Address various crucial topics, such as tracking and reducing CO_2 emissions, optimizing packaging, promoting remanufacturing and recycling practices, and optimizing logistics.

To accomplish these objectives, the FAAS prioritizes the critical concerns surrounding independent aftermarket operations. Consequently, four distinct working groups were established, each with its specific focus:

- **Working Group 1** is preparing a study to assess the CO_2 footprint of the automotive aftermarket sector and the CO_2 impact of maintaining a vehicle on the road for longer vs. replacing it with a new vehicle.
- **Working Group 2** is addressing existing obstacles that hinder the remanufacturing of components through improved knowledge and understanding of this practice while emphasizing its significant contribution to the circular economy, with the ultimate goal to increase the market share of remanufactured products.
- **Working Group 3** is developing a Product Carbon Footprint methodology that can be applied to all players in the automotive aftermarket sector and be consistent with other initiatives in Europe.
- **Working Group 4** is focusing on optimizing supply chain logistics by developing relevant partnerships between suppliers and wholesalers and preparing recommendations for reducing the environmental footprint of shipping.

The FAAS is open to all players in the automotive aftermarket. For further information, write to laurence@ext.figiefa.eu

Signed by Álvaro de la Cruz, Head of Communications of FIGIEFA.

Alvaro.delacruz@figiefa.eu

Álvaro de la Cruz Tomás is a political scientist by training who has worked in communication and marketing for more than 10 years. He worked for Spanish president Mariano Rajoy's office from 2012 to 2017 and after he was Communications Lead at EIT Digital. He led the agency's team working on the EPP's 2019 European Elections, then joined the Martens Centre in March 2020, and FIGIEFA in 2022. Álvaro holds a BA in Political Science by Universidad Complutense de Madrid, a Certificate on European Political Studies by Sciences Po Strasbourg, and a Master on Digital Marketing by EAE Business School. The content of this chapter and situation of the legislative dossiers forementioned were written in July 2023.

Part II

Perspectives on Knowledge, Education and Talent Development

Marketing Information Systems in the Automotive Aftersales Market

Zoran Nikolic

Abstract

This chapter provides a comprehensive overview of sectoral information and explores examples of marketing information systems (MIS) for managerial decision-making in the automotive marketing industry. It sheds light on the significance of accurate, trustworthy, and comprehensive data evidence required to support professional managers and entrepreneurs in their assessment, planning, and decision-making processes concerning strategies, products, customers, and new markets. By highlighting the importance of reliable data and evidence, this chapter contributes to the understanding that effective decision-making relies on a solid foundation of accurate information. It emphasizes the role of MIS in providing managers with the necessary tools and insights to make informed choices that drive business success. The chapter underscores the value of broad data evidence in assessing market dynamics, identifying opportunities, and developing strategies that align with customer needs and market trends.

1 Introduction

The automotive aftersales market refers to the market for products and services related to the maintenance, repair, and retrofit of vehicles after their initial sale. It is highly competitive and constantly evolving as modern technologies and products are developing. Initially developed as the marketing instrument (service) of the automotive industry, the automotive aftersales business has become increasingly emancipated from the parent automotive business. This "second" market has

Z. Nikolic (✉)
Wolk & Nikolic After Sales Intelligence GmbH, Cologne, Germany
e-mail: Zoran.Nikolic@wolk-aftersales.de

M. Elo, F. Katsardis (eds.), *Automotive Aftermarket*, Management for Professionals, https://doi.org/10.1007/978-3-031-62419-3_7

become independent with its own rules, market environment, players, and logics. Figure 1 offers a more detailed look at the sales structures in the automotive aftermarket and positions the most important players.

To succeed in this market, companies need to leverage various marketing information to inform their strategies and decisions. And, to carry out their tasks along the strategic planning process, marketing managers need an instrument for collecting, analysing, and utilizing this information—**a marketing information system.**

The objective of this chapter is to illustrate the specific types and formats of market information that are essential for marketing managers operating in the automotive aftermarket.

The initial section provides an explanation of the theoretical foundation underlying marketing information systems (MIS). It delves into the concepts and principles that form the basis of MIS and its relevance in the field of marketing.

The subsequent segment focuses on a practical demonstration of the necessity for marketing information systems (MIS) in the aftermarket. Through a hypothetical case study, it aims to shed light on the information requirements and available sources that can assist marketing managers in making informed decisions within this industry.

In the process of curating this content, the specific and relevant consultancy practices have been carefully chosen. Additionally, some situations and tasks that specifically pertain to the role of an aftermarket manager are included. These selections have been made with the aim of providing valuable insights and practical guidance in this field.

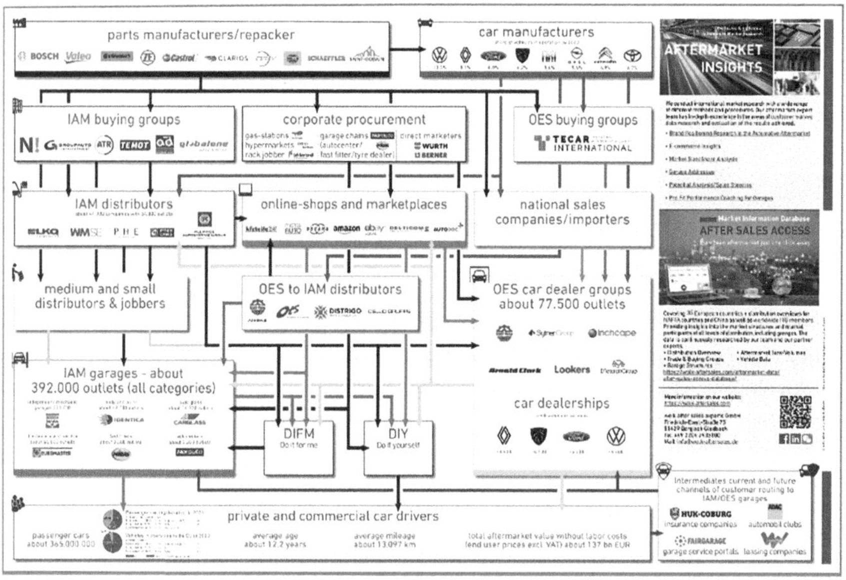

Fig. 1 Aftermarket structure in Europe. (Source: (Wolk After Sales Experts, 2021)

By examining the theoretical framework and presenting a practical application, this chapter provides a comprehensive understanding of the importance and utility of marketing information systems (MIS) for marketing managers operating in the aftermarket. It highlights the significance of having access to relevant information and data to drive effective decision-making and achieve business objectives.

2 What Are the Marketing Information Systems?

For the managerial and entrepreneurial decision-making context, **knowledge and information** hold significant importance in shaping the outcomes and success of businesses.

Knowledge refers to the understanding, skills, and expertise acquired through experience, education, and learning. It provides a foundation for decision-makers to comprehend various aspects of their industry, market dynamics, and organizational operations (Drucker, 2004).

Information, on the other hand, refers to data, facts, and insights that are collected, processed, and organized to generate meaningful and actionable knowledge. It serves as the raw material for decision-making, enabling managers and entrepreneurs to assess risks, identify opportunities, and make informed choices (Grant, 1996) To make well-founded decisions with far-reaching consequences and achieve maximum impact, a marketing manager must rely on solid information and facts.

In summary, knowledge and information are invaluable assets for managerial and entrepreneurial decision-making. They provide insights, understanding, and evidence-based guidance, enabling decision-makers to navigate complex challenges, seize opportunities, and drive the success of their organizations (Spender, 1996). Data-based decision-making is, therefore, an integral part of the planning process for a modern marketing strategy.

To effectively fulfil their responsibilities throughout the strategic planning process, marketing managers rely on a robust marketing information system. This system serves as a vital tool, equipping managers with the necessary data and information to steer their marketing initiatives in the right direction.

A marketing information system consists of three components (Kotler & Keller, 2012, S. 67–92):

- **Internal recording system**, which includes relevant in-house-generated information of operational company business, for example, product sales, prices, profit margins, etc.
- **Marketing intelligence system**, a set of sources to provide relevant information about developments in the marketing macroenvironment.
- **Marketing research system,** which allows for findings in the microenvironment relevant to a specific product, brand, or describes single relevant marketing situation.

The role of the marketing information system is to assess the managers' information needs, develop the needed information, and distribute that information in a timely manner. It is the best way of monitoring the external and internal information.

However, there is a clear and notable disparity in the management and utilization of management information systems (MIS) between small- and medium-sized enterprises (SMEs) and large companies. The difference is influenced by factors such as resources, technology adoption, and organizational structure. The financial capabilities of large companies enable them to afford advanced software, CRM tools, and market research capabilities. In contrast, SMEs may have limited financial resources, leading them to rely on simpler and cost-effective solutions for their marketing information gathering and analysis needs (Kotler P. & Keller K. L., Marketing Management 15th ed., 2016).

Here are some key distinctions:

Resource allocation: Large companies typically have more resources at their disposal, allowing them to invest in advanced marketing information systems. They can afford sophisticated software, customer relationship management (CRM) tools, and market research capabilities. In contrast, SMEs may have limited financial resources and may rely on simpler and more cost-effective solutions for gathering and analysing marketing information.

Data collection and analysis: Large companies often have dedicated marketing research teams or departments that focus on gathering and analysing market data. They may conduct extensive market research studies, customer surveys, and data mining activities to gain insights into consumer behaviour and market trends. SMEs, on the other hand, may rely on more streamlined approaches such as online surveys, social media listening tools, and utilizing third-party market research reports to gather marketing information.

Technology adoption: Large companies tend to have a higher propensity for adopting advanced technologies and marketing automation tools. They may utilize sophisticated analytics platforms, artificial intelligence (AI), and machine learning algorithms to process and derive actionable insights from vast amounts of marketing data. SMEs may have a more limited adoption of such technologies due to budget constraints or lack of expertise, resulting in a relatively simpler marketing information system.

Organizational structure: Large companies often have a more complex organizational structure with multiple departments and divisions. In such cases, their marketing information systems may need to integrate and consolidate data from various sources across different units. In contrast, SMEs typically have a flatter organizational structure, which may result in a more centralized marketing information system where data collection and analysis are relatively more straightforward.

Decision-making process: In large companies, marketing information systems often support decision-making at different levels of management, including strategic, tactical, and operational decisions. The availability of comprehensive marketing data enables managers to make informed decisions on product development, pricing strategies, promotional activities, and market segmentation. In SMEs, decision-making may be more agile and driven by the owner or a small management team, with a focus on more immediate and tactical marketing decisions.

It is important to note that these differences are generalizations, and the specific characteristics of internal recording systems can vary widely among SMEs and large companies based on industry, organizational structure, and individual practices.

But regardless of the size or type of business, the overall objective of a marketing information systems user is to leverage data and technology to improve marketing effectiveness, enhance customer insights, and drive business growth. The specific tasks they undertake will be tailored to the unique needs and goals of their organization within the marketing context.

2.1 Internal Recording System

Internal recording systems are an integral part of management accounting. They serve as the fundamental tools employed to gather, store, and organize internal data. By capturing financial and nonfinancial information, these systems establish the groundwork for analysis and decision-making processes. Management accounting, on the other hand, encompasses a broader scope. In essence, while internal recording systems focus on data gathering and storage, management accounting expands upon this foundation to encompass the analysis and utilization of that data for informed decision-making (Atkinson A. A., Kaplan R. S., Matsumura E. M., & Young S. M., 2013 sixth Edition, S. 25–38).

Internal recording systems typically include activities such as.

Recording financial transactions: This involves capturing and documenting financial transactions such as sales, purchases, expenses, and payments. It includes maintaining journals, ledgers, and other accounting records.

Inventory management: Internal recording systems help track and record inventory levels, including stock inflows and outflows, stock adjustments, and inventory valuations.

Payroll and employee records: Internal recording systems manage employee-related data, including payroll records, employee benefits, attendance, and personnel information.

Customer and sales data: Recording systems capture customer information, sales orders, invoices, and sales history, providing valuable insights into customer behaviour and sales performance.

Compliance and regulatory requirements: Internal recording systems ensure compliance with financial reporting standards, tax regulations, and other legal obligations by maintaining accurate and reliable records.

The information should serve to determine the advantages of the new strategy, record the company's capabilities, and map the process steps for the implementation of the strategy. Besides the already mentioned in-house-generated data for product sales, prices, profit margins, etc., other examples of the strategic internal information are capital, human resources, technology, organization, corporate culture, etc. (Bea & Haas, 2016, S. 327–352).

In this context, we assume the presence of internal information, acknowledging the quality variances between SMEs and large companies.

2.2 Marketing Intelligence System

Within the rapidly changing global picture, marketers must monitor **major environmental forces**: demographic, economic, social-cultural, natural, technological, and political-legal and find the opportunities for identifying trends and megatrends and changes that have long-lasting influence. (Kotler P., Keller K. L., Marketing Management 14th ed., 2012, S. 67–92).

The sources of information in a marketing intelligence system can include various channels such as market research reports, industry publications, government data, competitor analysis, customer feedback, social media monitoring, and economic indicators. The system involves the collection, analysis, and interpretation of this information to generate valuable insights and actionable intelligence. Marketing intelligence plays a crucial role in centralizing data sets from various systems into a single database. By consolidating data from different sources such as sales, customer interactions, social media, and market research, marketing intelligence facilitates a comprehensive understanding of the market landscape. Constantly evolving technologies and artificial intelligence make it possible to evaluate and use marketing data more and more effectively: the solution lies in marketing intelligence platforms. These bring together data sets from different platforms and sources. This diverse information is then organized, visualized, and evaluated within the software. These analyses result in a well-founded decision-making aid (Jaedtke, 2020). This centralized database allows for effective data analysis, trend identification, and performance measurement. By automating the data collection and analysis process, marketing intelligence systems streamline the decision-making process, providing timely and relevant insights into marketing professionals.

Marketing intelligence should act as the **guiding light** for marketing teams' decisions. By collecting and analysing contextual data about customers, industry trends, and behaviours, marketers can gain a holistic understanding of what is and is not working. This can give businesses an important advantage over competitors, educate them about their target audience, and evaluate insights into their various products (Evolution, Marketing, 2022).

With the aid of marketing intelligence, organizations can make data-driven decisions, optimize marketing strategies, allocate resources efficiently, and identify areas for improvement. By leveraging a comprehensive and centralized database, marketing intelligence empowers organizations to stay agile, adapt to **market** changes, and gain a competitive advantage.

2.3 Marketing Research System

Overall, market research is needed whenever businesses seek **reliable and evidence-based insights** to support decision-making, understand customers, assess markets, evaluate competition, and identify growth opportunities.

The marketing research system typically consists of the following key components:

- Problem definition: Clearly defining the marketing problem or research objective is the initial step in the marketing research system. It involves identifying the specific information needed to address the problem or objective.
- Research design: This step involves determining the appropriate research design and methodology to collect and analyse data. It includes decisions about the research approach (quantitative, qualitative, or mixed methods), data collection methods (surveys, interviews, observations), and sampling techniques.
- Data collection: Gathering data is an essential component of the marketing research system. Data can be collected through primary research (directly from respondents) or secondary research (existing data sources). Primary data collection methods include surveys, interviews, focus groups, and observations.
- Data analysis: Once the data is collected, it needs to be processed and analysed. This involves applying statistical techniques, qualitative analysis, or data mining methods to uncover patterns, trends, and relationships within the data.
- Insight generation: In this step, the findings from the data analysis are interpreted and transformed into actionable insights. The research results are used to answer research objectives, make informed marketing decisions, and develop effective marketing strategies.
- Reporting and presentation: The final step involves communicating the research findings through comprehensive reports or presentations. This includes summarizing the research process, key findings, and recommendations for management or stakeholders.

Market research can be classified into various types, depending on the specific objectives and areas of focus such as online–offline research, quantitative–qualitative research, descriptive research, and exploratory research. But all these types can be defined as meta-categories for the most common classification in primary and secondary research.

- Primary research involves collecting new data directly from the source. It can be conducted through surveys, interviews, focus groups, or observations tailored to the specific research objectives. Primary research provides firsthand and customized information for the research purpose.
- Secondary research involves utilizing existing data and information from various sources, such as government publications, industry reports, academic studies, and market research databases. It provides a cost-effective way to gather information and gain insights without conducting new research.

In conducting market research, companies must choose a research approach (observational, focus group, survey, behavioural data, or experimental) and research instruments (questionnaire, qualitative measures, or technological devices). In addition, they must decide on a sampling plan and contact methods (Kotler & Keller, 2012).

Most companies use a combination of marketing research resources to study their industries, competitors, audiences, and channel strategies.

In summary, it is important to emphasize that effective marketing research embodies the principles of the scientific method, creativity, the utilization of diverse research methods, precise model development, thorough cost–benefit analysis, a critical mindset, and an unwavering commitment to ethical practices.

3 Who Needs the Marketing Information Systems and Why?

Marketing information systems are essential for various individuals and entities involved in the field of marketing. The following are key stakeholders who benefit from marketing information systems:

- **Marketing managers:** Marketing managers require marketing information systems to make informed decisions and develop effective marketing strategies. These systems provide valuable data and insights regarding market trends, customer behaviour, competitor analysis, and product performance, enabling managers to optimize their marketing efforts and achieve desired business outcomes.
- **Sales teams:** Sales teams rely on marketing information systems to understand customer preferences, identify potential leads, and tailor their sales approach accordingly. These systems provide valuable customer data, such as demographics, buying habits, and purchasing patterns, which help sales teams enhance their targeting and improve customer engagement.
- **Product development teams:** Marketing information systems aid product development teams in gathering market feedback, identifying consumer needs and preferences, and evaluating the performance of existing products. By leveraging these systems, product development teams can make informed decisions regarding product features, improvements, and innovations, ultimately leading to the creation of successful products that meet customer demands.
- **Advertising and promotion specialists:** Marketing information systems offer valuable insights into target audience characteristics, media consumption habits, and advertising effectiveness. This information helps advertising and promotion specialists create targeted and impactful campaigns, optimize advertising budgets, and measure the return on investment (ROI) of their marketing activities.
- **Market researchers:** Market researchers heavily rely on marketing information systems to gather, analyse, and interpret data related to market trends, consumer behaviour, and competitive landscapes. These systems enable researchers to conduct in-depth studies, surveys, and experiments, facilitating the acquisition of valuable market intelligence.
- **Executives and decision-makers:** Marketing information systems provide executives and decision-makers with comprehensive reports and analysis, aiding them in strategic planning, resource allocation, and evaluating the overall performance of marketing initiatives. These systems offer key performance indicators

(KPIs), market forecasts, and other relevant metrics, enabling executives to make data-driven decisions.

In essence, marketing information systems benefit a wide range of stakeholders, empowering them with timely, accurate, and actionable information, enhancing marketing outcomes.

In the given context, the tasks of a **marketing information systems user** can indeed vary depending on the size and type of business or organization they are employed. The nature and scope of their responsibilities can be influenced by factors such as the organization's industry, target market, and overall marketing strategy.

4 What Kind of Data Does the Automotive Aftermarket Management Need?

The marketing strategy planning process requires a variety of information to be gathered. Marketing management directs its attention towards a specific product within a single market, aiming to develop a comprehensive marketing strategy that aligns with the product's objectives. This strategy serves as the primary instrument for effectively controlling and monitoring the marketing efforts deployed.

Next, we introduce a business case that illustrates how data is used for evidence-based decision-making.

5 Introducing a Descriptive Case from the Business

Company X, a manufacturer of maintenance products operating with its own brand in the aftermarket, is currently encountering various challenges. An analysis of internal data reveals two key findings. Firstly, all markets indicate a decreasing significance of the product, along with a challenging market position for the brand itself. Secondly, although brand recognition remains high, there is a noticeable decline, and brand awareness does not consistently yield positive outcomes.

To address these issues, the marketing management of Company X has made the decision to initiate a transformative process by reshaping the existing marketing strategy. The responsibility for creating a new marketing strategic plan has been assigned to the marketing manager. As part of this effort, the marketing management will launch a Pan-European brand repositioning project with the objective of revitalizing the brand. This project aims to redefine the brand's identity and rejuvenate its presence in the market.

The marketing strategic-planning process consists of the following steps (Kotler & Keller, 2012, S. 46–53):

- Development of a specific marketing mission which fits with the company vision and mission.

- Analysis of internal and external environment (SWOT).
- Based on that analysis, formulation of possible goals.
- Strategy finalization.
- Business plan as a detailed marketing programme.
- Implementation process.
- Control monitoring.

5.1 Mission

Defining a specific mission that aligns with the broader company mission is an essential initial phase in the strategic planning process. To establish an appropriate mission, both internal and external project teams engaged in brainstorming sessions. Following a thorough evaluation, the management reached a decision on the mission statement:

> Our mission is to enhance brand awareness and cultivate customer loyalty through the consistent delivery of innovative and exceptional-quality products. We are dedicated to empowering our customers by equipping them with the knowledge, tools, and unwavering support necessary for their success.

5.2 SWOT Analysis

The SWOT analysis is a strategic planning tool that provides a structured framework for assessing the internal strengths and weaknesses of an organization, as well as the external opportunities and threats it faces in its operating environment (Kotler P., Keller K. L., Marketing Management 14th ed., 2012, S. 48).

Internal sources for conducting a SWOT analysis refer to the information and data that come from within the organization itself. These sources provide insights into the organization's internal strengths and weaknesses.

The process starts with the description of the marketing situation with relevant background information.

As mentioned earlier, the management has obtained internal information through internal reporting.

To effectively execute the total marketing performance, project management requires additional external information pertaining to various key areas, including

- Market macroenvironment.
- Market development.
- Product insights.
- Competition insights.
- Distribution policy.

5.2.1 Macroenvironment Analysis

Based on the macroenvironment analysis of major forces demographic, economic, social-cultural, natural, technological, and political-legal, the market trends must be developed.

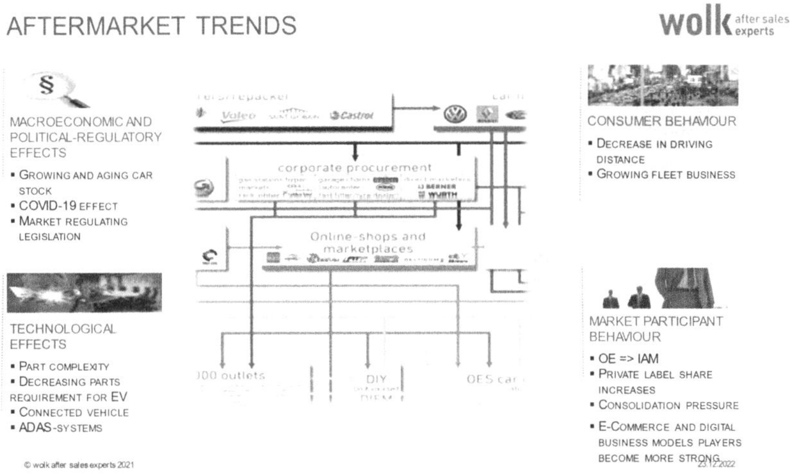

Fig. 2 Example: aftermarket trends 2021. (Source: (BCG, CLEPA, Wolk After Sales Experts, 2021)

In the automotive aftermarket, very often the branch associations such as **CLEPA, FIGIEFA, and ACEA** carry out this kind of analysis for the market participants.

An example of a trend development is shown in Fig. 2 (BCG, CLEPA, Wolk After Sales Experts, 2021).

The comprehensive analysis of external market forces beyond consumer markets, such as B2B trends and political influences, reveals valuable insights. The required data for this analysis are sourced and accessible from official registers, for example, Eurostat. Additionally, anecdotal insights and opinions from market leaders provide valuable information on technological effects and market participants' behaviour. A similar approach is applied to analyse the forces within consumer markets, ensuring a thorough understanding of the market dynamics.

Excurse: Consumer, Customer, Distribution Partner

The figure provided highlights the significance of consumer behaviour as a key trend in the consumer market. From the perspective of an IAM (Independent Aftermarket) parts manufacturer, it is important to distinguish between the consumer, customer, and distribution partner.

Although the aftermarket is primarily a business-to-business (B2B) market, B2B marketers face similar challenges to their business-to-consumer (B2C) counterparts. Understanding customers is crucial for both B2B and B2C marketers.

In the case of the IAM, the end consumer may be a car driver, but the target audience for marketing activities is the workshop (customer). The parts distributor in this hierarchy is the distribution partner.

On the contrary, in the original equipment (OE) channel, where the product and service are branded under the same name, the customer is indeed the end consumer. The relationship in the OE channel is primarily focused on serving the needs and preferences of the individual consumers, aligning with the dynamics of the consumer market.

Vehicle Fleet

In the context of aftermarket marketing, an essential aspect for managers is to comprehend the scale and composition of the vehicle fleet, which aids in understanding consumer behaviours. In Europe, there are only a few professional providers who analyse deeply the vehicle data and offer the solutions, for example, **Tec Alliance or S&P Global**. The official registers provide unfortunately only fragmented information on national vehicle parks.

Fig. 3 illustrates a segment from the After Sales Access, a specialized dashboard within the Aftermarket Intelligence System developed by Wolk After Sales Experts. This dashboard offers basic information regarding the European aftermarket. Specifically, the business intelligence (BI) dashboard presents current and historical data on the size and composition of the vehicle fleet across Europe. This knowledge equips marketing managers with valuable insights for formulating effective marketing strategies tailored to the specific characteristics of the European vehicle park.

Forecasting Vehicle Fleet

Alongside current and historical data, it is crucial to also have a grasp of the future trajectory of the vehicle fleet. Understanding the future trajectory of the vehicle fleet aids project managers in assessing potential risks and opportunities that may arise. It allows them to proactively adapt their plans, allocate resources efficiently, and mitigate potential challenges that could impact the project's success.

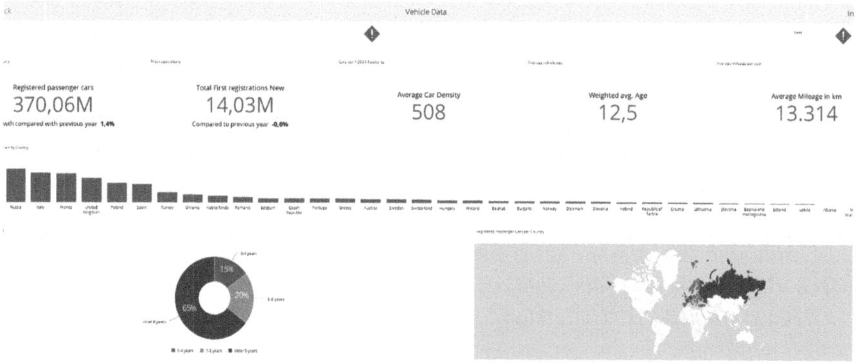

Fig. 3 Example dashboard vehicle data in Europe. (Source: (Wolk After Sales Experts, 2022a, 2022b)

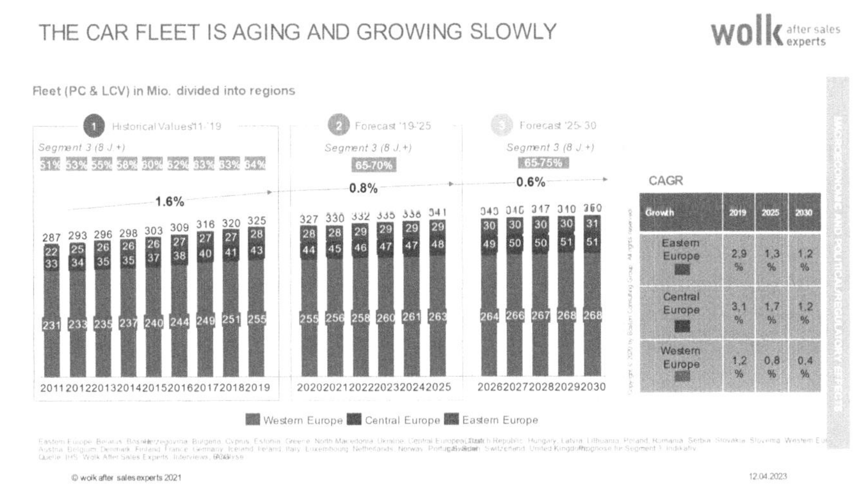

Fig. 4 Vehicle park forecast up to 2030. (BCG, CLEPA, Wolk After Sales Experts, 2021)

Achieving an understanding of the future trajectory of the vehicle fleet requires the use of various methods, with time-series analysis playing a pivotal role. This approach utilizes historical data on market demand to forecast future trends, assuming that past patterns will persist. By applying this method, businesses and analysts can gain valuable insights into the projected development of the vehicle park. This information assists project managers in formulating strategies that align with anticipated market demand.

Figure 4 shows a sample of vehicle data prediction from 2021.

Based on the understanding that there will continue to be a growing demand from the end consumer in the future, the project manager can confidently give the green light to proceed further with the project. This perspective acknowledges the market potential and the anticipated positive response from consumers. With the expectation of sustained demand, the project manager can move forward, knowing that there is a favourable market environment that supports the success and viability of the project.

5.2.2 Market Development

Understanding the current overall market development is probably the most important task for marketing managers. In addition to comprehending the vehicle park, relevant market data should encompass key aspects such as the market size, which represents the total sum of products or services purchased by consumers within a specific time frame.

Historical market development, based on past data, provides insights into the market's size and volume over time. This information enables the project manager to assess whether the market is expanding or shrinking, facilitating a deeper understanding of consumer behaviour, and supporting informed decisions regarding production and pricing strategies. Furthermore, it allows the manager and their team to evaluate the market's strength and make informed decisions regarding marketing policies tailored to that market.

Excurse: Aftersales ACCESS BI Tool

Our company offers a comprehensive BI Dashboard that provides standardized market volume data on an annual basis. This valuable tool aids management in understanding the performance of individual markets in terms of vehicle repairs and retrofits. Our team of dedicated market analysts closely monitors all European markets and gathers critical information that influences market development. This includes macroeconomic data, road conditions, investments in road infrastructure, road length, driven mileage, fuel consumption, and vehicle park data, among other factors. All this relevant information is collected and stored in a centralized database.

This aftermarket volume dashboard is shown in Fig. 5.

Market forecasts predict the future demand for a particular product or service based on various factors (Kotler P., Keller K. L., Marketing Management 14th ed., 2012, S. 85–91).There are several methods that businesses and analysts use to forecast market demand, including

- Time-series analysis: This method involves analysing past data on market demand to predict future demand. It assumes that past trends will continue.
- Survey research: This method involves collecting data from consumers through surveys to understand their future purchase intentions.

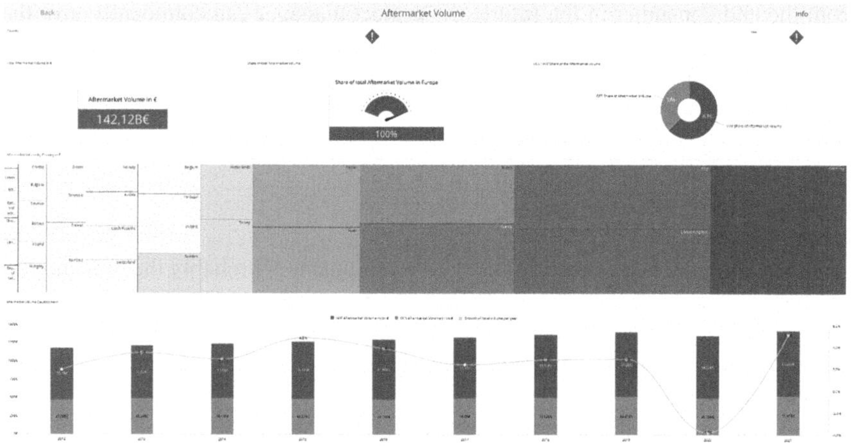

Fig. 5 Abstract BI dashboard aftermarket volume in Europe 2021 including aftermarket development. (Source: (Wolk After Sales Experts, 2022a, 2022b)

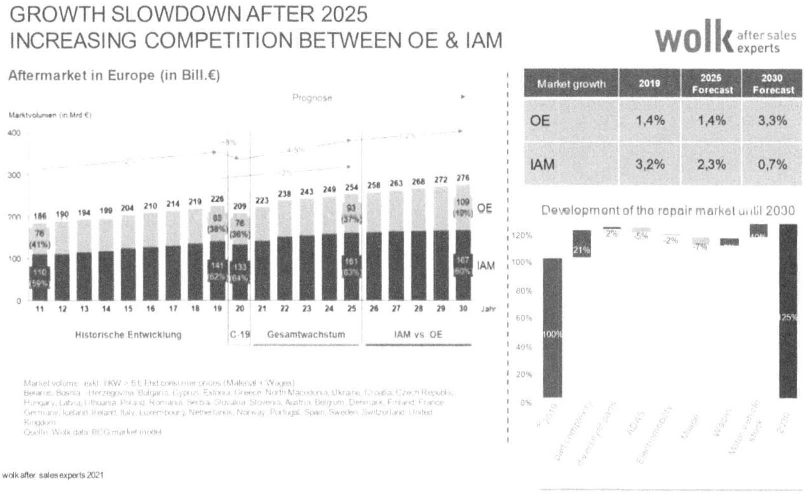

Fig. 6 Example: aftermarket value forecast up to 2030. (Source: (BCG, CLEPA, Wolk After Sales Experts, 2021)

- Expert opinion: This method involves consulting with experts in the industry or market to get their predictions about future demand.
- Econometric modelling: This method involves using statistical models to forecast demand based on economic and market conditions.

The aftermarket data should be generated through a specialized marketing intelligence system (Fig. 6).

Overall, the positive market indicators of a growing vehicle park and increasing aftermarket demand provide a strong rationale for the marketing manager to proceed with the brand relaunch project, taking advantage of the favourable business environment and seizing the business opportunity that lies ahead.

5.2.3 Product Insights

Within the SWOT analysis, the project manager requires information on the performance of the market for a specific product, such as maintenance products in our example. This entails understanding market dynamics, trends, and factors influencing the demand and competition within the market for maintenance products.

Product insights provide a comprehensive overview of various aspects related to the sales, prices, margins, costs, and profits associated with each product line. Additionally, the product policy encompasses critical elements such as product design, style, quality, packaging, size, brand name, range planning, product services, and warranties. These tasks involve collaboration with cross-functional teams to gather and analyse essential information.

Responsibility for managing these tasks lies with product, brand, and communication management. These teams work together to collect, synthesize, and interpret

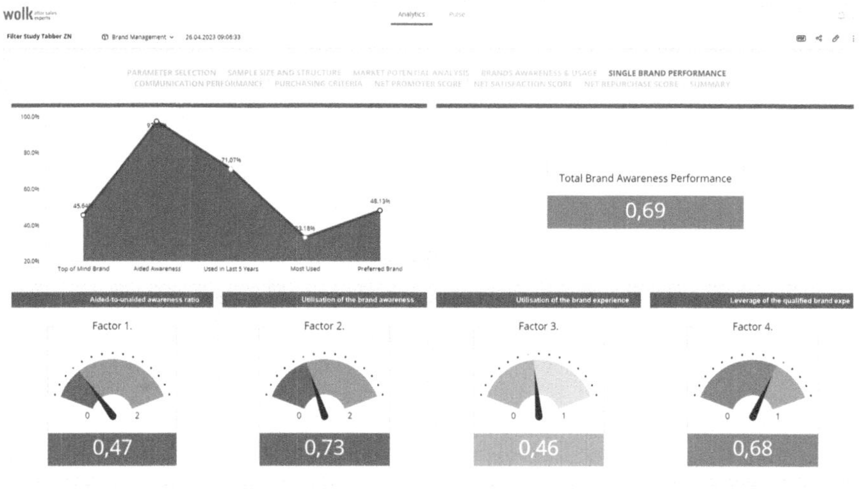

Fig. 7 Brand performance BI Tool: brand performance dashboard. (Source: (Wolk After Sales Experts, 2022a, 2022b)

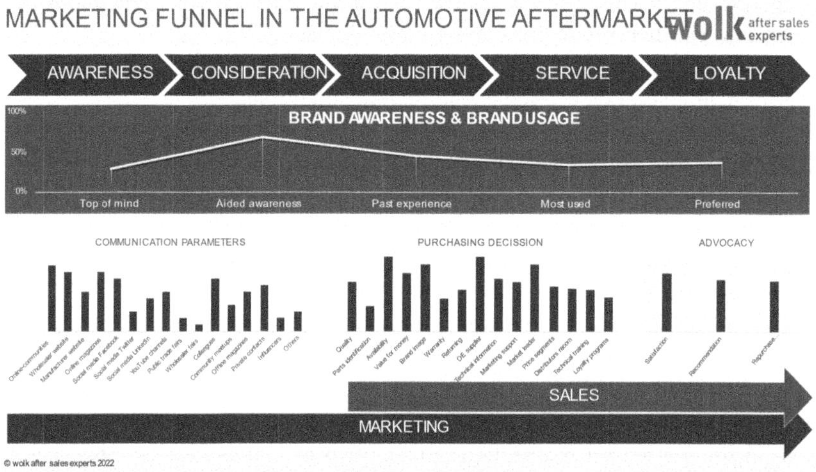

Fig. 8 Brand research structure of Wolk After Sales Experts

data to make informed decisions regarding product strategies, brand positioning, and effective communication approaches.

Product Management

Product management develops and observes the life cycle of a product.

Overall, the role of a product manager is to ensure that the product meets the needs of its customers and the business, and that it is developed and delivered efficiently and effectively.

A product manager must be able to track the performance of the product and make data-driven decisions to optimize its performance and meet business goals. This may involve analysing metrics such as customer satisfaction, user adoption, and revenue (Figs. 7 and 8).

Excurse: Brand Performance BI-Tool
Wolk After Sales Experts conduct a standardized workshop survey to evaluate brand performance, following the structure outlined in Fig. 8. This survey aligns with the general marketing funnel in the automotive aftermarket. The primary objective of this survey is to equip management with practical tools that enable a comprehensive comparison of our brand's performance with that of competitors across various relevant aspects. Communication aspects influence brand awareness, product features impact sales and usage, and advocacy measurements provide insights into customer loyalty. The results of this primary research are delivered as a multiclient product through our interactive BI-Tool. Below, you will find a few examples of the aspects covered in the survey.

Brand Management

Brand management is the process of building, nurturing, and protecting a brand. It involves creating and maintaining a consistent brand image and message across all customer touchpoints, including advertising, marketing materials, and the customer experience.

The key responsibility of a brand manager is monitoring brand performance. A brand manager must track the performance of the brand and analyse metrics such as customer satisfaction, brand awareness, and market share.

Overall, the role of a brand manager is to ensure that the brand is consistently presented in a positive and consistent manner, and that it effectively meets the needs and expectations of its customers.

Communication Management: Brand Marketing

Brand marketing is the practice of promoting a brand and its products or services to consumers to establish or increase brand recognition and loyalty. It involves developing and implementing marketing strategies and campaigns that are designed to communicate the value and benefits of a brand to consumers.

There are several different tactics that can be used in brand marketing, including advertising, social media marketing, content marketing, and public relations. To be effective, brand marketing efforts should be consistent, relevant, and aligned with the overall brand identity and messaging. Effective brand marketing can help companies establish strong, positive associations with their brand in the minds of consumers. This can lead to increased brand awareness, customer loyalty, and ultimately, higher sales and profits. It is an important part of the overall marketing process and can help companies differentiate themselves from their competitors in the market.

Fig. 9 Brand performance BI Tool: communication dashboard. (Source: (Wolk After Sales Experts, 2022a, 2022b)

At this stage, the management undertakes individual market research to obtain pertinent communication insights, specifically evaluating and assessing performance in both online and offline media channels. Besides our company, there are a few other specialized companies active in providing market research for automotive aftermarket, for example, **BBE Automotive and GIPA**.

The research aims to provide valuable information on the effectiveness of various communication strategies, advertising campaigns, and promotional activities across different media platforms. By analysing the data collected, the management can gain insights into the impact of their communication efforts, identify successful channels, and make informed decisions to optimize their marketing communications in both the online and offline domains (Fig. 9).

Product Value and Volume Size

It is crucial to determine the size of the product market, assess the brand's market share, and identify the potentially obtainable market share (SOM). The marketing management needs to comprehend whether the brand's sales are experiencing growth or decline. In case of growth, it is essential to determine whether the performance is higher or lower compared to competitors within the market.

By conducting dedicated primary market research, valuable insights can be gained regarding replacement frequencies within the relevant market. This information, when combined with data on the vehicle park or the total workshop population, can provide a comprehensive understanding of the product volume size. This combination of data sources enables marketing management to assess the scale and potential demand for their products within the market (Fig. 10).

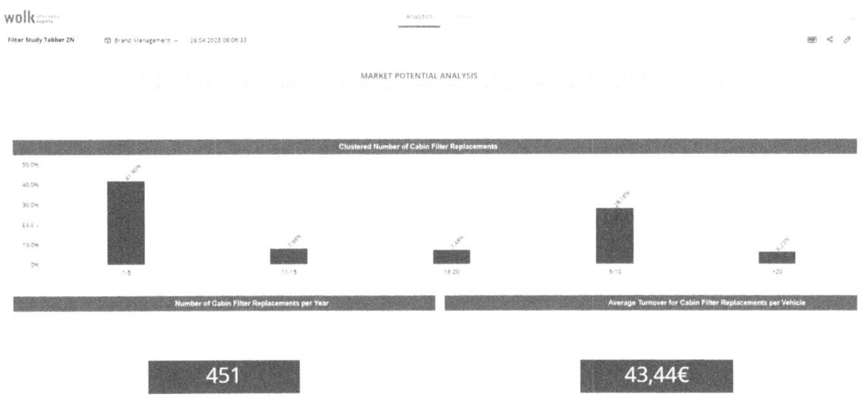

Fig. 10 Brand performance BI Tool: market potential dashboard. (Source: (Wolk After Sales Experts, 2022a, 2022b)

Managing Pricing Strategies and Programmes

Pricing is the process of setting the price for a product or service. It involves determining the value that the product or service provides to customers, and setting a price that reflects that value while also considering the costs of production and distribution.

Managing pricing strategies and programmes involves deciding which pricing approach is most appropriate for a business's products or services, and then implementing and monitoring the chosen strategy to ensure that it is effective. This may involve market research, data analysis, and ongoing adjustment of prices as needed.

Pricing strategies and programmes refer to the approaches that businesses use to set the prices of their products or services.

Factors that can influence pricing decisions are

- The cost of producing and distributing a product is a key factor in determining its price.
- The level of demand for a product or service can influence its price. If demand is high, a company may be able to charge a higher price.
- The prices of similar products offered by competitors can also impact a company's pricing decisions.
- Economic conditions, such as inflation or deflation, can also affect pricing decisions.
- The price of a product may also be influenced by how customers perceive its value.

To be able to meet the right pricing strategy, the product management will in a first step establish a price information system. The aim is to collect and report important price information such as competitor prices and cost prices.

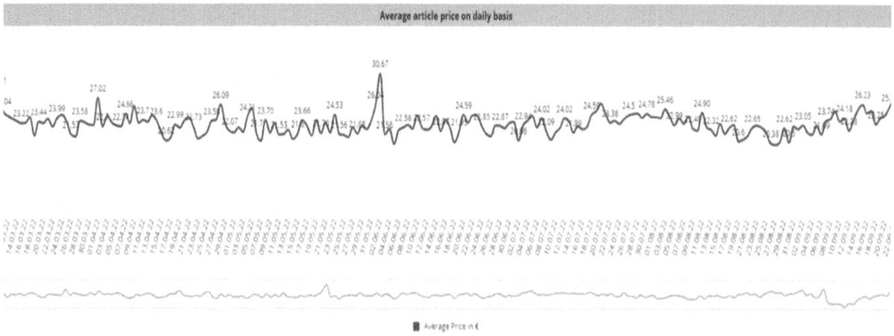

Fig. 11 Example online prices tracking

Due to the fact that the aftermarket components' prices change permanently, the ad hoc creation of the pricing data with the common market research methods is almost impossible.

The number of articles available in the market for one single OE reference number is increasing steadily.

That is why they developed the company **Eucon,** an industry system where each parts manufacturer reports the prices for all their products in a huge database where the data is processed and provided in a standard format to payable market players.

With this important information, the product management can understand the price position of its own products and relative position vs. competitors.

The company Wolk After Sales Experts has developed a pricing platform for B2C online parts prices on a monthly basis. An excerpt from the BI dashboard is displayed in Fig. 11.

The project management will use the abovementioned tools to understand what the current price corridor is and based on that information to develop the new brand pricing strategy.

5.2.4 Competition

Analysing competitors is an important part of business strategy and can help identify opportunities and threats in the market. To prepare an effective marketing strategy, the marketing manager must consider its competitors. That is especially necessary in slow growth markets such as saturated market for braking products because sales can be gained only by winning them away from competitors. The aftermarket needs to gather the information on competitors' strategies, objectives, strengths, weaknesses, and reaction patterns.

The project management will look at competitors' online presence, including their website, social media accounts, and online reviews. They promise to get a sense of customers' brand identity and how they are interacting with customers online. Also, by analysing the competitors, the management will gain valuable insights that can help better understand the market and better position their own brand for success. The abovementioned primary research will provide also valuable insights into the competitors' performance in the important brand evaluation aspects.

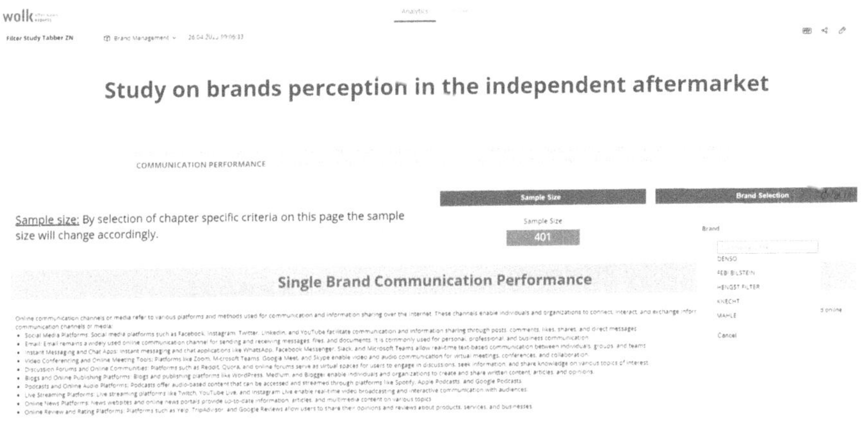

Fig. 12 Brand performance BI Tool: brand selection. (Source: (Wolk After Sales Experts, 2022a, 2022b)

Competitive intelligence needs to be collected, interpreted, and disseminated continuously. The management is sure that companies that manage a good balance of consumer and competitor consideration are practising a good market management (Fig. 12).

5.2.5 Distribution Channels

Distribution channel management is the process of managing the flow of a product from the manufacturer to the end customer. It involves deciding how a product or service will be distributed, and managing the distribution channels that are used to reach customers.

Some key considerations in distribution channel management include

- Direct vs. indirect channels: A direct distribution channel involves selling products directly to the end customer, while an indirect distribution channel involves using intermediaries, such as distributors or retailers, to reach customers.
- Channel partners: Distribution channel management involves choosing the right channel partners and developing strong relationships with them. Channel partners may include distributors, wholesalers, retailers, or online platforms.
- Channel strategy: A company's distribution channel strategy should align with its overall business goals and target market. For example, a company selling high-end products may choose a selective distribution strategy, using only a few carefully chosen retailers, while a company selling lower-priced products may use a mass distribution strategy, using a larger number of retailers.
- Channel management: Once the distribution channels have been chosen, a company must manage them effectively to ensure that products or services are delivered to customers in a timely and efficient manner. This may involve managing inventory levels, handling returns or complaints, and tracking performance metrics.

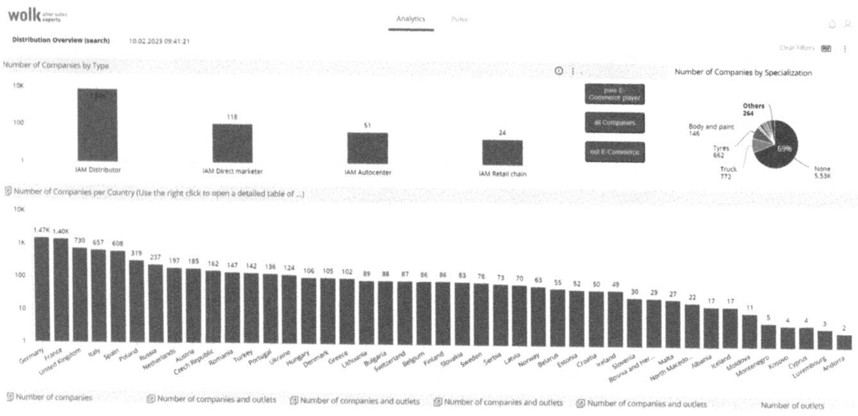

Fig. 13 Distribution overview of BI dashboard. (Wolk After Sales Experts, 2022a, 2022b)

Overall, effective distribution channel management is crucial for ensuring that products or services are delivered to customers in the most efficient and cost-effective manner.

The project management decides to review the existing distribution channels and analyse the potential in each possible channel. Therefore, they decide to get a full picture of possible customers in each channel. For this purpose, Wolk After Sales Experts provide in the After Sales BI dashboard entire European distribution data. The analysis of the database enables the management to select the distribution channels with the highest potential and prioritize them. The abstract from the BI dashboard is shown in Fig. 13.

5.2.6 Results of the SWOT Analysis

The growing vehicle park implies a larger customer base and potential market for aftermarket products and services. This increased demand indicates a positive market trend, presenting an opportunity for the marketing manager to capture a larger share of the expanding market.

With a growing vehicle park and increasing aftermarket demand, the brand relaunch project aligns well with the market dynamics. It allows the marketing manager to leverage the anticipated growth and position the brand to effectively cater to the needs and preferences of the expanding customer base.

Furthermore, the healthy business environment indicates a higher likelihood of success for the brand relaunch project. The growing market provides a receptive audience and potential customers who are actively seeking aftermarket solutions, creating a favourable landscape for the brand to make an impact and gain market share.

The project management has recognized that one of the most valuable intangible assets of the company is its brand, and it is incumbent on marketing to properly manage its value. Building a strong brand requires a good information base, careful planning, a deep long-term commitment, and creatively designed and executed marketing. Especially, as negotiations and sales in the B2B segment primarily take place with buying centres, it is always assumed that decisions are made primarily based on price. However, various studies have shown that buyers rely on other purchasing criteria.

Prior experiences of the marketing management are that the security function of the brand was often underestimated. Accordingly, brand management in B2B business was considered an unnecessary management discipline. Due to increased competition, market saturation, and increasing price pressure, the management has recognized that the brand plays a crucial role in success and are striving for professional and systematic brand management.

Based on the results of the analysis, the following key findings are recognized (Fig. 14).

To leverage its strengths and capitalize on opportunities, the company should focus on building a strong brand reputation, differentiating its offerings through unique features or value propositions, and developing strong supplier relationships. Adapting to changing consumer preferences and embracing sustainable practices can further enhance competitiveness. Additionally, staying abreast of technological advancements and exploring ways to integrate them into the aftermarket offerings can help mitigate potential threats and position the company for long-term success.

Fig. 14 SWOT analysis

5.3 Further Project Steps

After the SWOT analysis has been completed, the possible strategic goals for the brand in general but also specific details for each single market can be formulated as follows.

Based on the conclusion drawn from the SWOT analysis, the strategic goals for the company in the automotive aftermarket industry can be outlined as follows:

- The company should prioritize efforts to enhance its brand reputation by focusing on product quality, reliability, and customer satisfaction. Implementing strategies to improve customer loyalty and generate positive reviews can help build a stronger brand image.
- In a highly competitive market, the company needs to differentiate itself from other aftermarket suppliers. This can be achieved by identifying unique selling points, such as innovative features, superior customer service, or value-added services, and effectively communicating them to the target audience.
- As the company relies heavily on external suppliers, it should evaluate its supply chain to minimize disruptions and ensure consistent product quality. Building strong relationships with reliable suppliers, implementing quality control measures, and diversifying the supplier base can help mitigate potential risks.
- Company X should benefit from the growing demand for aftermarket automotive parts and accessories by considering expansion into new geographical regions or targeting untapped customer segments. Conduct thorough market research to identify regions or customer segments with high growth potential and develop tailored strategies to penetrate these markets.
- Offering personalized and eco-friendly aftermarket solutions. This may involve developing customizable products, integrating sustainable practices into operations, and communicating the company's commitment to environmental responsibility.
- Proactively explore opportunities to develop innovative aftermarket products that align with the emerging technologies, ensuring the company remains at the forefront of industry trends.
- Establishing processes to monitor and adapt to regulatory changes effectively, ensuring that the company remains compliant while minimizing operational disruptions.

By implementing these next steps, the company can address the weaknesses, capitalize on the opportunities, and mitigate the threats identified in the SWOT analysis. This strategic approach will position the company for growth, enhanced competitiveness, and success in the dynamic automotive aftermarket market.

Along the strategic redline, the business plan as a detailed marketing programme for brand relaunch can now be developed by identifying and establishing new brand positioning, planning new brand marketing, starting measuring and interpreting brand performance in a regular way, and growing and sustaining brand value deals with brand positioning.

The pricing strategy should be aligned with their overall business goals and the needs of their customers.

Besides the mentioned management teams, the creative teams, PR, and advertising specialists will join the project and help by developing the relaunch measures.

Finally, when all relevant measures are defined, the implementation of the new marketing programme can begin.

The brand plan will be executed in parallel so that the ongoing business within the category can be supported in the most sufficient way.

For the implementation of a new brand message, the communication manager needs a database with direct contacts of workshops in the target markets. This kind of information is actualized yearly and provided by our company to the market in a standard format. A sample of such data is displayed in Fig. 15 (Wolk After Sales Experts, 2022a, 2022b).

Once the implementation tasks have started, the monitoring of the entire process is needed. Very soon, the interaction with the market will show if in some process steps the rework will be necessary. If so, then the whole process step needs to be repeated.

With an efficient marketing information system, the project team will be enabled to relaunch the brand in the European aftermarket.

6 Discussion and Conclusion

Given the niche nature of the automotive aftermarket, obtaining relevant market data is challenging. It involves a continuous, intricate, and time-consuming process of data collection and analysis. Regrettably, the official registers and conventional sources lack the capability to provide the necessary market information that meets

Fig. 15 Abstract workshop database. (Wolk After Sales Experts, 2022a, 2022b)

the unique demands of the aftermarket. The specific questions that arise within this market require a more tailored approach to data gathering and analysis. Despite the difficulties, this effort is crucial because of the rapid evolution of the business landscape and utilization of knowledge and information.

By adopting data-driven decision-making processes, businesses empower themselves to navigate shifting market conditions, identify emerging trends, and gain a competitive advantage. In essence, the concept of management information systems (MIS) underscores the significance of data and evidence within the aftermarket. It highlights their pivotal role in supporting managerial decision-making, driving performance, and ensuring a competitive edge in the industry.

Aftermarket managers should ensure a comprehensive and systematic process for collecting relevant data from various sources within the aftermarket. They should develop analytical capabilities to derive meaningful insights from the collected data and align the MIS with the overall business strategy and goals of the aftermarket.

As the demand for data continues to rise in the future, advancements in technology will make data collection more accessible through alternative sources and direct data access. However, the importance of maintaining clear and high-quality data for effective management remains unchanged. To meet this need, organizations must establish a dedicated analytics system. The marketing information system (MIS) serves as a valuable platform to support this endeavour. By leveraging MIS, organizations can streamline data collection processes, ensure data accuracy, and uphold data integrity. MIS provides the necessary infrastructure and tools to effectively manage and analyse data, empowering managers to make informed decisions based on reliable insights. As data-driven decision-making becomes increasingly crucial, organizations that invest in a robust MIS will gain a competitive advantage in harnessing the power of data for future success.

References

Atkinson, A. A., Kaplan, R. S., Matsumura, E. M., & Young, S. M. (2013). *Management accounting: Information for decision-making and strategy execution* (6th ed.). Pearson education limited.

BCG, CLEPA, Wolk After Sales Experts. (2021). At the crossroads: The European aftermarket in 2030. Brussels.

Bea, F. X., & Haas, J. (2016). *Strategisches management*. UVK.

Drucker, P. F. (2004, June). What makes an effective executive. *Harvard Business Review, 82*(6), 58–63.

Evolution, Marketing. (2022, July 20). https://www.marketingevolution.com/marketing-essentials/marketing-intelligence

Grant, R. M. (1996). Toward a knowledge-based theory of the firm. *Strategic Management Journal, 17*, 109–122.

Jaedtke, K. (2020, October 21). https://blog.hubspot.de/marketing/marketing-intelligence.

Kotler, P., & Keller, K. L. (2012). *Marketing management* (14th ed.). Prentice Hall.

Kotler, P., & Keller, K. L. (2016). *Marketing management* (15th ed.). Pearson Education Limited.

Spender, J. C. (1996). Making knowledge the basis of a dynamic theory of the firm. *Strategic Management Journal, 17*, 45–62.

Wolk After Sales Experts. (2021). Poster: Aftermarket structure in Europe. Bergisch Gladbach.
Wolk After Sales Experts. (2022a). After sales ACCESS. Bergisch Gladbach.
Wolk After Sales Experts. (2022b). Workshop database Europe. Bergisch Gladbach.

Zoran Nikolic is the Managing Director and Co-Owner of Wolk & Nikolic After Sales Intelligence (former Wolk After Sales Experts), a renowned company specializing in automotive aftersales consulting. With over 20 years of experience, he has established himself as a trusted expert in the field. His career highlights include establishing a renowned study on the car aftermarket in Europe, which has become a standard in the industry. Moreover, he has successfully provided over 100 innovative consultancy services to aftermarket suppliers.

His consistent commitment to creating advanced solutions has bolstered the company's reputation as a dependable partner in the automotive aftersales industry.

Vocational Education Training (VET) for Electrical-Driven Cars: Development of a Training Concept for the Aftersales Market (Project DIAKOM-E)

Ralph Dreher, Jens Jüngst, and Fynn Komp

Abstract

The shift in mobility toward electrically powered vehicles (battery electric vehicle and fuel cell electric vehicle) also means a major change in the qualification requirements for workshop personnel. However, in the aftersales sector, there is currently a noticeable concentration on vehicle manufacturers (OEMs) to the exclusion of the independent repair and aftermarket. This applies both to the provision of measuring and testing technology and to vocational training measures, and leads to an anti-consumer market concentration, with a reduction in the range of tasks for independent repair shops and accessory dealers. The DIAKOM-E project presented here aims to counter the lack of manufacturer-dependent training by developing manufacturer-independent training measures that are then also available to the independent aftersales market. The project uses the project-based learning-like principle of complete action based on concrete, workshop-relevant test and repair orders. These are planned for real ("closed") vehicles, but with the help of an "open" vehicle on which all measuring points and aggregates are freely accessible. In order to be able to do justice to the heterogeneous groups of learners, who have not necessarily already been able to participate in measures of the manufacturers, an internal differentiation takes place via the principle of the learning situation matrix. The measures developed as a result of DIAKOM-E focus on the DACH (Deutschland Germany), Austria, Switzerland as common zone for Vocational Education) training system, but at the same time offer the possibility of an EQF (European Quality Frame)-compatible modularization through the learning situation matrix.

R. Dreher (✉) · J. Jüngst · F. Komp
Chair of Technical Vocational Didactics (TVD), Universität Siegen, Lehrstuhl für Technikdidaktik am Berufskolleg, Siegen, Germany
e-mail: dreher.tvd@uni-siegen.de; juengst.tvd@uni-siegen.de; fynn.komp@student.uni-siegen.de

M. Elo, F. Katsardis (eds.), *Automotive Aftermarket*, Management for Professionals, https://doi.org/10.1007/978-3-031-62419-3_8

1 Introduction

This chapter presents a vocational education training (VET) concept for battery electric vehicle (BEV) and fuel cell electric vehicle (FCEV), which was developed together with the "Competence Center of the Chamber of Skilled Crafts Bremen" within the framework of Project DIAKOM-E (Diagnostic Competence for Electromobility) funded by the German Federal Ministry of Research and Technology (BMBF). This happened against the background that on the part of the vocational training (institutionally represented by the vocational colleges and the training institutions of the skilled crafts) the development was recognized, so far incompletely accepted by the German motor vehicle trade, that the aftersales market will be completely transformed by the BEV/FCEV. The main reason for this is that many maintenance and repair tasks will simply no longer be necessary. Examples of this are the elimination of oil and spark plug changes or the exhaust system with exhaust aftertreatment (including the corresponding spare parts business). Diagnosis on EVs as well as repair of the expensive battery packs will, therefore, become a task to support the order volume of aftersales operations. In addition, it must be taken into account that the mixed calculation between new vehicle business, workshop operations, and aftersales business is no longer possible due to the Internet-based marketing of vehicles or sales via the manufacturer's own flagship stores.

The "DIAKOM-E" project was, therefore, supported by the Federal Ministry of Research and Technology in order to improve the hitherto inadequate initial vocational training of automotive mechatronics engineers in the use of EVs. This is because the introduction of electromobility for the automotive sector [electrical vehicle (EV)], which was foreseeable for a long time, but was not realized for a long time with regard to the broad market introduction, led to the effect that now suddenly a considerable need for training and further education has arisen. In addition, such qualification offers are intended to open up the market for repair services for independent (nonmanufacturer-specific) workshops, which will then also support the independent aftersales market as it can then once again take over the spare parts business in line with demand. The prerequisite for this, however, is that the corresponding spare parts, especially the battery cells, are also made available to the independent aftersales market.

The following should be noted here: vocational training enjoys a high social standing, especially in the DACH region (Germany, Austria, Switzerland), since countries poor in raw materials are always dependent on value creation shaped by innovation in the course of finishing or support processes. The primary example of this is the finding of the German mechanical and plant engineering industry, which, according to its own research, assumes that it can maintain its global market significance, on the one hand, through innovative products (which are predominantly the mental product of the interweaving of engineering and skilled labor), and above all through corresponding services in the area of maintenance and repair. In addition, with the emergence of the "key qualifications debate" at the end of the 1990s at the latest, it was recognized that vocational training or special education also plays a prominent role in social development (cf. fundamentally Blankertz, 1982). The more complex work processes become, the more they require independent action, which in turn must be justified in a reflective manner. This, in turn, can be understood as a universally usable ability in the course of action regulation.

In the continuous development of vocational training for vehicle repair, the duality of "combustion engine," on the one hand, and "electrical driven," on the other, was taken into account at an early stage in the "ordnungsmittel," that is, the curricula for vocational schools (KMK, 2013) and for in-company training (BMWi, 2013) through the possibility of specializing in "System and High-Voltage Technology," but the aforementioned specialization was chosen rather less, mainly because (as the Bremen Competence Center as the institution applying for the project puts it) the local companies were not able to afford such a focused initial vocational training given the long-standing "exotic status" of EV:

> Vignette 1:
>
> President of the VW and Audi dealer association Dirk Weddingen explains: "*We are at the beginning of the transformation. In the next ten years, we will still have a very brisk service business with combustion engines. After that, we will lose 70% of our maintenance revenues*" (Der Spiegel, issue 48/2022, p.76). Elsewhere, he states: "*Such an epochal transformation cannot be achieved if politicians change direction every few years. The government must make it clear to consumers that it is behind the technology turnaround. If the government continues to behave errartically, we will never become an e-car country like Norway.*" He refers here to subsidies for the purchase of EVs, but does not explicitly mention the problem that, in addition to financial incentives, service personnel are also needed in the workshops to give consumers the security of good vehicle maintenance and repair for the remaining 30%. However, in addition to the required political intervention in the new vehicle market, this is another essential building block for triggering the transformation conjured up in advance by Weddingen as a voluntary act on the part of the consumer (and not forced by "combustion forbidden"), notes Michael Brecht as a Tesla driver: "*Being an early adopter of electric vehicles myself, I am probably highly resistant to problems with this innovative technology. However, with the current mass adoption of electric vehicles in Europe, we have to cater for the needs of normal car buyers, less experienced drivers, and the elderly. The potential level of frustration of drivers less tolerant than me would make me nervous if I were a vehicle manufacturer like Tesla. There is work to be done for them when it comes to service. Building the most exciting electric car today is no longer enough - as other manufacturers are catching up*" (quoted from https://www.motion-mag.com/articles/can-tesla-do-service-my-experience-after-a-year-with-a-model-3; Last Call 2023-01-02).

DIAKOM-E aims to counteract this qualification gap that has arisen in this way in a manufacturer-independent way (since vocational training in the automotive trade in Germany is always focused on a manufacturer-neutral way) with its consequences for the automotive repair trade and is therefore to

- On the one hand, implement current vocational didactic concepts for this application [thus targeting train of trainers (ToT-)].
- On the other hand, to develop a realistic training concept for basic and advanced training in the field of "fault diagnosis on electrically driven vehicles." In this context, "realistic" means above all a vehicle-type-independent designation of typical diagnostic and repair tasks, which therefore had to be defined in advance as part of the "DIAKOM-E" project.

2 Overview

In developing this training concept for DIAKOM-E, the concept of work process orientation, which has been binding for German dual training since the mid-1990s, was used.

What is meant by this is a didactic orientation to real work tasks in the respective occupation determined by corresponding occupational science analysis. Instead of the previous principle of starting with theoretical instruction (in the automotive trades, this has so far mainly involved explaining the functions of assemblies such as the engine, transmission, brakes, and steering) and aiming to translate this knowledge into real diagnostic and repair activities as part of practical workshop training, work process orientation is intended to dovetail theory and practice from the outset by (re-)developing best-practice work processes. This is achieved primarily by means of a diagnostic or repair assignment at the beginning of a learning unit ("learning situation"), which is to be understood as a project task.

The real implementation of the work processes formulated by the learners, which is integrated into the learning situation, first serves as an evaluation phase (with the primary Frane, according to feasibility, responsibility, success, and meaningfulness). The resulting evaluation result then serves as the basis for a reflection phase that concludes the learning situation by answering the questions:

1. Based on the evaluation: What led in the work process defined by me to a technically correct problem solution in the sense of the task (e.g., was it possible to diagnose the fault on the vehicle reliably, quickly, and unambiguously?)?
2. In comparison with the solutions of other course participants: Where is my proposed solution better, why, where is it worse, and what is the optimal work process for this task?
3. What did I succeed in developing in my work process and why, what did I overlook and why? What does this mean for me and my personal development?

The reason for this definition was, in addition to the good fit with the existing concept in German vocational training, to ensure a high degree of realism in the training on EVs and at the same time meet the requirement that in-service measures by a nongovernmental training provider should also initiate vocational training processes ("education" instead of "training"). Following this definition, the following VET principles, which are described in more detail in Chap. 3, were defined as the framing for the training concept to be developed:

- The principle of design competence means that learners must be given the chance to formulate and try out troubleshooting routines on EV themselves in order to design their own solutions.
- The principle of holistic professional action means that errors identified as significant in the everyday work of the workshop are to be diagnosed in the context of learning situations (as a so-called "work process task" in the form of an assignment) and that the learners go through the phases of informing (type of

error and cause), planning (developing a diagnostic scheme as a work process), executing (implementing the work process they have found themselves), checking (success control for the correctness of the identified cause of the error) and reflecting (on the basis of the reflection questions mentioned above).
- The principle of internal differentiation via learning situation matrices means that in view of very heterogeneous learner groups, the same diagnostic task is differentiated in such a way that it is just as suitable for rather inexperienced learners as for learners who already have a great deal of experience. The goal here is to ensure that there is no individual over- or under-challenging.
- Principle of portfolio work by means of COMET: COMET is a procedure developed in the course of the introduction of work process orientation for the recording of vocational competencies. This instrument is used in two ways in the context of the DIAKOM-E project being worked on here: firstly, as a feedback instrument for the individual learner with regard to his or her design competence, and secondly, as an evaluation instrument to determine whether the instructional methodological approach of the project has an overall competence-promoting effect.

COMET is of particular importance in this context as a previously unused concept of teaching work with "open" and "closed" vehicles was tested especially for DIAKOM-E:

- The starting point of the project is an Artega sports coupe converted from combustion engine operation to electric drive. This vehicle offers full access to the measurement of relevant signals, electrical voltages, and electrical currents by means of break-out boxes and is therefore referred to as an "open vehicle." It thus enables, above all, the promotion of an understanding of the functioning of an electrically driven vehicle and the possible measurements with their justification as to what is useful when in the context of occupational safety and the containment of the fault on the vehicle.
- In contrast, "closed vehicles" were used in the test phase, on which the actual fault diagnosis is to be carried out as a respective work process task: an Audi e-tron quattro 55 and a Toyota Miral I—JTD10. These series production vehicles can be diagnosed exclusively via the OBD interface and using Y-cables such as those available to specialist workshops, in accordance with workshop reality, whereby basic functional understanding is then assumed.

3 VET Methods and Tools Used

3.1 Design Competence as a Goal of Vocational Training

The Standing Conference of the Ministers of Education and Cultural Affairs of the Länder in the Federal Republic of Germany (KMK) has defined the educational objective of vocational school curricula to be the promotion of competence to act in vocational and nonvocational life situations (KMK, 2021, p.10).

This requirement is currently summarized as "developing design competence" and is based on the following ideas:

1. In view of exponentially advancing technological development, work processes are highly volatile, that is, they have a low scope and duration of validity. It, therefore, makes little sense to learn work processes as procedures quasi-behavioristically (situation as stimulus, prefabricated work process as reaction) because they quickly become obsolete. Instead, specialized work means dealing with a problem and developing and (!) implementing an independent problem solution that can be considered sustainable. This must be promoted in every form of VET. The most important didactic question is always whether the selected diagnostic or repair task is sufficiently open to open up design potentials in the learning process. If it is not, it cannot be considered a learning situation in the sense of design-oriented VET.
2. However, it is precisely this goal of promoting creative competence for the further development of procedures and behaviors that shape people as individuals that is just as valuable in the nonprofessional sphere, be it in the context of social-political commitment (e.g., neighborhood or trade union work), settling the annoying neighborhood dispute, or, above all, developing perspectives for.
   - One's own life (what do I want, what do I want to achieve WITH my life partner, what do WE want as a mutual compromise why and what not).
   - In the education of one's own children (which should not be a copy of one's own education, but adapted to the technical-social reality of the present). The expectation is that the individual will succeed in such an adaptation or transformation—to actually initiate this is not clearly imposed on vocational training as a challenge. Instead, one relies here on effects such as those described in the "habitus theory" (Bourdieu 1978, p.164f), according to which professional action also very strongly shapes nonprofessional action.
3. For the assessment of competence, however, an additional requirement follows from what has been said above: In the simplest case, designing means the reflected adoption of what has been tried and tested, after it has been clarified why this represents an optimum. Far more often, however, it is a matter of meeting the situational challenge by adapting and recreating action in the face of the concrete problem (Becker, 2003). Such self-responsible work generates, on the one hand, the effect of humanization through self-determination, but above all a value-creating bottom-up innovation dynamic (Brödner & Oehlke, 2018, p.560ff) as a cross-professional challenge to skilled work. This, in turn, requires the presence of reflective capacity, on the one hand (Can I really do it this way—measured against the benchmark of accountability and sustainability versus the pressure for efficiency-based innovation?), and on the other hand, above all, the ability to be creative (to overcome familiar structures, thought patterns, and seemingly irrefutable guidelines that often proudly create professional identity as something learned and shaping habitus). Sustainability, responsibility, and

creativity are, therefore, recorded and reported as essential elements of "actual professional work" (which no longer imitates the familiar with perfectionist demands) within the COMET competence measurement on an equal footing as criteria (Rauner et al., 2009,p.91ff).

3.2 As a Consequence: Focusing on the Principle of Holistic Professional Action

Vocational education with the educational goal of promoting action competence is carried out by means of work process-initiated action to be planned, which in turn requires dealing with corresponding specialist theory as part of the planning.

A concrete cause for action in the form of a typical, open-ended task is taken as the starting point for a learning situation, so that trainees, in the course of mastering the task, that is, the work process for mastering this task, can both

- Develop a specific, self-designed, and reflected work process knowledge as well as.
- Acquire the necessary specialist knowledge as justification knowledge for the self-directed and open design of their work process.

Learning situations are, therefore, always created in such a way that "open tasks" with a solution space (and thus precisely not a solution to be clearly achieved) arise, giving the trainees the opportunity to develop their own, subjectively optimal solution and implement it in a testing manner. The principle of holistic action thus strongly resembles the principle of project-based learning (PBL) described in academic engineering education, which also calls for the self-designed and self-responsible solution of work-related development tasks (De Graaf & Kolmos, 2013 and Ulseth, 2016). In relation to vehicle repair, "self-responsible" includes the independent development or reflection of existing diagnostic and repair procedures while observing the criteria of social responsibility and environmental compatibility (see Vargas et al., 2019 for further details), which is also captured as a processural level by the COMET procedure for recording the professional competencies described below.

The common structure for processing is, therefore, oriented to the "principle of holistic action," which means to

- **Inform** themselves about the actual work task, the expected work result, as well as the necessary technical specifications for the task solution (including already existing work process descriptions, e.g., in the form of assembly or diagnostic instructions).
- **Plan** a concrete work action based on this.
- Execute or **implement** them directly as part of the instruction (ideally on the real object or else by means of a simulation medium).

- **Check the** result achieved in the implementation for actual feasibility, technical correctness, functionality, and accountability.
- Finally question through a moderated **reflection,**
  - Why this result was achieved in this quality?
  - Which strategies for success should be retained and what needs to be optimized with regard to the personal ability to design such work processes (cf. the following Sect. 4.2. in this document)?
  - Whether the solution found was designed precisely from the point of view of sustainability in such a way that it is economically feasible, ecologically sensible, and socially responsible, and which of these criteria was given which priority and why?
  - What this means for my personal ability to **shape** my life both professionally and nonprofessionally (thus named as a goal by the KMK in accordance with KMK 2021, at the same time identified in the company AO Kfz-Mechatroniker/in 2013 as "vocational training" with the goal of professional ability to act, BMWi 2013)?
  - Which consequences result from this for the task to be worked on next (mostly documented via a **personal agreement on objectives as the actual objective of a learning situation**)?

Vocational training structured according to this principle of a solution to a work task for which the learner is responsible and which progresses in phases must take two things into account:

- The individual phases must not be understood as clearly definable (progress) steps, but are worked through by "iteration loops." For example, there is frequent switching back and forth between informing and planning or planning and informing (because information is missing for planning); likewise alternating between planning, implementing, and checking (because it becomes apparent in the implementation or checking phase that the planning does not work/does not lead to the desired result). The use of these iteration loops by the trainees is then also always included in the reflection process (Why was this iteration necessary? What was not worked out or considered from the beginning and why?).
- The reflection phase represents the actual vocational training moment in this learning process and must be designed accordingly with care and above all in dependence on what has actually happened in the instructional work process. Vocational training, according to the principle of complete action, thus requires vocational training personnel who are able to design the reflection phase in a *situationally flexible manner in* such a way that questions that also directly reveal personal deficits are moderated in a correspondingly constructive manner in such a way that the result is an agreement on objectives between the teacher and the trainer. Such a target agreement then in turn represents the first element of a new information phase with which the next learning situation is initiated.

3.3 Principle of Internal Differentiation through "Learning Situation Matrices"

3.3.1 Why Internal Differentiation?

Vocational education is characterized by educational processes in and with very heterogeneous learning groups (ability potential, previous educational biography). Therefore, it is indispensable to take a closer look at the prerequisites of the individual students already during the conception of a vocational training measure and, above all, never assume that a learning situation with a work assignment is equally demanding for all students in the sense of promoting competencies.

For this purpose, the initially "factually-based" learning situations, that is, the learning content as a determined reflection of professional reality, must be internally differentiated in such a way that they offer moments of inclusion.

"Inclusive moment" here means that a basic task that is uniform for all students is varied to accommodate different levels of prior education, prior experience, and prior knowledge in such a way that neither over- nor under-challenging occurs.

It should, therefore, be possible to identify different levels of difficulty for each learning situation. This should ensure that all trainees can develop these further according to their existing competence in such a way that the technical objective of this measure is always achieved: the ability to diagnose and rectify faults on e-vehicles.

3.3.2 Internal Differentiation Via Complexity Level

One possible strategy to level learning situations is to increase or decrease the complexity level of a task via adjusting the salience of the task.

"Salience" means, on the one hand, to what extent which (helpful) guidelines have already been given in the task, and, on the other hand, which guidelines are

	Novice	Competent	Expert
OBD error	NTC resistor Coolant temperature defective	Supply voltage interrupted Camshaft speed-sensor	Resistance increase in the circuit of the injection valves
Lambda control loop	Interruption Heater circuit Lambda sensor	Lambda sensor defective	Resistance increase in Signal circuit Idle speed controller
Subsystem error	Clogged fuel filter	Pressure regulator defective	Clutch sensor defective

Fig. 1 Learning situation matrix with three learning situations ("motor management")

already available with recourse to existing experiences, so that ultimately the known best-practice imitating the learning situation can be mastered. For illustration (cf. Fig. 1)

- Engine control lamp indicates a defect.
- OBD readout shows: coolant temperature sensor defective.
- Verification of the sensor as NTC resistor indicates infinitely high resistance.
- Consequently, the following applies: the sensor is to be replaced, the fault memory is to be deleted, and then a test drive is to be carried out.

A large number of specifications are already included here: clear error description (engine control light), clear cause of error via OBD (sensor defective), and clearly detectable error (by the resistance measurement at the sensor). Such a task has a comparatively low level of complexity, but can be increased by, for example, simulating a break of the sensor line at an active sensor (speed sensor) (how to locate it?) or even a fault in the optocoupler or A/D converter in the ECU (how to check? how to repair in a youngtimer vehicle?).

What becomes clear from these simple examples: even the actually simple diagnostic task of a sensor stored in the fault memory as a defect can be varied in its complexity by the cause of the fault, the associated troubleshooting strategy, and the necessary technical knowledge (passive and active sensor).

3.3.3 Internal Differentiation Via Additional Tasks

In addition to the variation of salience, a further possibility for internal differentiation can be an additional task integrated into the assignment.

Related to the previous example of the defective temperature sensor, additional tasks could be here (for example!):

- Processing as a warranty case.
- Customer meeting to explain the error and its cause.
- Need to deal with English-language service literature.

3.3.4 Working with the Learning Situation Matrix

In order to achieve and document this internal differentiation in a safe and comprehensible way, the principle of the "learning situation matrix" (Fig. 1, exemplary for "troubleshooting the engine management of combustion vehicles") was developed, in which the individual learning situations were divided into different complexities (vertically) and additional tasks (horizontally):

In the development of such a learning situation matrix, a determination of the basic learning situations (here: OBD error, lambda control error, subsystem error) is first made on the basis of "macroanalytical work process investigations" as instruments of occupational science (Bremer, 2006, p.592ff), which means that It is determined in the companies, for example, through job analysis or employee interviews, which work tasks

- Occur frequently and are therefore relevant.
- How much design, that is, individual problem solving, for example, through a diagnostic plan to be developed very individually, the tasks required in each case, from which in turn it can be estimated what level of complexity this task will have as an instructional learning situation.

As a rule, a learning situation can be realized by means of the principle of complete action (cf. Sect. 3.2) in approximately 20 teaching hours. For a week of instruction in the context of a training course, this means a maximum of two learning situations, if phases of instruction (safety instructions) and final certification (cf. the explanations on the implementation of COMET, Chap. 7) are also provided.

Since more than 3–4 relevant work tasks are usually found in the occupational macroanalysis, a didactic selection must be made when choosing the tasks to be considered for the learning situation matrix. This can be guided by the following questions:

- Which learning situations can be realized particularly well at this learning location due to personnel and equipment? This is (unfortunately) the most important didactic question in real terms since otherwise the principle of complete action with implementation in practice cannot be realized.
- Which learning situation offers which design potential? Tasks that follow a highly standardized procedure (e.g., a guided visual inspection) do not represent an independent learning situation because these standardized work activities are not intended to deviate from the specifications for reasons of quality assurance.
- Which learning situation found in this way offers good opportunities for internal differentiation close to professional reality?

Based on such considerations, an initial definition is made, for example, in three basic learning situations (as vertical in Fig. 1).

The classification of the learning situations from 1 (entry) to 3 (measure goal) is based on an assessment of the salience of the learning situations found. The salience is determined by

- Reflective work process observation with concluding professional interview and/or.
- The teacher's documented and self-reflective self-experimentation on how to manage the learning situation.

Especially the latter (self-experimentation) is an indispensable element of instructional preparation since only in this way can the teacher develop a solution space for himself, which he can in turn use to moderate and support the work of the learners.

In order to designate the requirement level of learning situations, the division into learning situations is often used here in analogy to the developmental logic model according to Dreyfus/Dreyfus (Dreyfus/Dreyfus, 1980).

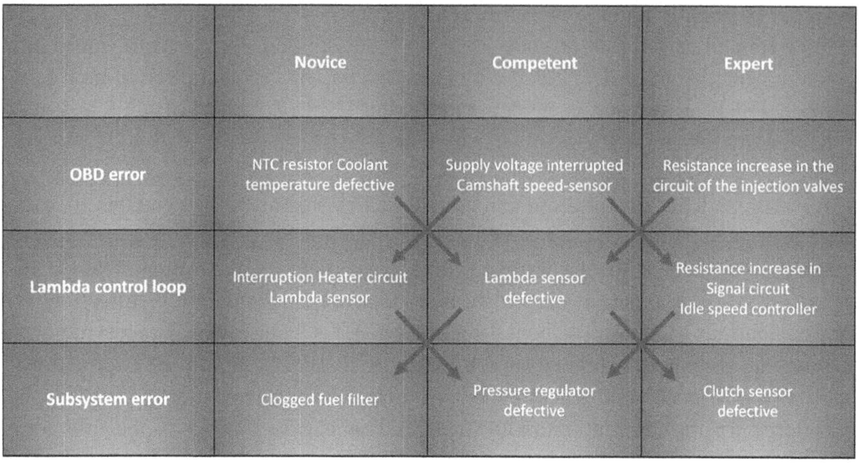

Fig. 2 Switching possibilities within different complexity levels

- For "novices" with few possibilities to shape the plot ("know that").
- For skilled workers who find that there is no regular procedure for solving tasks ("know how" with the use of situationally branching work actions).
- "Experts" who are confronted with tasks that they have not previously known from the professional requirements practice and, therefore, have to develop and take responsibility for completely self-developed work actions ("know why").

used.

In order to arrive at an internal differentiation within a learning situation, this is again varied in its scope or complexity—in Fig. 1 by the necessary measuring tool and the number of contacts as possible measuring points: ohmmeter with one measuring point for continuity testing, voltmeter with two contact points voltage measurement, and oscilloscope with switched ground for voltage curve measurement with the calculation of the internal resistance of the injection valve).

In order to create the necessary potential for inclusion, it can be decided for each learning situation identified in the matrix that each trainee will be assigned to a different group with a different level of differentiation for the following learning situations (Fig. 2). In this way, good novices can be assigned to the group of skilled trainees in the next learning situation.

Experts (and are thus already directly confronted with the application of the oscilloscope in the example in Fig. 2) or experts for whom the task was too complex in the first run go through the learning situation in the next run (lambda control) as a skilled person (who, again using the example in Fig. 2) and thus get another chance to practice the basic recording of measured values on the oscilloscope instead of additionally dealing with the principle of pulse width modulation (Fig. 2).

3.4 COMET Competence Measurement as an Evaluation Instrument

Picking up on the KMK's guiding principle of "enabling students to help shape the world of work and society in a socially, ecologically and economically responsible manner" (KMK, 1996; Rauner, 2018, p.20), the COMET process according to Rauner (Rauner, 2018) is based on the need to evaluate task solutions holistically with reference to the eight overarching criteria of vividness/presentation, sustainability/use value orientation, economic viability/efficiency, business and work process orientation, and social and environmental compatibility and creativity (Rauner, 2018, p.21 f.).

The COMET measurement method, according to Rauner et al., thus offers a possibility to evaluate the abilities of students to solve open and at the same time authentic (real vocational) tasks in a model-like manner. These tasks must be both challenging and qualifying in character—and presuppose at least the qualification level "skilled worker" as an essential characteristic of skilled worker status.

Due to the open task, the learners must not only create and present their solution, but also be able to take responsibility for it in a holistic and reflective way, in order to prove their competence.

In order to demand a holistic task solution that takes into account aspects of economic efficiency, environmental compatibility, and utility value orientation, COMET tasks must be carefully selected under the quality criteria of a design-open task and legitimized on the basis of a systematized selection process (Rauner, 2017, p.101ff).

The task solutions are evaluated by specially trained raters who did not participate in the task conception. Thus, a high interrater reliability is achieved (Finn coefficient of 0.7–0.85, which means a consistent score with two independent raters of 70–85% on the same task solution).

4 Development of the VET Training Concept "EVs"

4.1 Development of the "EV Diagnostics" Learning Situation Template

4.1.1 Vocationally Based Matrix Design

The need for internal differentiation is seen as particularly meaningful for the learning situations on the EV in this concept as the vehicle manufacturers have a highly different degree of brand-specific vehicle range and vehicle diagnostic capability (general/specific error message). Consequently, the learning groups with their previous experience regarding diagnosis and repair will also be very heterogeneous in this training measure.

Since the maintenance and repair of EVs is a completely new field of activity, a careful analysis of the occupational science is required to determine which work tasks are to be performed in this area.

- Due to their frequency.
- Their formability.

can be the basis for a vocational VET.

This is because tasks that are either unrealistic or can be performed independently of the manufacturer by means of guided diagnostics or corresponding manufacturer instructions in the form of a detailed, vehicle-specific work plan do not, in principle, require any fundamental training and further education measures.

An exception to this are the instructions

- In the handling of high-voltage systems on EVs.
- Especially for the FCEV in handling the tank system and the hydrogen under high pressure (700 bar).

It is assumed here that the pre-qualifications essential for the work on the vehicle as a "Qualified person for work on live HV systems (FuS, level S3, according to DGVU, 2021) as well as "Qualification for work on vehicles with gas drive" (level 2S, according to DGVU, 2018) have already been acquired in advance of this VET measure.

There is also a fundamental discussion to be had here about how and with what institutional responsibility such certificates of competence are to be integrated into initial vocational training because it does not seem very future-proof to implement initial vocational training for automotive mechatronics technicians that teaches the functioning, diagnosis, and repair of (FC)EVs in a competence-oriented manner, but at the same time does not ensure that they are allowed to work on the corresponding vehicles after successfully completing the training because the corresponding certificates of competence were not acquired at the same time.

Under this assumption, following the principles of macro-analytical occupational research, a multi-stage research concept was used to find and define the tasks with the elements

- Cross-brand order analysis (to find the frequency of errors).
- "Store floor" interview (information from shop floor management on orders and their actual processing in the shop floor without rejection or third-party contracting)
- Interview "workshop" (information of the skilled workers about the design potential of the respective order).
- Expert-skilled worker workshop (summarizing and evaluating the survey results together with the skilled workers).

planned.

Vignette 2:

The principle, which has so far been presented as very positive and sensible, of promoting both realism and the development of design competence by means of a scientifically

> supported definition of real, proven work process tasks and their transfer into learning situations, reaches its limits here. This is because the occupational macro-analysis mentioned as a precondition is particularly difficult to carry out for the "EVs" application case since there are precisely no results that can really be identified with certainty as to which faults occur on EVs, to what extent, and with what diagnostic and repair effort. Instead, findings like this circulate: "*Owners of BEVs and PHEVs cite more problems with their vehicles than do owners of vehicles with internal combustion engines (ICE). ICE vehicles average 175 PP100, PHEVs average 239 PP100 and BEVs-excluding Tesla models-average 240 PP100. Tesla models average 226 PP100 and are shown separately from the BEV average because the predominance of Tesla vehicles could obscure the performance of the legacy automakers that have recently introduced BEVs.)*" (J.D.Power Report 2022). Unfortunately, the report does not name the exact types of defects, neither for Tesla vehicles nor for the other manufacturers; therefore, it cannot be excluded that, for example, the frequently publicized malfunction for autonomous driving and the poor body quality of Tesla vehicles are decisive for the high defect rate for EVs.Better assurance would be possible here through interviews with the workshops. However, all of the workshops surveyed currently only provide information on manufacturer instructions in individual cases and on a basis of trust. For example, it can be stated that this often applies, to determine the SOC status (state of charge and residual charge capacity of the battery cells in BEVs) for used vehicles, that the temperature management of the battery module is a frequent source of error (solvable by guided software update or poor leakage of the cooling circuit), but that it would be mainly about solving problems in the noncharging of BEVs: "*If someone now tells me what is to be assigned as a task to the motor vehicle mechatronics technician what to the electronics technician for house and building technology who ensures the function of the charging station or wallbox, I would already be one step further. Are motor vehicle people also supposed to fiddle around with the high voltage of the charging points in the future? And furthermore, is it the job of the workshops to find out if and when the power supplier has capped the wallbox because it had to limit locally?*" (question from a trainer).

In concrete terms, this means for the current state of knowledge: at the moment, it cannot be determined with the usual and necessary precision in the development of a learning situation matrix,

	Novice	Competent	Expert
Security Interlock	Security Interlock interruption	Short circuit Security Interlock to ground	Short circuit Security Interlock to positive potential
Isolation error	Isolation error HV-line to ground	Isolation error HV-line to positve potential	Insulation fault HV-Plus via power resistor to HV-Minus
High-Voltage Interruption	Line interruption in theService plug	Interruption of the HV-line in an auxiliary unit	Interruption of a winding in the 3-phase-motor

Fig. 3 Draft learning field matrix "EV"

Table 1 Results for learning situation "security interlock"

<table>
<tr><td>Learning situation</td><td colspan="2">Interruption security interlock</td><td colspan="2">Short circuit security interlock against ground</td><td>Short circuit security interlock to positive potential</td></tr>
<tr><td>Fault connection</td><td colspan="2">Feasible</td><td colspan="2">Feasible</td><td>Not tested, because interlock is run by 2 V
Possible damage to CPU</td></tr>
<tr><td>Diagnosability</td><td colspan="2">Feasible</td><td colspan="2">Feasible</td><td>Not tested</td></tr>
<tr><td rowspan="2">Design potential</td><td colspan="2">Measuring points must be determined by the user</td><td colspan="2">Measuring points must be tested by the user</td><td rowspan="2">Not tested</td></tr>
<tr><td>Toyota:
Error message: "Fuel cell system malfunction"
OBD error code P0A0D: "High voltage in the interlock circuit of the high voltage system"</td><td>Audi:
Error message: "On-board power supply fault"
OBD error code P0A00: "Battery monitoring signal interruption, disconnected connector"</td><td>Toyota:
No error reported.
If you interrupt the security interlock based on the cable you connect to ground, you have either no error or the same as in the first earning situation</td><td>Audi:
Error message: "On-board power supply fault"
OBD error code P0A0C00: "Battery monitoring ground fault"</td></tr>
</table>

- Which work tasks have which relevance.
- There are no already developed sample solutions with which the lecturers can prepare themselves and (in the case of a reversed solution path) realize the error connection, which also means that.
- It is unclear how designable they are (except).

The latter criterion of designability meant that the SOC determination frequently mentioned in the course of the order analysis was not taken into account, nor were software updates, since these are carried out in an absolutely guided manner via OBD across all manufacturers and do not include any design options. Excluding these tasks, the learning situation matrix shown in Fig. 3 was developed on the basis of exemplary statements.

Table 2 Results for learning situation "isolation error"

Learning situation	Insulation fault HV line to ground	Insulation fault HV line to positive potential	Insulation fault HV line positive to HV line negative
Fault connection	Feasible, install separate relay circuit if necessary	Not feasible; Control unit damage foreseeable when exposed to HV during measurement	Initially not feasible; damage to HV battery fuse or battery itself foreseeable; if necessary, retrofit a fast-acting fuse
Diagnosability	Feasible	Not tested	Not tested
Design potential	Measuring points must be determined by the user, but: Audi: Measurements hardly possible due to plug insulation; relay circuit and break-out box must be installed beforehand. Toyota: The vehicle will not detect or report any error	Not tested	Not tested
	High design potential on the "open" vehicle without OBD diagnostic capability		

The learning situation matrix developed under these preliminary assumptions then led to the following results in the course of an initial actual implementation (Tables 1, 2, 3).

The critical review of the learning situation matrix developed in Fig. 3 documented in Tables 1, 2, and 3 thus shows, on the one hand, the poor practicability of the learning situations described there, but above all fundamental problems with instruction in the diagnosis and repair of EVs:

- With the exception of faults in the pilot line, fault connection requires the cooperation of specially qualified high-voltage technicians since separate relay-controlled circuits usually have to be implemented for this purpose.
- It must be checked to what extent manipulation of the inverter (on the input/direct current side as well as on the output/three-phase current side) is possible without using quickly detectable manipulated intermediate connectors.
- The designability appears too high (at least according to the results on the Toyota vehicle) since only an unspecified error message is generated, which does not permit the targeted development of an error-specific diagnostic scheme. Instead, a diagnostic scheme is created by self-development on the try-and-error principle, which is then based purely on empirical knowledge in practice.
- The "open" training vehicle can only serve as a medial aid if it has both the aforementioned (relay-controlled) switching options for faults in the high-voltage range and an extensive break-out box for hazard-free measurement. However:

Table 3 Results for learning situation "high-voltage interruption"

Learning situation	Line interruption in the service plug		Interruption HV-line into an auxiliary unit		Interruption winding drive motor
Fault connection	Toyota: Very well possible due to manipulated service plug	Audi: Poorly feasible; permanent manipulation of the inverter output through manipulated connection cable	Toyota: Well feasible (three-phase fan for air supply fuel cell)	Audi: Poorly feasible; permanent manipulation of the inverter output	Sealed drive units at Toyota and Audi; therefore, only possible on the "open" vehicle by manipulating the drive motor in advance
Diagnosability	Toyota: Feasible	Audi: Not tested	Toyota: Feasible	Audi: Not tested	Not tested
Design potential	Measuring points must be determined by the user		Measuring points must be tested by the user		Very high, provided break-out box with full range of measuring points is available
	Toyota: General error message: "HV system malfunction" OBD error code P0A95: "High voltage fuse" & P3004: "High voltage supply"	Audi: Not tested	Toyota: General error message: "HV system malfunction" OBD error code P1D7F: "Performance of FC air compressor speed"	Audi: Not tested	

these measurements thus increase the understanding of their feasibility and thus promote the understanding of the structure and functioning of an EV drive train with 12 V consumer circuit, 2–3 V pilot line, HV DC circuit, and HV three-phase circuit, but do not map the actual educational goal of designing real (!) work processes on series vehicles in a self-responsible manner since the possibility of a break-out box installation for diagnostic processes in the workshop is not provided.

4.1.2 Revised Learning Situation Matrix

The critical (self-)reflection of the learning situation matrix shown in Fig. 3 shows that the idea of defining vocational and internally differentiated design tasks on the basis of real operating faults occurring in vehicle operation and simulated by fault connection can currently only be achieved very incompletely.

For the creation of a revised learning situation matrix, it was, therefore, determined together with the workshops, independently of customer complaints or documented operational faults as the starting point of work process descriptions, which types of electrical measurements must be mastered across faults for fault diagnosis and vehicle inspection before and after repairs have been carried out. This focus on the electrical measurement procedures (instead of on the work order for fault rectification) departs from the principle of work process-oriented vocational training, but has the great advantage that the workshops surveyed were able to provide more precise information on what has to be done when and why. It was thus possible to determine that three types of measurements are used in the diagnosis and inspection of EVs, mainly before or after the resulting unit replacement: continuity measurement, insulation measurement, and potential equalization measurement.

Figure 4 shows that in this learning situation matrix, compared to the original proposal shown in Fig. 3,

- The faults in the pilot line could be retained for the purpose of diagnosing circuit interruptions. However, the learning situation "short circuit pilot line against 12V plus" was replaced by a combination of interruption and short circuit against 12 V ground....
- The insulation measurements were also kept, and it was accepted that the insulation fault HV plus against 12 V plus is still not treated (the die therefore has only two fields in this line).
- Checks for equipotential bonding were introduced as a third learning situation, a measurement that is always necessary when a component has been replaced. In workshop practice, this is a frequent repair measure for EVs, since assemblies such as charge control/transformation, inverter, HV battery, or drive motor cannot currently be repaired in the workshop. This learning situation is varied via the

	Novice	Competent	Expert
Continuity measurement	Security Interlock interruption	Short circuit Security Interlock to ground	Interruption and short circuit Security Interlock to ground
Insulation measurement	Interruption of one of the HV lines	Isolationsfehler einer HV-Leitung gegen 12V Masse	Short circuit of one of the HV lines to positive potential: possible damage to CPU HV+ to HV-: possible damage to HV-battery or fuse
Potential equalization measurement	Assembly is known. Resistor is OL. Line is missing(nominal values and test points are available)	Assembly is known Cable has a break(Points to be tested are present, nominal values are not available)	Assembly unknown (only limitation to charging infrastructure)Potential equalization line on charger not connected (No specified test points, no nominal values)

Fig. 4 Revised learning situation matrix EVs

Table 4 Results to the "potential equalization measurement" learning situation

Learning situation	Assembly is known. Resistor is OL Line is missing (nominal values and test points are available)		Assembly is known. Cable has a break (points to be tested are present, nominal values are not available)		Assembly unknown (only limitation to charging infrastructure). Potential equalization line on charger not connected (no specified test points, no nominal values)	
Fault connection	Toyota: Not feasible; all units are directly connected to the chassis	Audi: Feasible	Toyota: Not feasible; all units are directly connected to the chassis	Audi: Feasible	Toyota: Not feasible; fuel cell vehicle; therefore, no charger	Audi: Feasible
Diagnosability	Toyota: Feasible	Audi Feasible; no OBD error entry	Toyota: Feasible	Audi: Feasible; No OBD error entry	Toyota: Not feasible	Audi: Feasible; No OBD error entry
Design potential	Toyota: Only security measurement possible, no guided diagnosis	Audi: Measuring points must be determined by the user	Toyota: Only security measurement possible, no guided diagnosis	Audi: Measuring points must be determined by the user	Toyota Only security measurement possible, no guided diagnosis	Audi Measuring points must be determined by the user

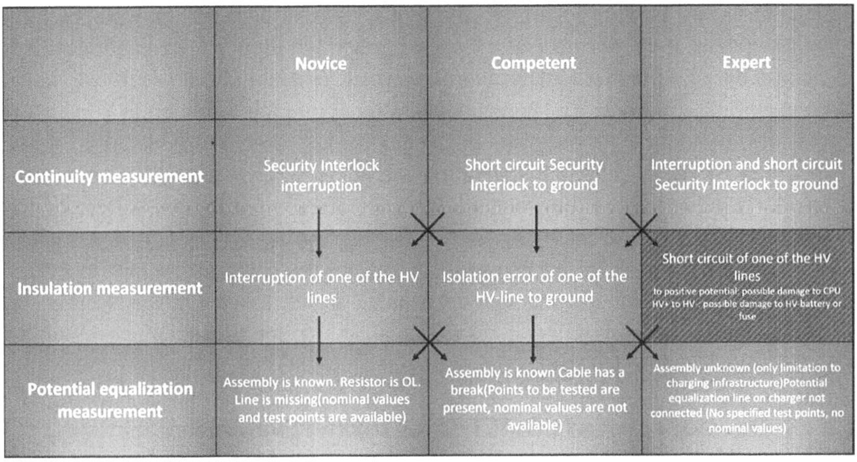

Fig. 5 Learning situation matrix EVs with internal differentiation

available information, so that one or more potential equalization measurements must be taken.

For this revised learning situation matrix according to Fig. 4, the third learning situation was evaluated again separately with the following result (Table 4).

The learning situation matrix in Fig. 4 allows the following jumps (cf. here the possible internal differentiation according to Fig. 5).

The learning situation matrix according to Fig. 5 is based on the following considerations regarding the possibilities for internal differentiation:

- At the novice level, only faults that can be traced back to a direct electrical interruption (due to cable breakage or missing connection) are dealt with. This means that the most common cause of faults in EVs according to workshop information is addressed throughout because the electrical cables in EVs are insulated in such a wear-resistant manner that short-circuit contacts due to damaged insulation are rather rare (exception: damage caused by animal bites).
- Due to the nonexclusion of faults caused by insulation damage, these faults are defined to the level of skilled and expert.
- In the second learning situation, both an internally differentiated upgrading (instead of cable breakage insulation fault for previous novices) and internally differentiated downgrading (same type of fault/repetition for adepts) are possible. Due to the lack of learning situation 2 for experts, a downgrading to the skilled level occurs here; a downgrading from the expert level for learning situation 1 to the novice level for learning situation 2 (instead of short-circuit faults, easily identifiable faults due to cable breakage) is also possible, but rather unlikely.
- For the third learning situation, all up- and downgrades are possible since this learning situation, similar to the example according to Fig. 2, has no direct con-

nection to the previous learning situations. This is because the equipotential bonding measurements are a specific test method whose negative result is always caused by an unrealized electrical/incomplete connection (no or loose screw connection), corrosion at a screw connection or complete cable break (i.e., interruption). In addition, it must be equally realized in all learning situations that a complete electrical potential separation can never be measured due to the (electrically conductive) cooling water for temperature management (effect of the residual leakage current). The only possible variation for a differentiation between novice and skilled expert is, therefore, the specification/nonspecification of the setpoints and the specification of which assembly was worked on beforehand.

4.2 Development of COMET Tasks for the Evaluation of Measures

4.2.1 Development of COMET Tasks

Based on the result according to the revised learning situation matrix in Fig. 4, a 2-week training concept was developed on the basis of the learning situations mentioned there, which in turn addresses two learning situations in a shortened form:

- The troubleshooting regarding the pilot line according to learning situation 1 in Fig. 4 (since frequently occurring).
- Carrying out a potential measurement in accordance with learning situation 3 in Fig. 4 (as this has hardly been relevant as a measurement method in automotive engineering to date and is therefore of particular importance as learning content).

Table 4 shows which design options should be available within the COMET tasks. Together with an experienced rater, these specifications for a competence-promoting learning task were analyzed on the basis of the problem "Measurement on an electric vehicle." The problem was systematically differentiated in a joint discussion between the subject matter experts with the help of the item catalog (Is the item present? How else can the task be designed so that it satisfies the respective item?). This resulted in tasks with a solution that encompassed all items. These tasks, in turn, were evaluated by raters who had not initially participated in task conceptualization and were now discussed with this team of raters. In this way, a solution space was formulated as a basis for rating solutions of course participants. This solution space is then at the same time a didactic basis for the conceptual design of the further teaching-learning process with regard to the thematic input and the learning support.

4.2.2 Input Task

In the context of the first COMET test, the students have 60 min to work on the task individually, without the specific criteria and requirements of the possible solution space having been addressed in advance in the school context.

The description of the situation serves in particular to familiarize the students with their function as experts or specialists for diagnoses in a PA. At the same time, the level of the expected solution is emphasized (maximum between novice and expert level).

The following COMET task was formulated as the initial task for the start of the course, taking into account the fact that the participants have already completed a safety briefing on unlocking and working on EVs.

Vignette 3:

Situation Description:

As part of your training at Autohaus Muster GmbH, your journeyman asks you to help troubleshoot a customer vehicle with an e-drive: After you recently replaced the air-conditioning compressor on the vehicle and performed a general inspection during which everything seemed fine, the disgruntled customer is now on site. The vehicle will not start and displays the error message: "Error high voltage system—No start possible."

Work Order:

Develop justified solution proposals and solution variants of your diagnostic plan for the troubleshooting that is now pending.

You must be able to present and explain your innovative proposals to the responsible foreman at the end of the planning process.

Aids:

For planning and development, the workshop equipment of Autohaus Muster GmbH is at your disposal.

Note: Autohaus Muster GmbH is represented here by the inter-company training center "handwerk bremen."

4.2.3 Initial Task

The second COMET test at the end of the course serves to check the competence gain and, as it were, evaluate the project. The task is currently still under development; the overall evaluation of the DIAKOM-E project based on this is planned for mid-2023.

5 Preliminary Evaluation Results and Conclusion

5.1 Preliminary Evaluation Results

Regardless of the concrete COMET test results, which are currently being evaluated, the instructors commented in the current course of the project that the students had problems with the unfamiliar confrontation with open real-life situations. The challenge of not reproducing a predefined work plan, but of creating and taking responsibility for it themselves in a fault-specific manner, was (correctly) perceived by the students as a major change in the learning and working culture. This is because the imitatio principle (pretending-after-doing with classic master–apprentice role distribution) that often still dominates in vocational training in reality is very largely broken through here, especially since the "best practice" example as an orientation framework does not exist in the social media channels either. It is true that student research was conducted on the Internet and examples were also found here—but unfortunately without any indication of whether what is shown there is really optimal and why.

5.2 Discussion of Results

A key conclusion about the project is illustrated by the following example.

Vignette 4:

Measuring success in projects of this kind is very well secured and scientifically respectable with the help of empirical methods. COMET, as an internationally recognized method of competence measurement, will therefore be able to deliver the corresponding results as a valuable instrument.

But success can also simply be perceived, which often has a much more convincing effect beyond all the measured values: "*You literally have to tie the students to the chairs so that they don't go to the open vehicle and get started as soon as they enter the room.*" This is how the project coordinator of the DIAKOM-E project describes the success of focusing on independent "*trying, planning and not least solving*" of diagnostic problems on the "open" training vehicle.

At the same time, the observation described above illustrates a current dilemma in vocational training in automotive engineering: there is a lack of essential information and opportunities to implement maintenance, fault diagnosis, and repair of *real* EVs in a didactically sound manner:

1. This is because experience with the typical causes of faults in EVs is scarcely available or poorly documented. It is doubtful whether the strategy of the manufacturer plants of not naming precisely these is actually helpful in pushing the mobility shift associated with EVs. This is because customer skepticism about technology increases if there is precisely no high-quality on-site maintenance and repair for these vehicles, which in turn can only be performed by appropriately competent, locally available service employees. Promoting this competence, however, can only succeed if didactically relevant work processes, such as those developed here in the DIAKOM-E prototyical framework, can be worked through. This then includes, above all, that instead of the path now chosen via measurement procedures, the requirements-based error analysis and correction again forms the starting point of a learning situation.
2. However, knowledge of typical faults and their causes is only one side of the coin, especially since it must be possible to demonstrate these faults in the course of qualification measures. The other side is the problems described with fault connection, which can only be carried out by special specialists for HV technology and can often only be realized in a less realistic way via additional relay circuits. In the medium term, the problem of fault connection can probably only be solved by appropriately prepared and thus expensive training vehicles, whereby it must be examined whether connection can also be realized in the future via software bugs.

3. Likewise, the test technology is still very much determined by manufacturers' own test systems. For example, the Bosch ESI system is hardly capable of diagnosing the Toyota vehicle; here, as an alternative, a "gray" solution was used with an adapter cable from Chinese production and with diagnostic software of unknown (probably American) origin. In contrast, the Audi vehicle can be diagnosed better; however, the exact localization of the fault requires additional test cables, which are also not or poorly available for the manufacturer's own workshops.
4. This means that there is a lack of such test cables as adapter and Y cables for realistic root cause analysis as part of the VET concept as these are also not (yet) available on the "free" workshop market. Here, too, manufacturers will have to act more openly and reliably for maintenance and training purposes in order to simultaneously strengthen the end customer's confidence that he has a vehicle that can be diagnosed and repaired without any problems (!) because test equipment and employee expertise are available.
5. It also needs to be clarified what should be part of the job description of an automotive service technician in the future. On the one hand, a decision must be made as to whether and to what extent the diagnosis and repair of wallboxes and charging columns should be part of the job description. Secondly, it must be clarified which degree of component repair (above all: drive battery, inverter, drive motor) should become part of the job description. Only after these decisions have been made can training concepts be further developed. From an ecological point of view, however, it would not make sense to dispense with on-site repairs or at least repairs at regional bases.
6. What remains unquestionable, on the other hand, is the certification required to work on high-voltage and gas tank systems must be integrated into initial vocational training.

6 Conclusion

The experience gained so far in the "DIAKOM-E" project allows the following conclusions to be drawn:

1. The general service and spare parts business is increasingly facing a crisis situation due to the introduction of EVs with their reduced service effort.
2. This appears to be particularly problematic for the manufacturer-independent workshops.
3. This negative development is triggered by four factors:
   - The OEMs are very reluctant to provide the independent repair and aftersales market with testing and diagnostic tools (adapter cables, testing software), contrary to the BER for the automotive trade that applies throughout Europe.
   - The same applies to the test data required for diagnosis, insofar as these could be determined with commercially available testing and measuring tools.

- It is also questionable whether smaller companies in the independent aftersales market are even in a position to make the necessary investments for the diagnosis and repair of EVs.
- Essential spare parts for carrying out repair work, especially on the battery pack and fuel cell, are not made available. Instead, the premise of expensive component replacement applies, which is then again preferably carried out by the OEM-linked workshops, precisely because there are no inexpensive spare parts alternatives or any spare parts at all for the independent repair market.

4. It is, therefore, all the more important to establish a vocational training and continuing education system that enables those working in the aftersales market to work on EVs independently of the OEM's qualification measures. "To work" means more than just complying with the relevant safety regulations; it means, above all, the ability to diagnose and repair on one's own responsibility. These teaching contents, identified here in the "DIAKOM-E" project, must therefore become the curricular standard of initial vocational training (and thus also initially in continuing vocational training) 0.5.
5. With regard to the structures of vocational training described in the DACH region, this appears to be quickly possible through dual training with the support of vocational colleges and inter-company training centers; however, it also became clear that for countries that have a more modularized training system at EQF levels 1–5, it is important to develop vocational training offers as quickly as possible. This is because the tendency of OEMs to seal off the market for repairs and maintenance from the free aftersales market leads to market distortion to the detriment of the consumer.
6. But there are also indications that this will happen: The many new OEMs of Asian origin entering the market are initially dependent on the availability of appropriately qualified aftersales service personnel, precisely because they have not yet built up their own workshop network—but at the same time offer inexpensive and therefore mass-compatible vehicles. The existence of a correspondingly manufacturer-independent aftersales service thus supports the rapid switch to EV vehicles, which must be completed quickly for ecological as well as economic reasons, especially for private customers.

References

Becker, M. (2003). *Diagnosearbeit im Kfz-Handwerk als Mensch-Maschine-Problem.* Bertelsmann Verlag.

Blankertz, H. (1982). *Die Geschichte der Pädagogik. Von der Aufklärung bis zur Gegenwart.* Wetzlar: Verlag Büchse der Pandora.

BMWi: Bundesministerium für Wirtschaft: Verordnung über die Berufsausbildung zum Kraftfahrzeugmechatroniker und zur Kraftfahrzeugmechatronikerin* vom 14. Juni 2013. Last call January 17, 2023, from https://www.google.com/url?sa=t&rct=j&q=&esrc=s&source=web&cd=&ved=2ahUKEwjC6didopf9AhV07LsIHVwiAKkQFnoECA4QAQ&url=https%3A%2F%2Fwww.bibb.de%2Fdienst%2Fberufesuche%2Fde%2Findex_berufesuche.

php%2Fregulation%2Fkraftfahrzeugmechatroniker2013.pdf&usg=AOvVaw1djbrc2OSgFEMKTF5roTu8
Bourdieu, P. (1978). *Entwurf einer Theorie der Praxis auf der ethnologischen Grundlage der kabylischen Gesellschaft. Aus dem Französischen von Cordula Pialoux und Bernd Schwibs*. Suhrkamp. (= suhrkamp Taschenbuch Wissenschaft. 291.).
Bremer, R. (2006). Zur Implementation grundlegender Methoden in der Berufsbildungsforschung (beobachten, experimentieren, befragen, Inhaltsanalyse). In F. Rauner (Ed.), *Handbuch Berufsbildungsforschung*. wbv-Verlag.
Brödner, P., & Oehlke, P. (2018). Gestaltung von Arbeit und Technik. Einführung. In F. Rauner & P. Grollmann (Eds.), *Handbuch Berufsbildungsforschung. 3. Aktualisierte und erweiterte Ausgabe*. wbv (utb-Verlagsgemeinschaft).
De Graaf, E., & Kolmos, A. (2013). Characteristics of problem-based learning. *International Journal of Engineering Education, 19*(5), 657–662.
DGVU: Deutsche Gesetzliche Unfallversicherung e.V. (2018, ed.): AKTUELL FB HM 099 „Gasantriebssysteme in Fahrzeugen – Qualifizierung für Arbeiten an Fahrzeugen mit Gasantrieb. Last call January 17, 2023, from https://publikationen.dguv.de/widgets/pdf/download/article/3528
DGVU: Deutsche Gesetzliche Unfallversicherung e.V. (2021, ed.): Qualifizierung für Arbeiten an Fahrzeugen mit Hochvoltsystemen. https://publikationen.dguv.de/widgets/pdf/download/article/3982
Dreyfus, H. L., & Dreyfus, S. E. (1980). *A five-stage model of the mental activities involved in directed skill acquisition*. ORC-80-2, Operations Research Center, University of California.
KMK: Sekretariat der Kultusministerkonferenz (2013): RAHMENLEHRPLAN für den Ausbildungsberuf Kraftfahrzeugmechatroniker und Kraftfahrzeugmechatronikerin (Beschluss der Kultusministerkonferenz vom 25.04.2013). Last call February 15, 2023, from https://www.kmk.org/themen/berufliche-schulen/duale-berufsausbildung/downloadbereich-rahmenlehrplaene.html?type=150&tx_fedownloads_pi1%5Bdownload%5D=12779&tx_fedownloads_pi1%5Baction%5D=forceDownload&tx_fedownloads_pi1%5Bcontroller%5D=Downloads&cHash=5b9f6d508b9d6f6f934f9250879d5397
KMK: Sekretariat der Kultusministerkonferenz Referat Berufliche Bildung, Weiterbildung und Sport (2021): Handreichung für die Erarbeitung von Rahmenlehrplänen der Kultusministerkonferenz für den berufsbezogenen Unterricht in der Berufsschule und ihre Abstimmung mit Ausbildungsordnungen des Bundes für anerkannte Ausbildungsberufe. Last call February 15, 2023, from https://www.kmk.org/fileadmin/veroeffentlichungen_beschluesse/2021/2021_06_17-GEP-Handreichung.pdf
Rauner, F. (2017). *Methodenhandbuch. Messen und Entwickeln beruflicher Kompetenzen (COMET)*. wbv-Verlag.
Rauner, F. (2018). *Berufliche Kompetenzmessung mit COMET. Erfahrungen und Überraschungen aus der Praxis*. wbv-Verlag.
Rauner, F., Haasler, B., & Heinemann, L. (2009). Berufliche Kompetenzen messen. In F. Rauner, B. Haasler, L. Heinemann, & P. Grollmann (Eds.), *Messen beruflicher Kompetenzen. Band I: Grundlagen und Konzeption des KOMET-Projektes. 2.Auflage* (pp. 23–50). LiT-Verlag.
Ulseth, R. (2016). *Self-directed learning in PBL. Disseration*. Aalborg University Press.
Vargas, N. V., Ortiz, J. L. A., & Rodriguez, A. R. L. (2019). Project-based learning to enhance environmental education through automobile mechanics. *Journal of Problem-Based Learning, 6*(2), 76–84.

Ralph Dreher, born 1965, vocational training as a car mechanic, master's degree for teaching at vocational schools (mechanical engineering/German), doctorate in didactics of automotive engineering, 15 years of teaching at vocational colleges, lecturer at the Technical University of Hamburg-Harburg and the University of Flensburg. In 2008, he was appointed to the chair of technology didactics at vocational schools at the University of Wuppertal. Since 2013, he has held the newly founded chair of technology didactics at vocational colleges and is the founder of the vehicle didactics laboratory at the University of Siegwen. Vice President of the IPW (Scientific Society for Engineering Pedagogy) and recipient of the Adolf Melezinek Medal of the IGIP (International Society of Engineering Pedagogy).

Jens Jüngst, born in 1974, successfully completed an apprenticeship as an automotive electrician after graduating from high school. After several years in vehicle maintenance with a focus on system diagnostics, he became a master automotive technician in 2006. He moved to a car dealership in 2011 as a service consultant & deputy location manager. Through regular further training, he is up-to-date in the future-oriented and safety-relevant areas such as high-voltage, air-conditioning, or airbag technology. Since 2017, he has been supporting Prof. Dr. Ralph Dreher at the Chair of Technology Didactics at the University of Siegen as a workshop master in the training of teachers for the vocational college.

Fynn Komp, I started my professional career after graduating from high school in 2016. First, I completed an apprenticeship as an industrial mechatronics technician, i.e. in the industrial-technical field. This gave me my first insights into the dual system of vocational education in Germany. After completing the apprenticeship, I decided to study to become a teacher at the University of Siegen, with the subjects of mechanical engineering technology and Catholic religious education. Through my work as a student assistant at the Chair for Didactics of Technology, I dealt with the didactic development of BEV, among other things, and also integrated this into my bachelor's thesis. I completed my bachelor's degree with this thesis at the beginning of the year 2023.

Gender Equality at Work: Tackling Gender Inequality in Senior Leadership in the Automotive Aftermarket Industry

Amanda Katsardis and Raushan Aman

Abstract

The primary aim of this chapter is to explore what measures companies in the automotive aftermarket are employing to achieve more gender diversity in senior leadership positions. Gender equality in leadership has been stated as a sub-goal of the UN's Sustainable Development Goal number 5, emphasising that businesses have to respond to gender inequalities immediately. To explore what actions the automotive aftermarket is currently taking and what can be improved to alleviate the challenges that women are facing in career advancement, qualitative semi-structured interviews were employed. A sample of 10 female employees in high-ranking positions in different European automotive aftermarket companies yielded findings which show that there is a gender bias in the automotive aftermarket, which might impede career advancement for women. Another main challenge proves to be work–life balance, which shows that increasing job flexibility and female empowerment would support women and enable them to advance to higher positions. Importantly, several companies are already increasing their focus on gender equality by implementing specific diversity policies and mentoring programmes. Despite these initial efforts made by companies, experiences from women in the aftermarket show that there are still significant hurdles which the industry has to overcome.

A. Katsardis (✉)
University of Aberdeen, Aberdeen, UK

R. Aman
Matti Ahtisaari Institute, University of Oulu, Oulu, Finland
e-mail: Raushan.Aman@oulu.fi

M. Elo, F. Katsardis (eds.), *Automotive Aftermarket*, Management for Professionals, https://doi.org/10.1007/978-3-031-62419-3_9

1 Introduction

Economies worldwide are still far from achieving gender equality. Reports show that 'it will take 135.6 years to close the gender gap worldwide' and 'it will take another 267.6 years to close the gender gap in economic participation and opportunity' (WEF, 2021, p. 5). In 2021, only 27% of all manager positions worldwide were occupied by women (WEF, 2021). Only 5% of Fortune 500 CEOs are women, and the number of women in leadership of a wide range of fields remains stuck at 5–20% (CAP, 2018).

The automotive aftermarket is a male-biased industry, women are significantly underrepresented, and specifically, senior leadership positions are predominantly occupied by men (Horak & Cui, 2017; Kjellson, 2017). Only 20% of the industry's workforce and less than 10% of its executives is female (Deloitte UK, 2020). A majority of women believe that the lack of diversity in leadership in the automotive industry can be explained by an industry bias towards men in leadership positions (Deloitte US, 2020). To grow further as a sustainable and profiting sector, the automotive aftermarket needs to focus on involving more women in high-ranking positions, as gender diversity in senior leadership positions has proven to bring multiple benefits (Monteiro et al., 2021; Opstrup & Villadsen, 2015). Despite positive development in getting more women into company boards, the number of women in leadership positions remains limited (Hoyt, 2010; Klettner et al., 2016; Taneja et al., 2012). The automotive aftermarket as an industry often gets overlooked. However, it is a crucial industry which has a significant impact on people's everyday lives and a major employment effect. Once cars get manufactured and sold, they enter the automotive aftermarket, which deals with these cars for many years. An official definition given by Figiefa (2020, p. 45) states, 'The Automotive Aftermarket is a business sector centred around the servicing and maintenance of vehicles after their production and initial sale. It includes the manufacturing, remanufacturing, distribution, retailing, repairing, maintaining and installing of all vehicle parts, chemicals, equipment and accessories and the provision of these respective services to private, corporate or public users. It covers the entire life cycle of the vehicle after its production.'

There are potentially many explanations for the lack of women in leadership positions, including women's leadership in the automotive aftermarket industry. Previous studies have emphasised the importance of female leaders as companies with higher numbers of women in leadership positions experience increased efficiency (Adams & Ferreira, 2009; Adams, 2016; Horak & Cui, 2017). Economies and businesses prosper when women are able to live out their full potential, unfulfilled female potential hinders economic growth (Adams, 2016; Taneja et al., 2012). Women tend to be transformational leaders, who are more effective when they drive organisations forward (Adams & Ferreira, 2009; Adams, 2016; Boulouta, 2013; Ely & Thomas, 2001; Randsley de Moura et al., 2018). One common issue impeding female careers is called the 'glass ceiling' (Hoyt, 2010). This describes how women are struggling to advance in their career due to diverse hindrances. When appointed to a job, they get less promotions and chances to develop in their career, which consequently makes it harder to gain a leadership position (Hoyt, 2010; Kjellson, 2017). Leadership positions that women are likely to be employed in are usually connected to higher risk and more criticism, which is described as 'glass cliffs' (Hoyt, 2010). There is a clear need to understand company-level practices related to female management as the previously mentioned percentages show significant gender inequality. The specific problem is that the progress in increasing female leaders is moving slowly or hardly at all, which exemplifies that companies need to update their responses and strategies (Demartini, 2019). This study is looking at the responses of the automotive aftermarket as this is a heavily male-biased sector which has remained underexplored. Despite the importance of this industry and its traditional gender bias, the automotive aftermarket's response to increase the number of women in senior leadership is surprisingly under-researched. It is important to research this as the automotive aftermarket is an industry of a significant size and employment effect. Increasing the number of female leaders in this sector would bring a meaningful advancement to gender equality in line with the UN gender equality goal number 5 in the Sustainable Development Goals 2030 (United Nations, 2022). However, the information presented in the Global Gender Gap Report 2021 shows that gender gaps will not be filled for more than 100 years. Thus, industry sectors and companies need to take rapid action to achieve the UN's goals. Achieving more gender equality and increasing female leadership would help companies in the automotive aftermarket develop a competitive advantage, making the industry more attractive for a diverse workforce, driving sustainability.

The aim is to investigate industry responses of the automotive aftermarket to tackle the problem of gender inequality in senior leadership. To address the research aim, three key research questions were developed and are outlined as follows:

1. Why is there a lack of female management representation in the automotive aftermarket, and how is this connected to the male bias in this industry sector?
2. What challenges are women facing in career advancement in automotive aftermarket companies?
3. What strategies do companies employ in order to recruit, retain, and advance female managers to senior leadership positions?

This research contributes to the extant literature by studying women's underrepresentation in this sector, and how this can be changed within companies by advancing their human resource and management practices. Investigating this contributes to discussions on gender equality, as stated in the UN's Sustainable Development Goals, which declare that each person should get the same chances at jobs, despite their gender. The automotive aftermarket has shown attempts at rethinking its practices and developing more female talent as it is facing challenges being a less attractive and unknown industry. A recent initiative is 'Talents for the Aftermarket', which specifically requests a higher participation of women in the industry (Talents4AA, 2022). It is important to focus on strategies which improve gender diversity as women form a significant source of management talent.

The rest of the chapter is structured as follows: we first start by presenting the literature review on women's leadership and the importance of gender-diverse teams in management; female management teams supporting UN SDGs; female talent empowerment and glass ceiling; stereotypes about women in leadership and management and previous suggestion to increase the number of female leaders and manager; and the gender pay gap. We then describe implemented research methodology, data collection method, and data analysis. Following this, the findings section demonstrates the seven main themes we came up with after our data analysis. Our research findings are then discussed in relation to extant research, and, finally, conclusions are provided.

2 Literature Review

This section covers the main points which have been discovered in previous research regarding female leadership and women's role in the automotive industry. It presents key arguments from existing literature. These include findings from scholars that investigate effective leadership, the advantages of gender-diverse management teams, connections to the Sustainable Development Goals, the male bias of the automotive industry, the gender pay gap, and suggested ideas of how to achieve more gender equality in senior leadership.

2.1 Women as Effective Leaders and the Importance of Gender-Diverse Teams in Management

Research on the role of women in management emerged in the early 1980s (Demartini, 2019). However, there has been more attention on the topic in recent years. Research on gender inequality and the role of women in business has increased significantly as the positive effects and contributions of female leaders have become apparent. Multiple studies over the years show that women are effective leaders. The term 'female advantage' describes the transformational leadership style of women, driving companies forward in efficiency (Randsley de Moura et al., 2018). Women tend to lead in a democratic style; they encourage participation, resulting in higher productivity (Hoyt, 2010; Randsley de Moura et al., 2018). An example is Hoogendoorn et al.'s (2013) experiment, which illustrated that gender-equal teams have better sales and profit performance compared to male-dominated teams. Cook and Glass (2018) explored data from Fortune 500 companies over several years, finding that gender-diverse boards and female CEOs lead to stronger business and equity practices, so diversity brings efficient outcomes for businesses overall. Studies by Monteiro et al. (2021) exemplified that gender diversity in management teams leads to more focus on labour and human rights in the company. Diverse management teams with more women on board pay more attention to corporate social responsibility (ibid.). Other advantages of having gender diversity in management are that more diverse leadership teams can understand the work environment better and relate to more diverse topics (Opstrup & Villadsen, 2015). Additionally, they offer more varied social and human capital, leading to higher creativity and innovation. Moreover, diverse teams solve problems better as they can think of various alternatives and consequences to issues. Gender-diverse teams work more efficiently and show better financial performance (ibid.). Men are more likely to take high risks and as women avoid taking excessive risks, gender-diverse teams might be perfectly balanced (Kiefner et al., 2022; Opstrup & Villadsen, 2015). Confirming the advantages of gender-diverse teams, study findings by Opstrup and Villadsen (2015) show that gender-diverse teams can outperform gender-homogeneous management teams. However, they emphasise that the crucial point in making gender-diverse teams effective is the right structure in a company (ibid.). Biswas et al. (2021) examine how more women on boards increase the number of women in senior management in firms, showing a trickle-down effect. So, improving gender board diversity affects overall diversity positively (ibid.).

2.2 Female Management Teams Supporting the SDGs

In addition to economic benefits, Kiefner, Mohr and Schumacher (2022) found that women on management teams increase multinational's support of the SDGs. So, companies that have gender-diverse top management teams (TMTs) are more likely to advance their business sustainably. Demartini (2019) suggested similarly, stating that women in business help advance sustainable growth. Achieving gender equality in TMTs advances the goal of gender equality, SDG number 5, and other sustainable development goals (Kiefner et al., 2022). Women on boards seem to increase companies' attention to sustainability, especially corporate philanthropy, as more attention is paid to CSR. Companies with gender-diverse management show better corporate governance and increased social engagement. Higher support of the SDGs is due to the different beliefs and behaviour of women compared to men. Another contributing factor is social expectations towards women, predicting that women should support the SDGs, as goal number 5 specifically talks about female leadership. Women face social pressure to adhere to these standards and pay attention to sustainable practices, which is not as pronounced for men. The representation of female managers has an impact on group decision-making as team members are more open to change (ibid.). Diverse TMTs are more open to take multiple perspectives into account which leads to higher CSR, which is one contributing factor for the support of the SDGs. Women seem to have higher ethical standards; thus, companies with female executive leaders have smaller chances of being involved in financial fraud. Despite clear findings showing that women advance their business towards the SDGs, Kiefner et al. (2022) only investigated multinational corporations, leaving unclear what results would be like for smaller firms with female managers. The contributions of female managers to the SDGs are still under-researched, Kiefner et al.'s (2022) findings might be limited as no other scholars have researched this specific topic. Additionally, Carmona et al. (2018) emphasise that as most existing literature on female management focuses on a Western context, it is not possible to generalise these findings for every situation. A gendered management style is dependent on the socio-economic and political context, which varies significantly across different settings (ibid.).

2.3 Female Talent Empowerment and the Glass Ceiling

Other research in the field of gender equality in the workspace has explored why companies should focus on female talent empowerment. Coleman (2010) indicated that companies reinforcing female empowerment show increased productivity through their labour force whilst also increasing the quality of their supply chains. Enhancement of women's economic power is good for business as the company's talent pool increases, resulting in higher efficiency and retention (Coleman, 2010). However, the problem is that women tend not to get employed into leadership positions (Cook & Glass, 2018; Ely & Thomas, 2001; Hoyt, 2010; Randsley de Moura et al., 2018). The phenomenon of women not rising into high-ranking positions has been described by multiple scholars using the term glass ceiling (Coleman, 2010; Opstrup & Villadsen, 2015; Siniscalco et al., 2014). Despite working hard, there seems to be an invisible barrier for women, keeping them from reaching leadership positions. Another issue facing women is the 'glass cliff'. This means women are prone to being promoted to riskier leadership positions than men (Hoyt, 2010). The organisations promoting them are facing higher risks to fail or are already struggling (Cook & Glass, 2014). Siniscalco et al. (2014) explain that this underrepresentation of women can be seen in high ranks of business, finance, government, and academia. The automotive aftermarket is one of these sectors showing a clear underrepresentation of women in senior leadership. After explaining common stereotypes about gender and leadership, the following will go into depth to explain the specific issues of gender in the automotive aftermarket.

2.4 Stereotypes about Women in Leadership and Management

A common issue confronting women in the workplace is stereotypes, specifically that management is a male task (Opstrup & Villadsen, 2015). Men are seen as 'better leaders', organisations tend to welcome an androgynous leadership style (Vasconcelos, 2018). Women are seen as less skilled in leading an organisation compared to white men. Women working in management are thought to lack experience of formal management systems (Carmona et al., 2018). Scholars agree that one very common stereotype is that 'prototypical leaders are white and male' (Cook & Glass, 2014, p. 1082). When appointed to high-ranking positions, minorities tend to face more objections than white men. Nevertheless, Cook and Glass (2014) argue that these challenges that minorities, such as female leaders, have had to overcome make them outstanding leaders. The discrimination that women face whilst climbing up the career ladder makes them accustomed to the challenges they have to face as leaders, which might be surprising for white male leaders (ibid.). In a critical review, Vasconcelos (2018) pointed out that despite organisations preferring a male leadership style, stereotypes about leadership have shifted slightly. Hoyt (2010) indicates similarly, stating that the image of women as leaders is seen to be more accepted; however, it is still highly stigmatised. Vasconcelos (2018) suggests that organisations should stop focusing on the gender of leaders as this discussion leads to a growing distinction between people of different genders and increases rivalry and prejudice at the workplace, enforcing a gender bias.

2.5 Gender Pay Gap

Another crucial topic when it comes to gender equality and work is the gender pay gap. Studies regarding the gender pay gap show mixed results, some findings show women earn significantly less than male colleagues, whilst others point out that this inequity has been overcome and female workers earn the same as male. For instance, a study conducted by Bugeja et al. (2012) focused on analysing whether there is a gender pay gap at the level of CEOs. However, findings from almost 300 US firms did not show significant results. The study concluded that women and men receive similar payments, so there is no gender bias in pay at CEO levels (Bugeja et al., 2012). Opposingly, Vasconcelos (2018, p. 1349) stated that there is an 'undeniable salary gap between men and women', which companies need to eliminate to overcome disparities at work and offer fair opportunities for everyone. Through qualitative research, Bullock discovered that the automotive industry in the United States does show inequitable pay (Bullock, 2019).

Other research by Stojmenovska (2019) investigated the effects on earnings when the number of female managers in a company increases. Findings showed that the gender pay gap decreased when more female managers were hired. Still, Stojmenovska (2019) points out that a gender pay gap does exist widely. Reasons for this are that men in leading positions are used to the advantage in payment and will not challenge this. Additionally, in most cases women are unable to get into these higher-ranked positions. This issue has been described earlier by studies that focused on exploring the glass ceiling. In her paper, Stojmenovska (2019) discusses various reasons why female managers reduce the gender pay gap, including role modelling and not being biased against other female employees. Women are more likely to implement gender equity-supporting policies at work due to the fact that they have experienced unequal treatment themselves. The gender gap decreases due to an increase in women's average earnings, so men's earnings do not suffer from this (ibid.). As no official information is given on employees' pay in the automotive aftermarket and no research has covered this subject, the results from the qualitative interviews conducted in this research project will try to offer new insights into whether a gender pay gap does exist in the automotive aftermarket, which may also serve as a challenge to women working in this field.

2.6 Previous Suggestions to Increase the Number of Female Leaders

To bring change into the underrepresentation of women in senior leadership, several approaches to increase gender equality exist or have been proposed as solutions for more diversity. These may stem from companies and their need for qualified managers developing, for example, diverse traineeship programmes, but also from society and institutions imposing regulations on gender issues. The United Nations agreed to work together on multiple goals by 2030, gender equality is listed specifically as goal number 5. One of the targets, number 5.5, states that women should have 'equal opportunities for leadership at all levels of decision-making in political, economic and public life' (UN, 2022). A key indicator listed for this is 5.5.2 'Proportion of women in managerial positions' (ibid.). As the SDGs are commonly agreed on goals for continuous growth until 2030, when they are set to be achieved, businesses should strive to reach this target soon. Some ideas that have been discussed by scholars and governments to advance towards SDG number 5 are listed below.

To reduce the gender gap in leadership, companies need to act swiftly. Industry responses include equal opportunity strategies, opportunity-enhancing strategies, reporting requirements, and setting gender targets (Sojo et al., 2016). Biswas et al. (2021) suggest that promoting women's empowerment would help increase the representation of women in top management. Implementing a policy on female empowerment would improve women's access to senior leadership positions. They further suggest an employment policy observing gender representation to ensure gender equality (ibid.). Taneja et al. (2012) propose a similar approach to Biswas et al. (2021), emphasising that organisations should focus on women's empowerment, to retain female talent and help women build careers. They propose implementing flexibility and gender equality strategies into the organisational culture and policies (ibid.). Focusing on empowerment and flexibility would increase productivity in organisations as employee and customer satisfaction increase. Taneja et al. (2012) claim that highly skilled new talent gets attracted to organisations with such a good reputation. Other specific recommendations are that companies develop specific mentoring and sponsorship programmes or utilise social media and provide leadership resources to women (Taneja et al., 2012).

Another proposed solution is quotas, officially adapted by several countries, a prominent example is Norway (Seierstad et al., 2021). The presence of a female board leads to better overall performance in a company as it is more supportive and less discriminating (Cook & Glass, 2015). These are some strategies that previous literature has examined and provided evidence on. However, research on strategies employed by companies in the automotive aftermarket is lacking. Hence, this chapter aims to bring new insights into this topic by employing explorative qualitative interviews investigating women's experience of discrimination and career progression in this field.

After examining existing literature, a lack of research on the role of women in the automotive aftermarket is noticeable. Despite many researchers investigating the underrepresentation of women in the automotive industry and numerous papers describing the benefits of female leadership and gender-diverse management teams, similar research has not been done in the context of the automotive aftermarket. Thus, this chapter will try to fill that gap in the literature and explore whether the findings for the automotive industry are applicable to the automotive aftermarket. It would seem likely that assumptions and findings of existing literature, such as on the male bias, will be supported by the results of this research.

3 Research Context

3.1 The Automotive Aftermarket: An Industry for Men? Issues Facing Women in a Male-Biased Workspace

The automotive aftermarket as an industry often gets overlooked. However, it is a crucial industry that significantly impacts people's everyday lives and a major employment effect. Once cars get manufactured and sold, they enter the automotive aftermarket, which deals with these cars for many years. An official definition given by Figiefa (2020, p. 45) states that 'the Automotive Aftermarket is a business sector centred around the servicing and maintenance of vehicles after their production and initial sale. It includes the manufacturing, remanufacturing, distribution, retailing, repairing, maintaining and installing of all vehicle parts, chemicals, equipment and accessories and the provision of these respective services to private, corporate or public users. It covers the entire life cycle of the vehicle after its production.'

There may be sector-specific issues attributed to female leadership in the automotive aftermarket; due to lacking previous research, findings from the automotive industry are presented. These may be related or similar to women's experiences in the automotive aftermarket. The automotive industry can be defined as a predominantly male industry, emphasised by the low percentage of employed women (Horak & Cui, 2017). The setting around this industry is male-centred, for instance, cars are seen as masculine leadership symbols, with pictures showing men driving cars and strict domestic gender roles on car usage. The workspace is gendered, with organisational hierarchies favouring men in leadership positions (Bullock, 2019; Kjellson, 2017). Gendered substructures make it harder for women to excel career-wise and be hired into high positions (ibid.). Gender bias is one main contributing factor in restricting women's career progression (Biswas et al., 2021). Inequalities are seen in corporate settings, where women get paid less, having to fight for equal pay (Bullock, 2019). Organisational structures push men more, they advance in their career based on gender bias, whereas it has been witnessed that some women are advancing based on physical appearance. Women struggle to overcome these gendered pressures and experience social isolation at work or 'gender othering' when occupying leadership positions. Hostile work environments and others challenging their authority negatively affect women's ability to lead, caused by high pressure and scrutiny (Cook & Glass, 2015; Hoyt, 2010). Challenges at work include sexual harassment; almost all women who participated in a recent study answered that they had experienced sexism at work (Bullock, 2019). Biased hiring processes restrict women's access to leadership positions, impeding them from showing their full potential at work (McKinsey, 2021). Some barriers to career progression that Kjellson (2017) identified in the automotive industry include sexist comments, and most importantly gendered practices and substructures. Sexual harassment has also been mentioned by Foley et al. (2020). Their research highlighted gender harassment, which is demeaning behaviour based on women's gender, including sexist remarks. The high rate of discrimination women face presents a burden to women's career progression (Foley et al., 2020; Kjellson, 2017; Lloyd & May, 2007). Significant barriers to leadership positions for women include unconscious bias, stereotyping, and ineffective management of diversity at a corporate level (Hoyt, 2010; Randsley de Moura et al., 2018).

Discrimination and stereotypes are connected to negative expectations and reactions in the workplace. Another prominent issue presenting a challenge for women's career advancement is work–life balance (Demartini, 2019; Taneja et al., 2012; Vasconcelos, 2018). As long as women have to take a break from work to take care of their children and without receiving support from work, women will not be able to release their full potential in business (Demartini, 2019). Female leaders with families need more support in terms of childcare services (Kjellson, 2017; Kowaleska, 2021). Tanjea et al. (2012)) suggest that organisations should focus on flexibility to support women and minimise the challenges of balancing family life and their career.

3.2 Research Methodology

This study employs qualitative research methods, using interviews and observations. Qualitative research tries to unveil new or previously less researched phenomena, including investigating people's experiences and behaviour (Ghauri & Gronhaug, 2010). Qualitative research methods capture differences and dissimilarities in data, providing a broad picture of the studied phenomenon, involving nuances and unexpected observations, as qualitative data are likely to be more varied and complex than quantitative data (Saunders et al., 2019). Business studies often use qualitative data collection methods, which include interviews (structured, semi-structured, or unstructured), surveys, or observations (Ghauri & Gronhaug, 2010). These qualitative methods help researchers understand how organisations work or understand the behaviour of groups and individuals (ibid.). Qualitative interviews and observations may involve longitudinal time horizons instead of snapshots in time, which 'provide the best opportunities for the study of processes' (Gummesson, 2000, p. 35).

This study builds on qualitative methods, employing semi-structured interviews that allow for comparability across respondents and safeguard that adequate and relevant data is received (Blumberg et al., 2008; Hussey and Hussey, 1997; O'Gorman & MacIntosh, 2015). They also enable, when personal or mediated by teams, observations in parallel that serve as field notes. Qualitative interviews may discover how organisations work in terms of gender equality and help understand behaviour of groups and individuals. This form of qualitative research aids in finding out about women's experiences working in the automotive aftermarket as it enables them to explain their experiences, observations, and opinions which could be more complicated to capture through survey questions. Hence, their reflections in the interviews will bring new insights into an unexplored area of research as primary data.

3.3 Sampling and Sampling Procedure

The sampling procedure for this project is purposive-snowball sampling, a type of non- probability sampling, meaning that participants are selected strategically having the research questions in mind. People relevant to the field and able to answer the interview questions, contributing to new insights into the field are chosen as the sample. Snowball sampling means some participants were recommended to the researcher by other participants chosen for the sample from a personal network of the researcher. Qualitative research is primarily associated with purposive sampling, as researchers select the sample knowing their questions can be answered (Saunders et al., 2019). As it is non-probability sampling, results from the research cannot be generalised for the whole population. Sample members differ, they have various key characteristics, so there is variety in the resulting sample. This ensures some heterogeneity in the sample, providing richer findings (Bryman & Bell, 2015). Different characteristics such as age, background, and employment position might impact the respondent's experiences and answers.

This research draws on the recommendations of Saunders et al. (2019) to explore a sample size of 5–30 respondents. Due to time and resource constraints, this research project chose a sample size of 10 respondents. Through a personal network, it was possible to get contact information from people that were approached and then selected for interviews. All participants are women, employed in high-ranking positions (from sales director to CEO) in the automotive aftermarket industry in Europe (Germany, Italy, Denmark). The selection process tried to focus on various job positions in terms of area of expertise; however, a majority of participants work in sales. Every participant works or has been working in a different company of the automotive aftermarket, except for two women who were employed in the same company at some point during their career. Most respondents have worked in the industry for more than 20 years, with only two women being employed for 10 plus years and one person having retired from the industry after 3 years of working in it. The majority of the sample is working in companies located in Germany; however, their cultural background differs. So, there is some geographical variation in the companies and participants.

3.4 Data Collection

This study builds on qualitative methods as they allow accessing the information needed and offer insights that are necessary to address the research questions. The research seeks to understand women's lived experiences in the automotive aftermarket. As these can be best unfolded during interviews, data was collected through semi-structured qualitative interviews. The semi-structured qualitative interviews in this research comprise a standard set of questions on the demographic data and less structured open-ended questions. This leaves room to explore people's experiences, while providing factual information such as the respondents' gender (ibid.). The interviews were carried out in English and were recorded and transcribed. Every participant was asked 16 questions, the interviews were scheduled to last for approximately 30 min. They were person-to-person mediated interviews due to COVID-19-related safety. The interviews took place on Microsoft Teams, a well-suited medium for an interview situation as it is largely used in the business context.

3.5 Data Analysis

The data obtained from the interviews was transcribed and analysed. For transcribing the transcription software, Otter was used. The transcripts were then double-checked and corrected by the researcher as there were some misunderstandings in the transcripts. Everything said in the interviews was written down in separate word documents. The next step was organising the materials. The transcripts were organised per work position and perspectives given by respondents. Next up, written information was transferred into the software NVivo, offering a more organised depiction of the interview answers. The software aids in storing all transcripts in the same place and organising the data, assisting in sorting information into codes (called nodes on NVivo) and respective thematic groups that support the theorising. Such thematic groups include, for instance, women's experiences of gender bias or discrimination in the automotive aftermarket, or advancing to getting a promotion. There are many possibilities for analysis, here thematic analysis is employed (Braun & Clarke, 2006, 2012). After creating first-order codes, a back-and-forth movement between the data and literature made it possible to combine and collapse some codes and create new ones, to reach second-order codes. After this thematic grouping, it was reflected what these answers mean, and how the results tie into research. So, what are women's experiences and what are companies doing regarding increasing gender diversity? Understanding these themes and linking them to theory helped get to the relevant findings and analysis of the research, which then moves on to the discussion. Lastly, the validity of research had to be evaluated, as well as the relevance and rigour in terms of the whole industry sector's response. Limitations of the research are discussed in the end (Saunders et al., 2019).

4 Findings

The analysis of our interviews revealed seven main themes: the underrepresentation of female managers in the automotive aftermarket; the existence of male bias; equal treatment and discrimination at work; challenges to career advancement; gender and perceptions of family commitments; male bias and image building; and lastly, company strategies for improving gender equality.

4.1 Underrepresentation of Female Managers

To investigate the lack of women in the automotive aftermarket, and specifically in leadership positions, respondents were asked if they have ever had a female manager whilst working in the industry. The findings show that half of the women have had at least one female manager during their career in the automotive aftermarket. Noticeably, these were all respondents from Germany or Denmark. When asked if this impacted their ambition for career development, a few women answered that it seemed normal to them to have a female superior, so there was no impact. One person did state that this might, however, be more normal because of the geographical location she was in, which is a Nordic country, and these countries tend to show more gender equality in the workplace. Having a female manager seems to have a positive effect on women, as one respondent stated:

> '*I definitely know that it motivates people... this belongs to leading by example.*' (Respondent 8)

Interestingly, all respondents who never had a female manager were either working in the biggest company out of the sample or in a company located in Southern Europe. Having never had a female manager showed to have a stimulating effect:

> '*Probably a positive effect, because it made me even more angry... so probably it stimulated me.*' (Respondent 9)

When asked about how many women are currently employed in management positions in the company, the findings support the statement that women are underrepresented in the automotive aftermarket as a majority of respondents affirmed that there are significantly more men in leadership positions. One response noted that despite an increasing share of women working in sales, there are still no women in the management of the company. Three women who were interviewed stated that they are the only females in the management of the company. Only one woman said that there are more women in leadership than men, with three out of five department leaders being women. So, the overwhelming majority still shows an underrepresentation of women in management positions.

4.2 Existence of a Male Bias

Participants were asked if they feel like there is a gender bias in the automotive aftermarket. The answers to this question are particularly relevant to the study as the male bias was stated numerous times, and despite answering no to the question in the first instance, some women explicitly stated situations in which a gender bias becomes obvious in later phases of the interview.

> *'The men's club is there, the glass ceiling somewhere exists.'* (Participant 1)

Mentioning that the glass ceiling exists confirms findings from previous literature which stated that it is harder for women to advance further in their job once they have reached a high- standing position.

Other answers pinpoint to the fact that there are regional-geographic and cultural aspects at work, which show a male bias. For example, women felt like the Arab automotive aftermarket treats women and men differently, although the respondents acknowledged that this is not direct gender discrimination, it is due to cultural differences. Participants shared that it is clear that the share of men is still much higher in the industry and.

> *'there is a gender gap in the industry. Absolutely.'* (Participant 5)

Despite women feeling like they have gotten used to it and do not notice it in their daily work, answers showed that at bigger international meetings, there is a much larger number of men represented.

> *'If you look at international meetings, there are always many, many men and much less women.'* (Participant 5)
>
> *'There are always 500 people, and 480 or 470 are male and 55 plus, so, in some of them, especially the older ones, they do not have the necessary respect for female leaders.'* (Participant 2)

One participant shared her experience of a time when a man asked her if she is customer service, instead of recognising her actual position at a larger convention. Despite initially saying that there is no gender bias, one participant then explained that it takes longer for women to show customers that they are competent. Interestingly, there is a gender bias on the customer side too, not only within an employer-organisation. For example, experiences show that customers would rather go to a male employee working in export than to a female one. These examples do show an ingrained male gender bias in the automotive aftermarket.

However, a few answers allude to the fact that there is no gender bias as people shared that the daily work situation does not show male bias. Their teams work well together, and no one is looking for differences, or paying attention to the gender of others. Nevertheless, overall most women—except one—imply that the automotive aftermarket is a male-centred industry, which can be seen by the fact that more men are employed in the industry. Some specific examples do show the more direct gender bias. For instance, one person observed that another female employee was judged based on her appearance, which she said would not have happened to a man in that situation. Additionally, a man was employed for a vacancy, despite not having the required language skills, which female candidates did have. These situations show a clear discrimination towards women and contradict the statements made by other women who claimed that jobs in the automotive aftermarket are filled up based on the person's skills and not their gender.

Looking at the topic of equal pay, there is no clear finding on whether women in the automotive aftermarket receive equal pay compared to male colleagues. As pay is something which is typically not talked about at work, only two women knew that they do indeed receive equal pay. Three assumed that this would be the case, and four assumed that their pay is not equal to male colleagues. Interestingly, one person stated that there is an inequality in manual labour:

> '*What I experience is often that they really enjoy taking the women because they pay them less.*' (Participant 4)

This implies that women in higher positions of the company might receive the same pay as men, but it could be that there is a clearer gender pay gap in lower-ranking positions in the industry. This assumption could be supported by two women stating that at an earlier stage in their career they received a smaller pay than their colleague.

4.3 Equal Treatment and Discrimination at Work

Investigating how women feel in the industry reveals that half of the sample felt fully recognised as an equal member of the organisation, whilst a few of them indicated that it took some time to feel this way. One person stated that she feels fully recognised by her colleagues but not from the top level. Opposing the statements of others, one person explained:

> '*I felt like a number.*' (Participant 10)

Incidentally, a person who worked in the same large-scale company felt fully recognised until she got pregnant, and when she returned to work, she started feeling some limitations. The organisational context and the size of the company influenced their perceptions.

Looking at gender-based discrimination, most of the answers allude to the fact that women tend to not get their competencies downplayed based on their gender. However, even though some women did not experience a situation in which they felt that their competencies were downplayed based on their gender, they feel like women in the industry are not taken as seriously as men. Numerous quotes indicate the challenges and feelings that women have in this industry:

> '*For women, it is harder to show that you are competent, it takes a longer time.*' (Participant 6)
>
> 'As a woman, you have to say five more sentences to convince them that you are competent.' (Ibid.)
>
> '*I always felt that I worked harder to get the same level of credibility.*' (Participant 9).
>
> '*In order to be taken seriously I really need to be stainless.*' (Ibid.)
>
> '*I think that it's quite typical, that women kind of feel that they need to commit much more, and to work harder to get to certain results.*' (Ibid.)

Having numerous statements which show these challenges that women face suggests that women indeed must put in more effort than men to have their skills taken seriously.

While the majority of women replied that they have never faced direct discrimination in the job, there were some notable cases of discrimination found in the industry. Even if some women have not faced discrimination themselves, some heard from colleagues or others who have faced discrimination in the automotive aftermarket industry. This discrimination occurred, for instance, based on sexuality or nationality. These cases did not necessarily take place inside the organisation, the discrimination also occurred from customers. Surprisingly, answers differed greatly as some women shared that they feel comfortable with customers, even saying that being a woman can be something positive as a customer might remember her better. For some, the discrimination they felt has come from within the organisation and not from customers. Contrasting this, other women explained that customers discriminate them, for instance, they stated that some customers expect a man to work in a car-related industry. In this situation, women need to work harder to show that they are as competent as men. Additionally, there were cultural issues where a party felt discriminated. Another finding is that women have felt discriminated due to their family commitments, which emphasises that there should be more attention on supporting flexibility and family-friendly policies at work.

4.4 Challenges to Career Advancement

As the sample only included women in high-ranking positions in the automotive aftermarket, it is not surprising that all of these women have received promotions in their current company, or the one they were in before. To understand if the industry provides sufficient opportunities for women to grow, every participant was asked if they feel like they could advance further in their company. Due to some participants already being in the highest management position in their company, they unsurprisingly said that it is not possible to step up further. The other views illustrated differing perceptions. Some other participants do feel like further promotions are possible and that the organisation they are in offers good opportunities.

> '*(company name) is open for good managers to develop them, to promote them, independent if they are women or men.*' (Participant 7)

Opposingly, some respondents did not see opportunities for further career growth:

> '*No, I have to step out there.*' (Participant 3)

> '*I don't see a step ahead.*' (Participant 9)

These findings show that the glass ceiling might exist in some companies in the automotive aftermarket, as women do not have the possibility to progress in their career. For the industry as a whole, the opinions on whether the automotive aftermarket provides career opportunities were very split. Half of the women agreed with the more negative standpoint, the industry is not offering career growth for women.

> *'No, you need to push it.'* (Participant 2).
> *'I think we are not on the promotion agenda.'* (Ibid.)
> *'Especially if you're talking about women between 25 and 35... the risk to get pregnant is there.'* (Ibid.)

A negative point which was brought up by one participant was that

> *'Today, the automotive aftermarket provides really poor career opportunities for everybody.'* (Participant 9)

She argued that this is because the industry is not 'charming' (ibid.), which another person agreed with as she said that

> *'the product is not sexy for a woman.'* (Participant 10)

Despite not being an attractive industry for women, the same person said that

> *'the opportunities are there for everyone.'* (Ibid.)

Another participant also believes that.

> *'the opportunities are there but you need to have an interest in it.'* (Participant 8)

This implies that women might not be interested in the automotive aftermarket industry themselves, which could be a potential reason for the underrepresentation of female leaders and employees in the industry. Surprisingly, one person's view differed greatly from the other participants. She believes that the industry does offer opportunities for women as she can see women working in top positions in other companies of the automotive aftermarket that she is working with. Importantly, two women said that despite the opportunities being there, it is harder for women to advance because of a 'men's club'. Some companies.

> *'they try to push women ... but it's hard.'* (Participant 4).

It might depend on the supervisor,

> *'if they want to have this diversity on the management level or not.'* (Participant 7)

4.5 Gender and Perceptions of Family Commitments

To understand what the specific challenges are that women face in career advancement in the automotive aftermarket, each participant was asked to share one specific thing that came to their mind. An overwhelming majority of the sample answered that their biggest challenge in advancing in their career has been family.

> *'It was challenging to be a mother and a manager at the same time.'* (Participant 6)

> *'… to have this balance, work-life balance, as a woman is hard.'* (Participant 1)

Many women have the feeling that they had to give up time with the family to maintain their good position in the company. For instance, they had to spend time travelling, which proved to be 'stressful' and 'a big challenge'. Other women who shared that they took time off work to spend this time with their family felt like they had to regain the trust of the organisation and demonstrate their commitment repeatedly,

> *'every time I had the impression that I had to start from the beginning.'* (Participant 5)

4.6 Male Bias and Image Building

Creating a managerial identity and image as a woman seems to be challenging. In particular, some participants once again alluded to the male bias in the industry as their biggest challenge.

> *'To fight for something that, it was automatically granted to others, to my male colleagues.'* (Participant 9)

To stand out and advance to a management position, one person mentioned:

> 'I was working like crazy to create trust with customers and show I am a professional.'

(Participant 7)

The automotive aftermarket seems to be a challenging industry, as another person explained that

> *'You learn to fight for yourself', 'this is really helping a lot to … to survive in the automotive aftermarket.'* (Participant 2)

So, the industry shows numerous challenges, which might inhibit women's career growth.

4.7 Company Strategies for Improving Gender Equality

To understand what is currently being done in the automotive aftermarket to improve gender diversity, participants were asked about specific policies and mentoring in the companies that the participants are employed in.

Only two companies have their own policy in place which ensures gender equality in the company. These policies include, for example, having someone employed in HR who is responsible for gender equality and who organises specific activities on the topic of diversity, or women networks or meetings in which women of the company can discuss current issues and support each other. Another example is a female lunch organised by a company, which also sent out brochures and messages for International Women's Day. Having this in place seems to have a significant effect on women, as one participant mentioned that she feels appreciated by these messages. Other companies seem to emphasise the topic of gender equality by adhering to general policies made by institutions such as the Internationale Handelskammer (IHK) in Germany, which is Germany's International Chamber of Commerce. These policies focus on developing female talent in business. Another company adheres to policies set by the fund-owning company, which have set targets for gender equality. Those women who shared that there are no policies in place stated that it might be a good idea to implement some:

> *'I think it would make sense to keep very qualified women in the company.'* (Participant 3)

Interestingly, one woman who stepped out of the automotive aftermarket stated that while she was working in the industry, there were no policies in place, but she assumes that this would be the case now. Her assumption did prove to be true. Thus, it might be the case that many companies in the automotive aftermarket are working on developing and implementing more policies and are trying to achieve more gender equality, but their awareness of them is yet limited.

In comparison, more companies focus on mentoring, although only one organisation focuses this mentoring specifically on women. In this organisation, men have complained about the fact that they are not receiving the same resources as women. Most others have mentoring for every person in the organisation who wants to advance into a higher position, regardless of the gender of that person. The general feedback for this mentoring has been positive:

> *'I think mentoring programs is absolutely great stuff.'* (Participant 8)

But comments did imply that this mentoring should not be exclusively for women, it should focus on developing an individual's skills regardless of their gender. Surprisingly, one person said that she does not think that there is anything to help women advance in their career. Others stated that there are no programmes,

> *'Not especially for women and not especially for leadership.'* (Participant 6)

5 Discussion

5.1 Investigating the Gender Bias and Inequality Around Leadership

Existing literature and official reports on female leadership point out clearly that women are underrepresented in top management positions as these are typically occupied by men and it is harder for women to advance in their career (Siniscalco et al., 2014). Findings from the automotive aftermarket showed that there are significantly more male managers in the industry, although it is important to note that having a female manager is not unusual as some respondents claimed it was normal for them. Perhaps there is a geographical difference as the sample from Italy showed that having a female manager is rare. So, despite some women having the chance to advance to senior leadership positions, these positions are still dominated by men. Interestingly, one participant mentioned that the representation of women in the company's board has increased and it feels like diversity is improving in the company. This example confirms the findings of Biswas et al. (2021), who stated that board diversity improves overall diversity in a company.

Investigating the male bias and gendered discrimination in the workplace shows that the automotive aftermarket can be a gendered workspace, like Bullock (2019) and Kjellson (2017) examined, the organisational hierarchy tends to favour men in leadership positions. Importantly, experiences from participants showed that this male bias does not solely exist inside organisations, but it often comes from customers. This is an interesting point which exemplifies that society as a whole might think of the automotive industry as a male industry. This unconscious bias was identified as a barrier to leadership by Randsley de Moura et al. (2018), and the findings from this research do confirm their predictions. A male bias in the automotive aftermarket can be presumed to be connected to the underrepresentation of women. Investigating whether this bias exists in pay levels in the aftermarket did not lead to clear results. Similarly to existing literature, there were some opposing opinions, with some participants assuming there is a gender pay gap and some assuming there is none. It can be presumed that unclear information on pay levels is due to a lack of transparency on salary as participants stated that pay is not revealed to employees. Interestingly, Bullock's (2019) research in the US automotive industry discovered that there is a gender pay gap, so it could be possible that once again inequalities differ per geographical location. Nevertheless, to avoid misconceptions and emphasise that there is no distinction between gender, it might be beneficial to give official accounts of the pay that employees receive in the automotive aftermarket.

5.2 Examining Challenges for Female Career Advancement

A recent article by McKinsey (2021) pointed out how biased hiring processes are restricting women from getting into leadership positions and from showing their full potential at work. This biased hiring process might pose a challenge for women in the automotive aftermarket as well as a participant explained a case of clear gender discrimination in a hiring situation, where a man was chosen for the positions, despite not having the suitable qualifications, which female applicants did have.

Cook and Glass (2014) observed the objections that minorities have to face when advancing in their career and stated that minorities tend to face more objections than white men, which accustoms them to challenges they thus have to face as leaders. As the sample for this research paper only focused on women's experiences in the automotive aftermarket, it is not possible to say whether women face more challenges than men in their career advancement. However, statements from participants imply that women and minorities do have to face some objections in career advancement. The automotive aftermarket can be a challenging environment for women to build their careers in as 'you learn to fight for yourself' (Participant 2). It is possible that these challenges do indeed make women more accustomed to challenges which they face as leaders. Other objections such as sexual harassment or sexist comments were mentioned several times by other scholars, but findings from this research show that discrimination is not faced by women in the automotive aftermarket in Europe.

Multiple scholars have identified work–life balance as one of the main issues impeding women's career progression (Kjellson, 2017; Kowaleska, 2021, Taneja et al., 2012; Vasconcelos, 2018). Results from this chapter support previous findings as numerous women stated that they have struggled to balance family responsibilities with their work and named this as their main challenge in career progression. This emphasises that companies in the automotive industry and aftermarket should focus on improving their family policies to retain female talent in management.

5.3 Exploring Actions Taken by the Automotive Aftermarket to Retain Female Talent

Findings from interviews show that strategies implemented by companies in the automotive aftermarket differ slightly to those mentioned in the previous literature. Only one participant touched upon the idea of quotas to solve gender inequality in leadership as proposed by Seierstad et al. (2021). The participant mentioned that Germany has been considering implementing quotas as a solution to gender inequality, but she explicitly stated that she is 'not sure if this is the right process to do it' (Participant 2) due to the fact that a woman might be discriminated by other employees if she only got the position because of a certain percentage. So, quotas seem to not be a favoured solution to achieve gender equality. Meanwhile, research findings emphasised that mentoring programmes are a great approach to developing talent in the automotive aftermarket as long as it is open to everyone who wants to advance in their career. Several companies seem to invest into these mentoring programmes, whilst also paying attention to gender equality policies. Promoting diversity in the company makes women feel appreciated and could be an efficient strategy at retaining talent in the aftermarket, which could support to increase gender diversity in leadership.

6 Conclusions and Contributions

The research conducted in this chapter aims to investigate the responses of the automotive aftermarket to address the issue of gender inequality in top management positions. Three key objectives were examined: the underrepresentation of women due to the male bias in the industry, specific challenges women face in career advancement, and lastly, exploring strategies taken by companies in the automotive aftermarket to improve gender diversity and retain female talent, specifically in leadership. This chapter aims to contribute to existing literature on female leadership and gender equality in business, and tries to increase awareness of the importance of women in business as this is related to overall increases in development in the economic and social dimensions. Focusing on retaining female talent is particularly relevant to businesses as women help contribute to new and diverse business perspectives which increases an organisation's growth and profitability and thus creates a competitive advantage (Adams & Ferreira, 2009; Adams, 2016; Coleman, 2010; Horak & Cui, 2017; Taneja et al., 2012).

Investigating women's underrepresentation in the automotive aftermarket is crucial as there is no existing information on this industry sector, despite its significant size and influence on society. Advancing gender equality in leadership is a pressing issue, which is addressed in the UN's Sustainable Development Goal number 5 (UN, 2022). In order to obtain relevant insights into women's perspectives on the industry, qualitative semi-structured interviews were chosen as the most suitable research method as they help capture women's lived experiences. To understand specific challenges for female career advancement, the chosen sample consisted of 10 women from different automotive aftermarket companies, all of whom work in a high- ranking position. Findings from the 10 interviews conducted pointed out that there is in fact an underrepresentation of women in leadership in the industry sector, with a gender bias ingrained in some organisations, as well as customers. Nevertheless, several female employees believe that this gender bias does not exist and that it depends on women's ambitions and aspirations, whether they can make it to top leadership positions. Specific challenges in advancing to a higher level in the industry revolved predominantly around work–life balance as numerous respondents stated that it is hard to focus on their family commitments whilst keeping up with their work responsibilities. Another big challenge is that women tend to be taken less seriously than men, which results in them having to work harder to receive the same recognition. These are two issues that were identified in previous literature, so it could be presumed that they are common struggles women face in a male-dominated work environment (Bullock, 2019; Biswas et al., 2021; Kjellson, 2017). From the sample, it can be concluded that the automotive aftermarket as an industry is trying to improve its gender equality policy as several companies prove to pay a lot of attention to the topic and adhere to, or have implemented, specific policies for more diversity. Mentoring for employees who plan to advance in their career exists in a couple of companies and tends to be a popular strategy in the aftermarket. Interestingly, findings from the sample show that companies located in Southern

Europe are not as active in driving forward progress for gender diversity as some companies in Germany and the Nordic countries.

6.1 Recommendations

Based on these findings, a recommendation for companies in the automotive after market would be to focus more on female empowerment and increased flexibility, specifically in regard to work–life balance. An idea would be to implement more family-friendly policies, which would support women who struggle to divide their time between work and family. Moreover, as the feedback from participants regarding mentoring programmes was predominantly positive, a suggestion would be for more companies to introduce these programmes. As a participant pointed out, these should be open to anyone who wishes to advance in their career, regardless of their gender. This would avoid unfair treatment as mentoring, which is solely focused on advancing female talent, has been shown to give male employees the feeling that they are being treated unfairly. To avoid misconceptions around pay, there should be transparent criteria on the salary level and performance for the respective position in the company.

6.2 Limitations

The limitations of this research include the sample as having selected women in high-ranking management positions for the sample of this study provides a limited outlook on the entire sector. It would be interesting to investigate the experiences of women who are employed in lower positions in the automotive aftermarket in order to see what the main challenges and perceptions of the automotive aftermarket are like for them. One participant mentioned that there is a gender pay gap for women who work in the manufacturing part of the company she is in, so future studies could investigate the gender pay gap in this branch of the industry.

Choosing a different sample would likely yield different results; it might, for instance, be interesting to explore the experiences of men in the automotive aftermarket and see which issues can be generalised for career advancement in the industry sector. Having selected a sample of only 10 women makes it difficult to generalise the results for the entire industry sector. All participants are European white women. It is likely that women from different geographical locations and from different ethnic backgrounds have different experiences. So, it is not possible to say that these are the main challenges for women in leadership in the automotive aftermarket. One could say that these are likely challenges white women in the European automotive aftermarket face. In addition to this, these are company responses from the European automotive aftermarket, not the worldwide context. Future research could focus on companies from different geographical areas to see if the location has an impact on ingrained biases, challenges for career advancement, and company responses. It would be likely that this is the case as even Europe seems to differ in the Nordic countries, Central Europe, and Southern Europe.

References

Adams, R. B., & Ferreira, D. (2009). Women in the boardroom and their impact on governance and performance. *Journal of financial economics*, 94(2), 291–309

Adams, R. B. (2016). Women on boards: *The superheroes of tomorrow? The Leadership Quarterly*, 27(3), 371–386.

Biswas, P. K., Chapple, L., Roberts, H., & Stainback, K. (2021). Board gender diversity and women in senior management. *Journal of Business Ethics, 182*(1), 177–198.

Blumberg, B., et al. (2008). *Business research methods. 2nd European ed* (p. 193). McGraw- Hill Higher Education.

Boulouta, I. (2013). Hidden connections: The link between board gender diversity and corporate social performance. *Journal of Business Ethics, 113*(2), 185–197.

Braun, V., & Clarke, V. (2006). Using thematic analysis in psychology. *Qualitative Research in Psychology, 3*, 77–101.

Braun, V., & Clarke, V. (2012). Thematic analysis. In H. Cooper (Ed.), *APA handbook of research methods in psychology* (Research designs) (Vol. 2, pp. 57–91). American Psychological Association.

Bryman, A., & Bell, E. (2015). *Business research methods* (4th ed.). Oxford University Pres.

Bugeja, M., Matolcsy, Z. P., & Spiropoulos, H. (2012). Is there a gender gap in CEO compensation? *Journal of Corporate Finance, 18*(4), 849–859.

Bullock, C. L. (2019). Driving new narratives: Women-leader identities in the automotive industry. *Gender in Management, 34*(3), 200–216.

Carmona, S., Ezzamel, M., & Mogotocoro, C. (2018). Gender, management styles, and forms of capital. *Journal of Business Ethics, 153*(2), 357–373.

Center for American Progress (CAP). (2018). *The women's leadership gap*. Accessed November 23, 2021, from https://americanprogress.org/article/womens-leadership-gap-2/

Coleman, I. (2010). The global Glass ceiling: Why empowering women is good for business. *Foreign Affairs, 89*(3), 13–20.

Cook, A., & Glass, C. (2014). Above the glass ceiling: When are women and racial/ethnic minorities promoted to CEO? *Strategic Management Journal, 35*(7), 1080–1089.

Cook, A., & Glass, C. (2015). Diversity begets diversity? The effects of board composition on the appointment and success of women CEOs. *Social Science Research, 53*, 53137–53147.

Cook, A., & Glass, C. (2018). Do women leaders promote positive change? Analyzing the effect of gender on business practices and diversity initiatives: Women leaders. *Human Resource Management, 57*(4), 823–837.

Deloitte UK. (2020). 2020 women in automotive industry study. Driving greater diversity and inclusion within the automotive industry, [online]. Accessed November 26, 2021.

Deloitte US. (2020). 2020 women at the wheel. Key findings from the 2020 diversity, equity, and inclusion in automotive study, [online]. Accessed November 27, 2021.

Demartini, P. (2019). Why and how women in business can make innovations in light of the sustainable development goals. *Administrative Sciences, 9*(3), 64.

Ely, R. J., & Thomas, D. J. (2001). Cultural diversity at work: The effects of diversity perspectives on work group diversity and outcomes. *Administrative Science Quarterly, 46*, 229–273.

Figiefa. (2020). Definition, p. 45. Reference given in Appendix F.

Foley, M., Oxenbridge, S., Cooper, R., & Baird, M. (2020). I'll never be one of the boys': Gender harassment of women working as pilots and automotive tradespeople. *Gender, Work, and Organization, 29*(5), 1676–1691.

Ghauri, P. N., & Grønhaug, K. (2010). *Research methods in business studies* (4th ed.). Financial Times Prentice Hall.

Gummesson, E. (2000). *Qualitative methods in management research* (2nd ed.). Sage.

Hoogendoorn, S., Oosterbeek, H., & van Praag, M. (2013). The impact of gender diversity on the performance of business teams: Evidence from a field experiment. *Management Science, 59*(7), 1514–1528.

Horak, S., & Cui, J. (2017). Financial performance and risk behavior of gender-diversified boards in the Chinese automotive industry: Initial insights. *Personnel Review, 46*(4), 847–866.

Hoyt, C. L. (2010). Women, men, and leadership: Exploring the gender gap at the top. *Social and Personality Psychology Compass, 4*(7), 474–498.

Hussey, J., & Hussey, R. (1997). *Business research: A practical guide for undergraduate and postgraduate students*. Macmillan Business.

Kiefner, V., Mohr, A., & Schumacher, C. (2022). Female executives and multinationals' support of the UN's sustainable development goals. *Journal of World Business, 57*(3), 1–9.

Kjellson, J. (2017). *Barriers to female career progression in the automotive industry. – A qualitative study of how gendered substructures affect women's opportunities for career advancement in an automotive MNC*. Masters thesis. Gothenburg: University of Gothenburg [Online]. Accessed November 14, 2021, from https://gupea.ub.gu.se/bitstream/2077/53151/1/gupea_2077_53151_1.pdf

Klettner, A., Clarke, T., & Boersma, M. (2016). Strategic and regulatory approaches to increasing women in leadership: Multilevel targets and mandatory quotas as levers for cultural change. *Journal of Business Ethics*, 133, 395–419.

Kowaleska, H. (2021). Bringing women on board? Family policies, quotas and gender diversity in top jobs. *Work, Employment and Society, 35*(4), 753–773.

Lloyd, H. R., & May, M. R. (2007). Gender differences in perceptions of workplace progression; an automotive industry case study. *Southern African Business Review, 11*(3), 95–120.

McKinsey. (2021). *Women in the workplace 2021*. Accessed November 17, 2021, from https://www.mckinsey.com/featured-insights/diversity-and-inclusion/women-in-the-workplace

Monteiro, A. P., García-Sánchez, I. M., & Aibar-Guzmán, B. (2021). Labour practice, decent work and human rights performance and reporting: The impact of women managers. *Journal of Business Ethics, 180*(2), 1–20.

O'Gorman, K., & MacIntosh, R. (2015). *Research methods for business & management: A guide to writing your dissertation* (2nd ed.). Goodfellow Publishers Ltd..

Opstrup, N., & Villadsen, A. R. (2015). The right mix? Gender diversity in top management teams and financial performance. *Public Administration Review, 75*(2), 291–301.

Randsley de Moura, G., Leicht, C., Leite, A. C., Crisp, R. J., & Gocłowska, A. M. (2018). Leadership diversity: Effects of counter stereotypical thinking on the support for women leaders under uncertainty. *Journal of Social Issues, 74*(1), 165–183.

Saunders, M., Lewis, P., & Thornhill, A. (2019). *Research methods for business students* (8th ed.). Pearson Education Limited.

Seierstad, C., Tatli, A., Aldossari, M., & Huse, M. (2021). Broadening of the field of corporate boards and legitimate capitals: An investigation into the use of gender quotas in corporate boards in Norway. *Work, Employment and Society, 35*(4), 753–773.

Siniscalco, G., Damrell, L., & Morain Nabity, C. (2014). The pay gap, the glass ceiling, and pay bias: Moving forward fifty years after the equal pay act. *ABA Journal of Labor & Employment Law, 29*(3), 395–427.

Sojo, V. E., Wood, R. E., Sally, A., Wood, S. A., & Wheeler, M. A. (2016). Reporting requirements, targets, and quotas for women in leadership. *Leadership Quarterly, 27*(3), 519–536.

Stojmenovska, D. (2019). Management gender composition and the gender pay gap: Evidence from British panel data. *Gender, Work and Organization, 26*(5), 738–764.

Talents 4 AA. (2022). Talents 4 AA. Accessed April 11, 2022, from https://talents4aa.com/#manifesto

Taneja, S., Golden Pryor, M., & Oyler, J. (2012). Empowerment and gender equality: The retention and promotion of women in the workforce. *Journal of Business Diversity, 12*(3), 43–51.

United Nations. (2022). *The 17 goals—sustainable development goals*. Accessed March 27, 2022, from https://sdgs.un.org/goals

Vasconcelos, A. F. (2018). Gender and leadership stereotypes theory: is it reaching the boundaries? *Management Research Review, 41*(11), 1336–1355.

World Economic Forum. (2021). *Global gender gap report 2021* [Online]. Accessed November 14, 2021, from https://www3.weforum.org/docs/WEF_GGGR_2021.pdf

Amanda Katsardis is a young professional, who graduated with a Master of Arts in Business Management and International Relations from the University of Aberdeen in 2022. Her thesis focused on the topic of gender equality in leadership positions in the automotive aftermarket, which is also being discussed in the book chapter. She has gathered experience on the automotive sector through two internships in the aftermarket. She is currently working in the context of software sales and SaaS gaining insights to the startup context. Her interests also relate to women in Tech and sustainability issues.

Raushan Aman is a researcher at the Martti Ahtisaari Institute, Oulu Business School, University of Oulu. She holds a master's degree in project management from the Royal Holloway University of London in the United Kingdom. Raushan was a visiting researcher at the University of Southern Denmark and Shanghai University within the frame of her doctoral research. Her current research interests focus on female entrepreneurship, highly skilled female immigrant talent and their integration, immigrant entrepreneurship, migrant women entrepreneurship, minority entrepreneurship, gender diversity, diversity, equity and inclusion, female brain drain, brain gain and brain circulation, and entrepreneurial or entrepreneurship ecosystems.

Acquiring Talent for the Automotive Aftermarket: How Sustainability Is Becoming a Key Driver in Talent Acquisition and Development for the Sector

Gyula Kimpán and Maria Elo

Abstract

The war for talents fuels a competition among automotive aftermarket (AA) companies, big tech, consultancies and banks. Simultaneously, the demographic changes and ageing populations influence labour markets and talent potential. Specifically in Europe, the AA is facing difficulties with attracting talents and with its lacking sectoral awareness and positioning. As a response, a multi-stakeholder initiative "TALENTS4AA" was formed. We explored the problem through the views of the talent pipeline starting with Denmark and Danish students. The purpose was to generate a better understanding of their perceptions on the AA and its attractiveness. Our study found that Danish engineering and business students are largely unaware of the existence of the AA, while they are interested in contributing to sustainability. With new regulations, e.g. the EU Taxonomy Regulations and the Green Deal, companies require talent to comply with shifting institutional demands. By updating its strategies and meaningfully engaging with diverse talent, the sector as a whole may better tackle emerging resource needs.

G. Kimpán (✉)
TEMOT International Autoparts GmbH, Ratingen, Germany
e-mail: g.kimpan@pm.me

M. Elo
University of Southern Denmark (SDU), Odense, Denmark

Belt and Road Institute of International Business, School of Economics, Shanghai University, Shanghai, China

Abo Akademi University, Turku, Finland

Turku school of Economics, University of Turku, Turku, Finland

Institute Arbeit und Technik, Gelsenkirchen, Germany
e-mail: melo@sam.sdu.dk

M. Elo, F. Katsardis (eds.), *Automotive Aftermarket*, Management for Professionals, https://doi.org/10.1007/978-3-031-62419-3_10

1 Introduction

European labour markets are changing. There are two concerns for the labour markets and industry sectors: on one side, there is a fierce competition for diminishing supply of highly skilled talent, and on the other side, there is a societal issue of not all talent potential being in use or in development. The median age of the EU-27 population is rising, caused by decreased fertility rates and increased life expectancy (Eurostat, 2020a, 2020b). The "Being young in Europe today" report highlights the negative impact that demographic changes can have on the effectiveness of the green and digital transition, suggesting that strategic planning around policy-making is essential to address the issue. The slow recovery after the last global financial and economic crisis left young graduates in difficult positions when they entered the labour market, especially those with lower levels of education. Youth unemployment has remained a major problem in many countries in the EU, especially in southern Member States where young people likely face long-term unemployment. Within the EU-27, 17% of young people aged 25–29 years were neither in employment nor in education and training in 2019. There was an unemployment rate of 15% among the labour force of youth between 15 and 24 years in the same year. (Eurostat, 2020a, 2020b).

The Eurostat report states that higher levels of education resulted in better employment outcome, but nevertheless, the working-age population in Europe is shrinking and the European Union is in need of more effective ways to fuel economic growth by both reducing unemployment and increasing productivity. The European Union established a strategy that was renewed in November 2018 to address the problems that the youth face by focusing on three aspects found most important to aid them, being engagement, connectiveness and empowerment (Council of the European Union, 2018). The issues remain recognised and continuously monitored by the Council, granting an opportunity for interested stakeholders to shape the approaches taken to bolster the chances of young talent in becoming employed in meaningful positions. These concerns highlight the focus on youth as a source of talent.

Simultaneously, the concerns on climate change underline the aspect of sustainability. The European Union announced the "European Green Deal" with the goal of becoming the first climate-neutral continent (European Commission, 2021). Among the areas covered by the resulting action plan are key points that directly affect the transportation industry, both product and service providers. The overall objective of boosting the efficient use of resources by moving to a clean, circular economy is linked to investments in environmentally friendly technologies and the adoption of cleaner, cheaper and healthier forms of public and private transport. Sustainability has risen to the top of the agenda for the European Commission when discussing the future of the European Union in the coming decades, but it is also an issue for the youth that are concerned about the future and employment opportunities (Aceleanu et al., 2015).

The economic environment described above led scholars and industry stakeholders from the automotive aftermarket to form the "TALENTS4IAM" initiative in

September 2019 and address the challenge of scarcity of young talent interested in the industry, and the resulting implications on the future of the sector. The initiative was renamed to TALENTS4AA later. The stakeholders represented managers from firms such as ABR, ATR, Bosch, Delphi, Global One, Nexus, Schaeffler, SKF, TEMOT International, Zbeda and ZF and scholars from the University of Southern Denmark in Denmark and Karlsruhe Institute of Technology in Germany. The prominence of the sustainability agenda combined with the unfavourable demographic changes in Europe creates challenges to the talent development of the sector, partly due to lacking awareness of it that does not produce respective action (Łącka-Badura, 2015). Simultaneously, the industry is facing a paradigm shift as electromobility is gaining widespread acceptance as the most viable way into a sustainable future, affecting not only personal but freight transportation as well, which implies that all the business partners that the automotive aftermarket sector is currently working with are fundamentally affected. However, industry leaders, such as Bosch, have been able to achieve significant sustainability milestones, such as carbon neutrality, by 2020 (Bosch, 2020), but the notion of fully adopting the circular economy is proving to be challenging even for large multinational organisations.

In addition, there is also a potential gap regarding how smaller organisations in the automotive aftermarket with less resources should approach and execute a strategy that would assist them to adhere to the regulations proposed by the European Council and comply with not only the European Union's efforts towards a sustainable future but also the Sustainable Development Goals, for example, the Goal 12 "Responsible consumption and production", published by the United Nations (United Nations, 2015). Various methodologies to assess the current level of sustainability in a business exist, such as life cycle assessments and the triple bottom line, but knowledge within the field is constantly evolving, requiring experts to keep up-to-date regularly, but students in higher education institutes are already being equipped with the latest developments as they graduate, raising the value of such young talent significantly in the eyes of companies that currently lack such competences.

This makes students a relevant target group. The chapter contributes to the discussion on young talent, their perceptions, views and their values, offering a perspective from Denmark. It addresses the young talent by exploring opinions and perceptions of business and engineering students who study in Denmark (Elo, 2022).

The purpose of the chapter is to discuss the talent challenges with empirical insights from students, taking the view of one talent target group. As we are interested in exploring how automotive aftermarket companies can attract competent young talent which possesses the necessary knowledge to adhere with upcoming sustainability regulations, the role of students is of central interest representing the pipeline of talent. The chapter aims to increase the understanding of the student perspectives regarding the AA.

Our empirical study reflects themes like the following:

- What are students thinking about AA as a sector, employer, entrepreneurial partner and societal actor?
- What kind of perceptions, ideas and plans do they have?
- How can they help AA develop ways to communicate with them and build talent for future?

The chapter is organised as follows: first, it discusses the relevant theme, then it presents the study and the findings from the explorative study. Finally, it provides managerial implications for the automotive aftermarket sector and suggestions for education and training sector. It concludes with suggestions for future research and action.

2 Talent Development and Attraction—Sectoral Challenges

The automotive aftermarket is facing significant changes (McKinsey&Company, 2017). Such changes generate connected effects on the talent acquisition and development as well as on the overall human resource management, strategy and communication needs (DiFonzo & Bordia, 1998). The need for talent is aligned with the trends that shape the sector. For example, the COVID-19 pandemic offered a chance for the suppliers of both OEMs (Original Equipment Manufacturers) and the aftermarket to rethink, realign and renew their business models to emerge from the crisis as winners (Lazard, 2020), while pressure is increasing from the European Union for companies to adjust their business to fulfil the goals defined by the "Green Deal" (European Commission, 2021). Market-changing trends were already present before the COVID-19 pandemic and remain strong regardless of the ensuing economic crisis. The following trends have been, and will keep, shaping the future of the sector (Roland Berger, 2020):

- Mobility.
  - Consumer behaviour is shifting towards on-demand short-term vehicle rentals, ride sharing and new forms of personal mobility rather than owning and maintaining their own private vehicle.
- Autonomous.
  - Automotive OEMs are investing heavily in technologies that take the burden of driving away from humans, promising less stress and accidents on the roads.
- Digitalisation.
  - New forms of digital business models are emerging as a result of the increasing amount of entertainment devices installed in new vehicles.
- Electrification.
  - Electromobility received a significant push from European governments, including subsidies, electric charging infrastructure investments and strict vehicle emissions.

These four trends set requirements for particular skills and competences that educational sector is producing through programs and courses. These skills and competences—when in employment—contribute to the organisational talent pipeline. The types of professionals needed within the sector is rapidly shifting as skills within electromotors, batteries, electronics, and software become essential to designing and manufacturing new vehicles, especially electric-powered ones (McKinsey&Company, 2018). While companies in the sector are traditionally driven by financial results, and therefore strive for financial sustainability in their strategy-making, large multinationals such as Bosch have succeeded in achieving carbon neutrality within their organisation (Bosch, 2020). However, many supply chain partners are not yet in such a situation, and the notion of circular economy also requires additional talent. Hence, the newly emerging job positions offer excellent opportunities for suppliers to advance their diversity metrics and embrace a renewed mentality towards fulfilling global goals regarding social, developmental and environmental issues.

Sustainability is one of the sectoral challenges. Top consultancy firms and expert institutes have not listed an issue specifically around the attraction of young, competent and purpose-driven talent within the automotive aftermarket sector. In similar vein, many educational institutions lag behind in developing respective educational programs and courses to boost the needed talent pipeline. Yet, sustainability and the green transition have climbed to the top of the European Union's agenda (European Commission, 2021) with the formation of the "Green Deal", further supported by the Sustainable Development Goals (United Nations, 2015). The Green Deal specifically raises awareness to shifting towards "cleaner, cheaper and healthier forms of private and public transport", which directly affects the AA. This is a challenge that requires multi-stakeholder and institutional attention.

The European Union aims to achieve climate neutrality by 2050 and is expected to introduce climate laws to ensure that appropriate action is taken to achieve the goal. The EU Taxonomy Regulation was introduced with the intention to require companies to describe the environmental sustainability of their economic activities (Ramboll, 2021). The Taxonomy focuses on the following six environmental objectives (Ramboll, 2021):

- Climate change mitigation.
- Climate change adaption.
- Sustainable use of water and marine sources.
- Circular economy.
- Pollution prevention.
- Healthy ecosystem.

All non-financial companies, including those in the AA sector, had to adopt the objectives latest by the end of 2022. To avoid greenwashing, the guidelines specifically demand substantial contribution to at least one of the mentioned points, and no significant harm to the remaining ones.

Circularity is another challenge, but again a challenge that holds inherent potential. The concept of a circular economy is not recent, for example, many automotive aftermarket actors have been early adopters of reuse of materials. Circularity as a business concept is gaining increasing attention, for example, regarding operations related to remanufacturing and recycling. Circularity encompasses waste prevention, resource efficiency and dematerialisation of the economy (Geissdoerfer et al., 2017). In principle, circularity aims to eliminate the linear flow of goods and materials by creating a loop that allows end-of-life products to be returned to the beginning of the cycle with the purpose of retaining as much value during each loop iteration as possible. A regenerative system that incorporates maintenance, reuse, repair, remanufacture, refurbishment and recycling as means to minimise or eliminate environmental degradation from harmful emissions may foster AA as a sector. It may increase the attractiveness of the sector fostering the purpose that drives many young talents (e.g. Ronda et al., 2018).

A further challenge is the demographic change in many developed countries. The demographic changes in the European population have been recognised and researched extensively as the impact of the phenomenon is expected to substantially shape the future of the continent (see e.g. Zimmermann, 2005; Harper, 2014). The European Union Youth Strategy is continuously renewed to address the evolving challenges that young people face during their everyday lives, be it around education, employment or general social engagement (Council of the European Union, 2018).

Eurostat (2020a, 2020b) describe in their report "Being young in Europe today" the key issues revolving around the ageing population and the declining ratio of children and young people to elders. The member nations of the EU are found to be especially affected in the world, with the median age of the population in EU-27 rising faster than in any other region. This trend is further driven by decreasing fertility rates and the increased life expectancy of individuals. The overall effect of the situation is causing a decreasing amount of working-age people, but as an anomaly, unemployment rates among the youth is considered high. Reasons behind the unemployment are listed as the lack of higher-level education among affected individuals or a reduced need for the knowledge and skills they possess. These conditions are challenging from the sectoral perspective, as the emerging trends in the automotive industry demand highly skilled talent next to skills produced in vocational training. For example, digital skills at work become increasingly important due to digital devices and processes from diagnostics to business to cyber security.

Extant research suggests that by fostering human mobility and migration, the lack of talent can be at least deferred, if not solved. Zimmermann (2005, 425) refers to the rigid labour market frameworks as the cause for the perception on "an immobile labour force and the eurosklerosis phenomenon". He argues: "In such a situation, an economically motivated immigration policy at the European level can generate welfare improvements. A selective policy that discourages unskilled migrants and attracts skilled foreign workers will vitalize the labour market, foster growth and increase demand for unskilled native workers" (Zimmermann, 2005, 425). Particularly, policymakers have the power to create better policy incentives that would attract more highly skilled talent into Europe, tilting the global talent flow balance in favour of the region (Kerr et al., 2016). However, such policy programs can be counteracted by the talent sending nations that wish to delimit their brain drain or attract returnees with novel schemes.

Sustainable talent attraction and recruiting forms another challenge, particularly if talent is pulled from developing countries that lack the same talent (Emmanuel et al., 2019). The practices per se and this form of sustainability are enacted at the company level (Elo, 2021). Sustainable human resource management is introduced as an interesting approach to the increasing difficulties around attracting talent, emphasising not only financial rewards, but the contribution of the position towards better social and environmental welfare (Kramar, 2014). Financial incentives, however, remain as strong tools for motivating job seekers to apply and existing employees to perform better, regardless of their age, gender and race (Schlechter et al., 2014). Linley suggests (2015) that the best practice for finding hidden talent lies in the avoidance of arbitrary screening approaches and instead using assessment insights and data analytics to determine the suitability of candidates. Focusing on the future potential rather than the past experiences allows organisations to discover talent towards which they were previously blind. By engaging in candidate sharing across respective industries it becomes possible to create a business environment that the entire sector can benefit from. Sustainability efforts are often perceived by companies as costly activities without meaningful financial returns, but corporate sustainable business practices have been proven to improve the talent attraction capabilities of firms (bin Magbool et al., 2016). Klotz (2013) states that the young people hold the key to a sustainable future, calling them the "Sustainability Generation". She argues that the youth is geared towards making radical changes, which seems necessary and is essential to push the green transition agenda forward, even if there are not immediate and clear financial benefits.

The literature indicates that there is lack of research focusing on the automotive aftermarket and its talent attraction and development opportunities, for example, ScienceDirect finds zero articles on this theme. It illustrates that there are overlapping interests and developments that could benefit from deeper research insights to the different perspectives.

3 The Study

This chapter presents a study conducted in Denmark at the University of Southern Denmark (SDU). The study explores students' perceptions and views on the independent automotive aftermarket contrasting the usual views on fashionable employers representing multinational enterprises like Google, Facebook, Vestas, Ørsted or Maersk. Danish engineering and business students, both bachelor and master-level, constitute the kind of talent that is relevant as talent potential for AA. The study employs primary data collected by a research team at SDU surveys that include both exploratory qualitative open-ended and quantitative questions for assessing claims. The questions are designed at SDU and are built on related literature and have been validated by experts of TALENTS4AA. The method selected was mixed, using a structured questionnaire-survey combined with qualitative analysis of assignments.

The sampling strategy focuses on two types of students, those with technical study programs and those with business study programs. The data collection is carried out as a voluntary exercise in respective classrooms. The students did not have any exercise or prior readings on AA before answering the survey or completing a related small assignment. Our study consisted of total 139 participants. The online survey obtained 54 business student respondents: 41% male, 59% female and 25 engineering student respondents: 68% male, 32% female, the remaining 60 participants responded via an assignment.

The analysis of the data follows a twofold strategy. First, the survey software assisted in generating a visualised analysis of the quantitative responses. Second, the open-ended qualitative data was processed into an analysable format and coded based on recurring themes and keywords using the research analysis software "NVivo".

The survey contents followed the concerns that stem from the challenges related to the sector. The questions aimed to elicit impressions, associations and knowledge about the automotive aftermarket among the participants. Here, the design of the questions emphasises young people who represent potential as future talent from the perspective of the automotive aftermarket sector. The purpose of the study was to understand how students, i.e. young talent, perceive AA as a sector and what they think about it. The survey included questions around the following themes:

- Are the students aware of the automotive aftermarket?
- What do young people think that AA means / comprises?
- What kind of associations come up with the AA as a sector?
- How attractive the students view the AA as a potential employer/ a career option?

The survey study was conducted in March 2021 and it received 79 responses from business and engineering students at the University of Southern Denmark. We also received open-ended assignment responses that allowed deeper qualitative insights. Here, these were mainly used to triangulate the survey data.

4 Findings of the Danish Study

The findings are interesting and illustrate that many of the industry concerns exist in Denmark. For example, there is an issue of lacking awareness of the sector as 44% of business students and 62% of engineering students had never heard of AA before the study. "Because I have never heard of AA/IAM" was a typical response across students. The sector does not have a good reputation that would attract students, which refers to a problem that is larger than just some employer branding issue. The majority of all participating students see the AA as a not prestigious sector for work. Interestingly, only a minority of the students see the automotive sector in general as an attractive career option, which this comment explicates: "The image of the automotive sector has an influence on the image of AA with being stuck in the past, not being innovative enough regarding resolving global issues like climate change". The lack of positioning from the primary automotive sector reflects on AA as there is a lack of awareness and knowledge that would differentiate the sectors. Students mix secondary and primary automotive sectors and have paradoxes in their understanding due to lack of awareness. "Selling used cars" or "old male"- descriptions illustrate such perceptions next to the narrow views on mechanic repair activities.

We found that 88% of business students and 81% of engineering students have never thought of AA as a career option, which is rather alarming. But more importantly, there is a discrepancy of the overall concerns that the students have (such as sustainability or inclusion) and their perception of the AA. Such value-based perceptions are highly relevant for the decision-making and career plans (Greenbank, 2009). When addressing valuable themes, there were differences across study programs. Business students emphasised themes such as innovation, IT, data and technology, environmental sustainability and social sustainability. Engineering students emphasised innovation, environmental sustainability and circular economy. One business student illustrated an important dynamic in a comment on the competitive advantage of the AA: "This is technological-based, so as this is the era of technology thus young people will get more attracted to this sector and this will eventually flourish this sector even more in coming years." Another response called for "regenerative business models as this is increasingly important for future employees/ young talent".

We reflected some assumptions. Figures 1 and 2 show an extract of how the students responded to particular claims (on the left) that are critical for understanding the perception of the students regarding the AA.

In terms of the qualitative textual data from the open-assignments, we used different techniques to assess the findings. Figure 3 shows a world cloud that was generated from the most frequently mentioned words by the Danish students.

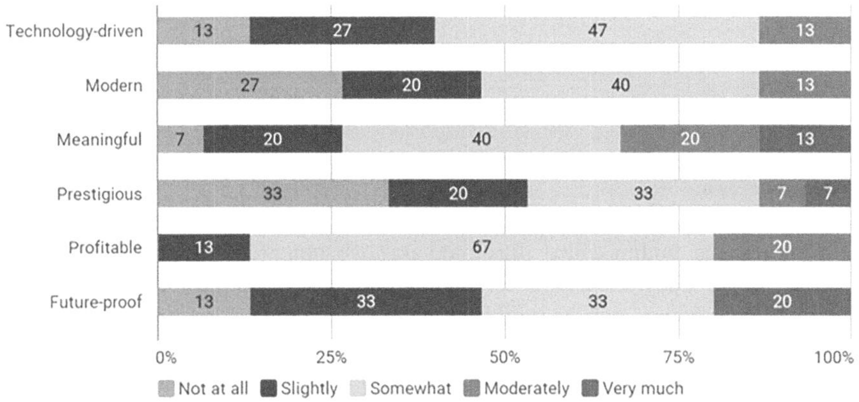

Fig. 1 Perception of the IAM from master's level engineering students at SDU. Source: own

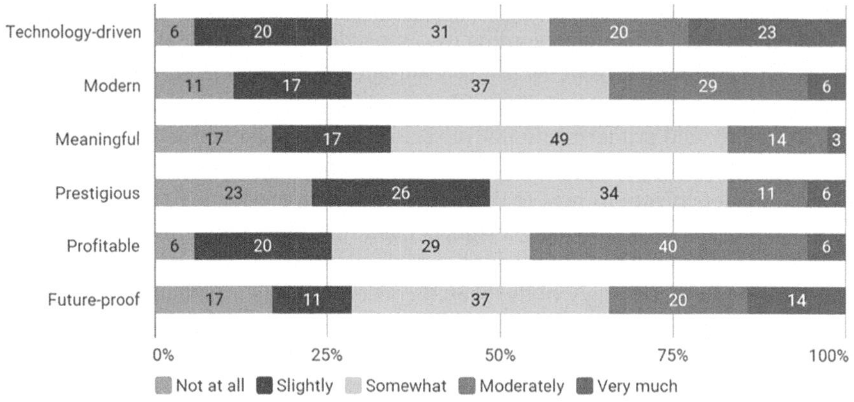

Fig. 2 Perception of the IAM from master's level business students at SDU. Source: own

Fig. 3 Most frequently mentioned words by SDU (Denmark) students regarding the IAM. Source: own

However, the findings also illustrated positive potential. Untapped resources appeared in the study, these refer to women, migrants, business students and other non-perceived groups, such as rural area talent.

For example, engineering students recommended that the AA informs them about the future and sectoral strategy in following ways for higher interest: "Tell me how I can use my degree to make a difference, maybe?", "Re-brand the industry toward being the sustainable choice. Maybe attract talent to lead this change.",

"Rebrand the entire industry". These comments show that the AA efforts must be sectoral; single organisations are not adequate to change the landscape of perceptions. Business students, and especially female students, were more interested in the sector than engineering or male students, which suggests that there is potential to be discovered. The value and interest perceptions of the students also overlap with many of the aspects that the AA is dealing with, showing a connection that deserves attention.

The student views can help AA develop ways to communicate with them. Their suggestions included the following:

- Engagement in the education system, universities, training centres, and schools.
- Involvement via social media, workshops, traineeships and others.
- Events and purpose-driven activities, e.g. sponsoring.
- Showcasing AA talents.

The students suggest using multiple channels to connect with them, particularly social media. These were universities and their channels, LinkedIn, job platforms, alumni pages, Facebook, YouTube ads and websites. They also called for information and collaboration via posters, presentations, projects, competitions, formal university communication, school courses, entrepreneurial course, innovation programs.

Interestingly, the students call for a new narrative that is more sustainable and inclusive highlighting empowerment, inclusion and environmental sustainability. The following narrative examples were suggested to attract talent: "The sustainable narrative", "an opportunity to gain experience within a large industry", "That is an interesting career path where you can impact while working with interesting technology", "We need you to solve this problem", "Sustainability, change, impact, jumpstart on a career, highlight the possibilities", "Help save the world by reducing automotive waste" and "Rebranded towards sustainability aspect and the changing world". The majority of suggestions implicitly underline the impact (cf. being a change agent).

5 Managerial and Policy Implications for the Sector

The AA sector expressed their concern about their perception of not being able to attract young talent within the TALENTS4AA initiative. This formed the research problem that the authors explored from the student perspective, being one key target group. Theoretically, it is known that values and perceptions are impactful in considering employment and career, but we find that they also influence sectoral attractiveness (Greenbank, 2009). Furthermore, instead of simple employer branding the study highlights the need for sectoral awareness campaigns and branding as the unawareness of the AA forms an impediment for interest. The classic AIDA (awareness, interest, desire, action) approach can be confirmed in this context

(Łącka-Badura, 2015) as the interest potential was explicated by the descriptive elements provided, while the students did not know that such elements exist in the AA sector.

The results support many of the concerns expressed by TALENTS4AA, but they also show that the automotive industry as a whole is not perceived as generally attractive by these students (cf. Gupta et al., 2018). This transformation triggers the need to re-theorize the sectoral development and the respective challenges (see e.g. Winkelhake, 2018). These findings show a lack of knowledge or wrong knowledge about the sector. This is detrimental for talent attraction because the assumptions include misunderstandings on the nature of the AA and its activities, e.g. on sustainability and innovation, creating hindering effects for interest.

Managerially, the findings show that there is ample room for advancement and development of potential. Although many responses revolve around negative phrases and misconceptions, those are often followed by recommendations and calls to action with ways to improve the young talents' perception of the sector. This can be conveying a message that there is an interest in AA, but it needs to improve its image and position in the economy to be seen as attractive by the desired young talent. Moreover, numerous suggestions on communication show that especially social media engagement and direct involvement would benefit the sector's talent attraction efforts. Communicating the challenges and possibilities to take action for

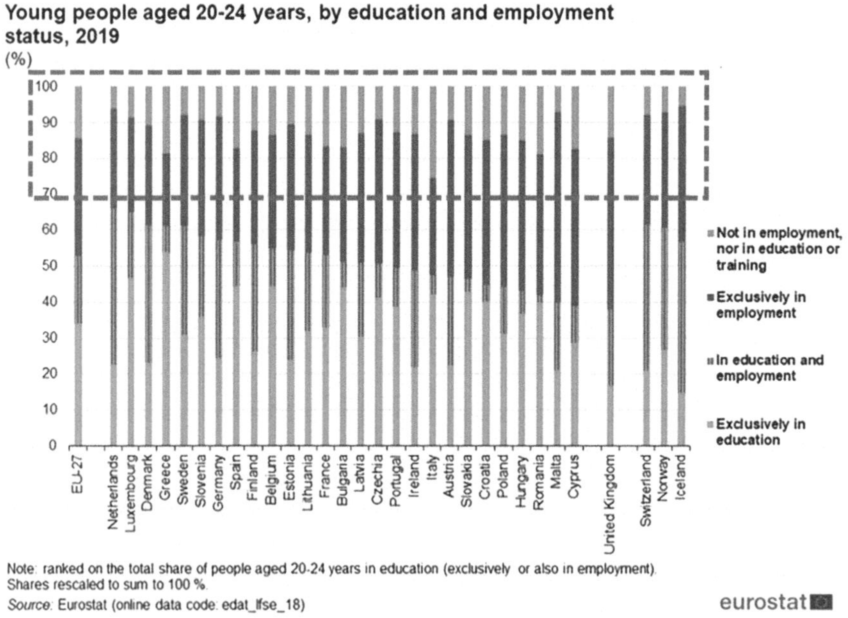

Fig. 4 Highlighted percentage of economically inactive youth aged 20–24 in the EU. Source: Eurostat, 2020b

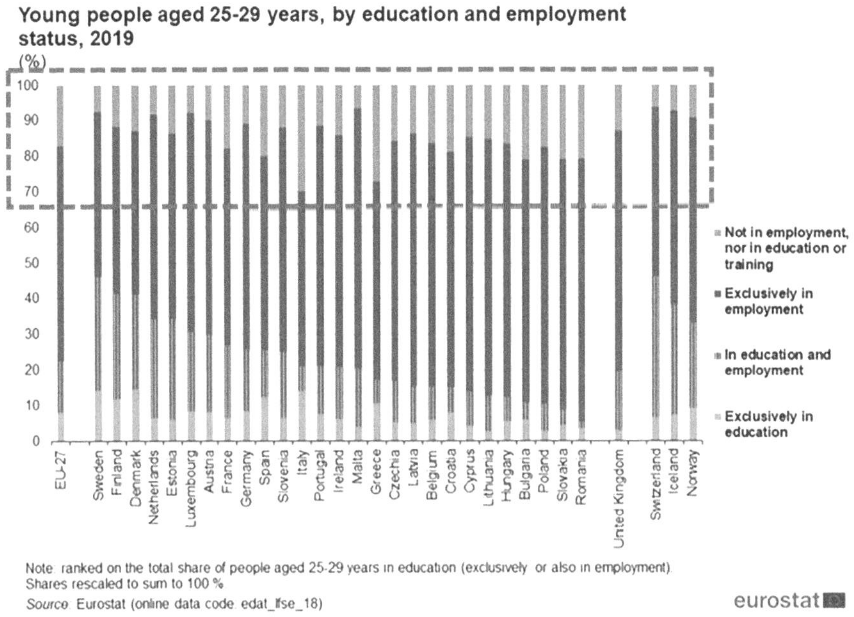

Fig. 5 Highlighted percentage of economically inactive youth aged 25–29 in the EU. Source Eurostat, n.d.

a more sustainable, digital and innovative future, and demonstrating purpose are among the managerial recommendations for attracting AA human resources.

If simplified, the Danish students have multiple options as privileged, educated youth. Hence, they rather represent potential for managerial positions or the so-called white collar talent. Yet, as the AA sector also requires talent from vocation training, for example, for garages, a broader audience is relevant. For this audience, we reflect statistical data and notice that the AA human resource management should look at the talent potential landscape in an updated manner. It is not just demographic factors or educational level that determines the level of potential, there are societal issues that generate concerns for the talent pipeline as a whole (see Figs. 4 and 5). The highlighted percentage of young people who are neither in employment, nor in education or training demonstrate a concern for underdevelopment and waste of talent potential. Particularly, countries with relatively high youth unemployment have environments that negatively affect young people, their inherent potential and their professional opportunities. As Linley (2015) suggests, difficulties around human resource management and recruitment of new talent can be traced to ineffective candidate evaluation approaches. By assessing the future potential of the young talent rather than fixating on their past experience and achievements, the AA sector could explore and uncover missing talent in contexts that they have not noticed before or in categories that have been previously marginalised.

Concerning the overall implications, the findings highlight the need to consider a multi-stakeholder view to the talent pipeline. Talent potential for the AA is co-created by several factors, such as demography, education, training and study programs, but also in the industry per se and its in-house sectoral efforts, and in collaboration with policymakers and local communities. The socio-economic setting needs to comply with the needs and preferences of the young people as well as those of the industry and the sector as a whole, but in a balanced and inclusive manner. Scarcity of talent can be alleviated by more inclusive and broader efforts in addressing it.

Societally, there are problems stemming from the un- or underemployed population of young people who remain economically and professionally inactive, and from the unattractive or unknown image of the sector. Partly, this may be a communication issue, a mismatch of information on the professional opportunities or the lack of it. As the "Being young in Europe today" report" (Eurostat, 2020a, 2020b) revealed, appropriate education is crucial towards meaningful employment among the youth, which presents a potential answer to the research question. Hence, also the educational institutions play a role in generating awareness for developing professional capacity and respective career pathways.

Counterbalancing the lack of interest among university students, the opportunity to reach out to the inactive young people with the purpose of training them to become change makers, such as sustainability heroes, should be investigated. As Klotz (2013) explains, the role of the youth is critical in pursuing substantial sustainability goals, and the notion of adopting the circular economy can be regarded too radical for some organisations, in which cases the young mindset could function as the change agent within the organisation.

The efforts could start with industry-education partnership piloting that uses existing learning programs and materials employing multimodal and accessible courses. The rewards for the efforts should be certifiable and lead to employability and work fostering self-efficacy and empowerment. The possibility of the AA sector to partner with educational institutions to offer improved apprenticeships and re-education programs as a form of reskilling and upskilling should be investigated as a diagonal option to enter as a talent for a new career. As Schlechter et al. (2014) alluded to, financial incentives are strong motivators among all individuals regardless of age, race or gender. Capturing those young people that are not in employment and education, and not allowing them to "drift" out of the job market would offer one societally interesting pillar for talent development beyond the classic degree education.

Another major group of interest identified is women. Female students in Denmark are more interested in good career opportunities in this sector than male students, but lack targeting by the sector. Due to gender bias, the male students tend to have more options leaving more room for success of the AA sector to develop strategies that are competitive particularly for female talent. Also many non-Danish students in Denmark found the AA as an interesting option, suggesting that there are target groups with different preferences and perceptions even in one university.

Such a factor that seems to be overlooked is the level of economically inactive young people in the European Union, who are neither in employment nor in education or training. If those young people would be meaningfully engaged and trained within the state-of-the-art sustainability knowledge, the AA sector would secure the supply of young talent they currently lack to execute sustainability strategies that are necessary to conform with the "Green Deal" and the EU Taxonomy Regulation. Learning materials are readily available, and with appropriate incentives the previously ignored or marginalised young talent could be reinstated into the job market.

6 Conclusion

We conclude that there is disconnection between the AA sector and the young talent in education. The image and reputation of the AA sector is largely missing or not positive among Danish students. The automotive aftermarket is misunderstood as used car sales or with the overall automotive industry and the service or innovation element is not perceived. As a result, the perceptions on elements that the students value, such as innovation, environmental sustainability and circularity, are incorrect or misunderstood. Yet, there is interesting potential in terms of the actual values and interests that needs to be directed to the automotive aftermarket with targeted campaigns and communication. Particularly women, migrants or other somewhat vulnerable groups represent untapped potential. Cross-dissemination of campaigns across disciplines and social engagement in education of all sorts is recommended. The study suggests following actions for the sector in line with prior findings (Elo, 2022):

- More direct communication and engagement.
- Updating the status and image of AA.
- New narrative, responsible, sustainable and circular AA.
- New narrative, modern AA with inclusion, equality and diversity.
- New strategy to connect and communicate with future talent.

The students wish to make a positive impact, but are not aware of the opportunities in the automotive aftermarket and call for a new narrative. There is a great willingness to contribute for the global challenges.

Naturally, this study is limited to one country and has a limited number of students. A selection bias of study programs may influence some views. Hence, it does not provide generalisable hypothesis per se, but it offers interesting food for thought for the talent development and understanding the talent pipeline. Future studies could investigate measures and campaigns that address particular target groups, such as inactive young people, female students or migrants as potential talent. Future studies assessing effective communication strategies for sectoral awareness are recommended.

Further research is needed to investigate the linkage between young talent and the transition towards a sustainable future for the automotive aftermarket sector to be able to formulate an exact action plan on how to best proceed with complying with the upcoming sustainability regulations. Finally, we suggest that the role of values and purpose in connection with enabling skillsets and education on sustainability issues are examined and further developed into managerial approaches that enhance equity, inclusion and diversity next to environmental concerns on the organisational level.

References

Aceleanu, M. I., Serban, A. C., & Burghelea, C. (2015). "Greening" the youth employment—A chance for sustainable development. *Sustainability, 7*(3), 2623–2643.

bin Magbool, M. A., Amran, A., Nejati, M., & Jayaraman, K. (2016). Corporate sustainable business practices and talent attraction. *Sustainability Accounting, Management and Policy Journal, 7*(4), 539–559. https://doi.org/10.1108/SAMPJ-06-2015-0042

Bosch. (2020). *Sustainability report 2020.* Retrieved May 18, 2021, from https://assets.bosch.com/media/global/sustainability/reporting_and_data/2020/bosch-sustainability-report-2020-factbook.pdf

Council of the European Union. (2018). *The European Union Youth Strategy 2019–2027.* Retrieved April 20, 2021, from https://eur-lex.europa.eu/legal-content/EN/TXT/?uri=uriserv:OJ.C_.2018.456.01.0001.01.ENG

DiFonzo, N., & Bordia, P. (1998). A tale of two corporations: Managing uncertainty during organizational change. *Human Resource Management, Fall/Winter, 37*(3–4), 295–303.

Elo, M. (2021). Den globale konkurrence om arbejdskraft modarbejder FN's mål om at reducere uligheden. *HR Chefen, Februar, 2021*, 40–42.

Elo, M. (2022). *Research report 1. TALENTS4AA join the future of mobility* (pp. 1–9).

Emmanuel, N., Elo, M., & Piekkari, R. (2019). Human stickiness as a counterforce to brain drain: Purpose-driven behaviour among Tanzanian medical doctors and implications for policy. *Journal of International Business Policy., 2*. https://doi.org/10.1057/s42214-019-00036-7

European Commission. (2021). *A European green deal.* Retrieved May 05, 2021, from https://ec.europa.eu/info/strategy/priorities-2019-2024/european-green-deal_en

Eurostat. (2020a). *Being young in Europe today.* Retrieved April 10, 2021, from https://ec.europa.eu/eurostat/statistics-explained/index.php?title=Being_young_in_Europe_today

Eurostat. (2020b). *Young people aged 20–24 years, by education and employment status, 2019 (%).* Retrieved April 20, 2021, from https://ec.europa.eu/eurostat/statistics-explained/index.php?title=File:Young_people_aged_20-24_years,_by_education_and_employment_status,_2019_(%25)_BYIE20.png

Eurostat. (n.d.). *Young people aged 25–29 years, by education and employment status, 2019 (%).* Retrieved April 20, 2021, from https://ec.europa.eu/eurostat/statistics-explained/index.php?title=File:Young_people_aged_25-29_years,_by_education_and_employment_status,_2019_(%25)_BYIE20.png

Geissdoerfer, M., Savaget, P., Bocken, N. M., & Hultink, E. J. (2017). The circular economy—A new sustainability paradigm? *Journal of Cleaner Production, 143*, 757–768. https://doi.org/10.1016/j.jclepro.2016.12.048

Greenbank, P. (2009). Re-evaluating the role of social capital in the career decision-making behaviour of working-class students. *Research in Post-Compulsory Education, 14*(2), 157–170. https://doi.org/10.1080/13596740902921463

Gupta, R., Sahoo, S. K., & Sahoo, T. R. (2018). Employer branding: A tool for employee retention. *International Journal of Business and Management Invention (IJBMI) ISSN, 7*(6), 6–12.

Harper, S. (2014). Economic and social implications of aging societies. *Science, 346*(6209), 587–591.

Kerr, S. P., Kerr, W., Özden, Ç., & Parsons, C. (2016). Global talent flows. *Journal of Economic Perspective, 30*(4), 83–106. https://doi.org/10.1257/jep.30.4.83

Klotz, L. (2013). The sustainability generation: Why young people hold the key to a better future. Clemson University TigerPrints. Retrieved May 04, 2021, from https://tigerprints.clemson.edu/cgi/viewcontent.cgi?article=1007&context=civileng_pubs

Kramar, R. (2014). Beyond strategic human resource management: Is sustainable human resource management the next approach? *The International Journal of Human Resource Management, 25*(8), 1069–1089. https://doi.org/10.1080/09585192.2013.816863

Łącka-Badura, J. (2015). *Recruitment advertising as an instrument of employer branding: A linguistic perspective*. Cambridge Scholars Publishing.

Lazard. (2020). *Global automotive supplier study 2020*. Retrieved March 16, 2021, from https://www.lazard.com/media/451494/global-automotive-supplier-study-2020.pdf

Linley, A. (2015). Best practices for finding hidden talent and empowering social mobility. *Strategic HR Review, 14*(5), 194–198. https://doi.org/10.1108/SHR-03-2015-0023

McKinsey&Company. (2017). *The changing aftermarket game—and how automotive suppliers can benefit from arising opportunities*. Retrieved March 15, 2021, from https://www.mckinsey.com/~/media/McKinsey/Industries/Automotive%20and%20Assembly/Our%20Insights/The%20changing%20aftermarket%20game%20and%20how%20automotive%20suppliers%20can%20benefit%20from%20arising%20opportunities/The-changing-aftermarket-game.pdf

McKinsey&Company. (2018). *Ready for inspection—The automotive aftermarket in 2030*. Center for future mobility. Retrieved March 15, 2021, from https://www.mckinsey.com/~/media/McKinsey/Industries/Automotive%20and%20Assembly/Our%20Insights/Ready%20for%20inspection%20The%20automotive%20aftermarket%20in%202030/Ready-for-inspection-The-automotive-aftermarket-in-2030-vF.ashx

Ramboll. (2021). *The EU taxonomy regulation: 5 reasons to start aligning today*. Retrieved May 12, 2021, from https://ramboll.com/ingenuity/the-eu-taxonomy

Roland Berger. (2020). Fast forward after the Covid-19 crisis. How suppliers can successfully adjust their strategies for the future. Retrieved March 16, 2021, from https://www.rolandberger.com/publications/publication_pdf/roland_berger_automotive_suppliers_2020.pdf

Ronda, L., Valor, C., & Abril, C. (2018). Are they willing to work for you? An employee-centric view to employer brand attractiveness. *Journal of Product & Brand Management, 27*(5), 573–596.

Schlechter, A., Hung, A., & Bussin, M. (2014). Understanding talent attraction: The influence of financial rewards elements on perceived job attractiveness. *SA Journal of Human Resource Management/SA Tydskrif vir Menslikehulpbronbestuur, 12*(1), 13. https://doi.org/10.4102/sajhrm.v12i1.647

United Nations. (2015). *Sustainable development goals*. Department of Economic and Social Affairs. Retrieved March 12, 2021, from https://sdgs.un.org/goals

Winkelhake, U. (2018). *The Digital Transformation of the Automotive Industry*. Springer International Publishing Cham.

Zimmermann, K. F. (2005). European labour mobility: Challenges and potentials. *De Economist, 153*, 425–450.

Gyula Kimpan is a recent graduate of engineering and business master studies, with interest and experience in the automotive industry, currently being active in the automotive aftermarket. He has lived in various countries around the world and participated in international talent competitions where his teams were awarded with the highest level of recognition for cutting-edge applications in areas such as biotechnology. Today, he provides consultancy for digital projects, product development, and innovation management. While Gyula is enthusiastic about all things virtual, he just as much enjoys cycling and spending time in nature.

Maria Elo full professor of international business and entrepreneurship at the University of Southern Denmark, professor at the BRIIB Institute of International Business at Shanghai University, Senior Research Fellow, University of Turku, Adjunct professor at Åbo Akademi

University, and Research Fellow at Institut für Arbeit und Technik, works in the area of international business, -entrepreneurship, and -migration with topics such as sustainability, business ecosystems, policy frameworks, networks, internationalization, mobility forms, human stickiness, migrant entrepreneurship, transnational and family businesses, diaspora investment, and remittances. She has published books and articles, for example, in Journal of World Business, Journal of International Management, Journal of International Business Policy, Industrial Marketing Management, Regional Studies, Journal of International Entrepreneurship, and International Journal of Entrepreneurship and Small Business. Maria is one of the founders of Talents4AA and a Talents4AA ambassador leading the collaboration with the education sector.

Part III

Perspectives on Regions, Markets and Internationalization

Automotive Aftermarket in the Western Balkans: Opportunities for the Region and the European Automotive Industry

Iris Koleša and Andreja Jaklič

Abstract

In this chapter, we address the strengths, weaknesses, opportunities and threats in the Western Balkans' automotive industry and its aftermarket sector. Based on desktop analysis and anecdotal evidence from interviews with business, cluster and support environment representatives, we first discuss the similarities in the region and how they are reflected in the development of the automotive industry and its aftermarket. We then address the specifics of individual Western Balkan countries in this respect. We conclude the chapter with policy recommendations, whereby we particularly stress the importance of (1) the modernisation of the scientific higher education system and investments in employee development in accordance with the global trends in the industry, (2) facilitation of partnerships between governance and business, (3) promotion of inter-business networking and (4) acceleration of internationalisation— especially in the form of inward foreign direct investment that promotes knowledge sharing and development as well as investments in infrastructure in the region (especially in terms of digitalisation, automation and technology upgrades).

I. Koleša (✉) · A. Jaklič
Faculty of Social Sciences, University of Ljubljana, Ljubljana, Slovenia
e-mail: iris.kolesa@fdv.uni-lj.si; andreja.jaklic@fdv.uni-lj.si

M. Elo, F. Katsardis (eds.), *Automotive Aftermarket*, Management for Professionals, https://doi.org/10.1007/978-3-031-62419-3_11

1 Introduction

The Western Balkan countries (i.e. Albania, Bosnia and Herzegovina, Croatia, Kosovo, Montenegro, North Macedonia, Serbia and Slovenia)[1] have experienced drastic systemic transformations at both national and regional levels throughout their history— including the transition from socialism to capitalism and slow integration into the European Union (EU). The continuous institutional, political and economic uncertainties have resulted in these countries' and their inhabitants' (businesses and citizens alike) greater flexibility, resilience and preparedness for and to change. However, these countries' responsiveness to change has been slower compared to the EU member transition countries from outside the region, which is why they are sometimes described as 'late reformers' (International Bank for Reconstruction and Development/The World Bank, 2017, pp. 123). This can be attributed to the persisting conflicts in the region combined with the limited resources (knowledge and capital alike) for reform. This also holds for actors in the automotive industry and its automotive aftermarket sector that are highly affected by the developments in global markets, yet often depend on foreign investments to be able to respond to them.

The automotive aftermarket in the region mainly consists of (1) for the most part export-oriented and often foreign-owned automotive components manufacturers[2] and (2) the largely domestically owned and locally oriented[3] repair shops (Agency for Foreign Investments and Export Promotion, Invest North Macedonia, Investor servicing Unit, Investor servicing and aftercare department, personal communication, August 24, 2022; FIPA, personal communication, August 10, 2022; OECD, 2008; The Croatian Chamber of Economy— Croatian Association of Car Importers and Distributors, personal communication, August 12, 2022; The Investment Promotion Section at the Montenegrin Investment Agency, personal communication, August 19, 2022). The latter service the aftermarket as their primary market, whereas the former often consider the aftermarket as a secondary area of their business. More specifically, these firms engage in the automotive aftermarket because they are contractually bound to provide auto parts they are manufacturing for new vehicles for the entire duration of the vehicles' lifecycle (SRIP ACS GIZ, personal communication, April 19, 2022). For the most part, they thus do not approach the aftermarket systematically and strategically, but rather responsively— according to

[1] We include the two European Union member states Croatia and Slovenia in the analysis due to the joint history and maintained strong regional connections between all eight countries even after the breakup of Yugoslavia in the 1990s. This connectedness is also reflected in the automotive industry and its aftermarket sector.

[2] The Western Balkans' automotive components sector is highly integrated in the European regional value chain (bne IntelliNews, 2022), which makes the development of the sector very much dependent on the developments in the European automotive industry, but also the normative framework in that region (e.g. transitioning to green mobility).

[3] This introduces the challenge of servicing a small and limited market (FIPA, personal communication, August 10, 2022).

the demands of the original equipment manufacturers (OEMs)[4] and the developments in the overall automotive industry (survey[5] among SRIP ACS GIZ members, 2022).[6]

Focused on these players, policy-makers tend to forget to develop aftermarket-tailored support instruments and mechanisms. They rather rely on spillover effects from the overall automotive industry to the sector. This is even more true for the independent aftermarket (IAM), dominated by small and medium-sized enterprises (SMEs)[7] that predominantly rely on imports. To the best of our knowledge (also based on the interviews with industry representatives and policy-makers), there are currently no automotive aftermarket-specific policies in the region.[8] In addition, data on the automotive aftermarket that could drive informed decision-making is practically non-existent. Data collection is made additionally difficult because companies from the automotive aftermarket can be registered in and often supply different industries (FIPA, personal communication, August 10, 2022).

The automotive aftermarket in the region nonetheless demonstrates great growth potential. This is because of the high average age of vehicles in these markets[9] that necessitates more repair work and other related services (FIPA, personal communication, August 10, 2022; The Croatian Chamber of Economy— Croatian Association

---

[4] Partnerships with OEMs are predominantly long-term and focused on the European Union (survey among ACS-GIZ members, 2022). These partnerships are crucial, as OEM actors often have greater potential to contribute to local and regional competitiveness through research and development (R&D) and innovation, whereas IAM actors (although numerous and successful due to the old carpark in the Western Balkan countries) often lack similar capacities (FIPA, personal communication, August 10, 2022).

[5] The sample was too small for statistical analyses. However, businesses' responses to the open-ended questions in the survey enabled qualitative analyses.

[6] This somewhat depends on the products they produce (SRIP ACS GIZ, personal communication, April 19, 2022).

[7] In response to the Covid-19 pandemic- and the war in Ukraine-related global, regional and local value chain disruptions, the increased demand for IAM products and services due to the effect of reduced new car sales and accelerated digitalisation (unequally distributed through the region) enabling access not only to customers but also business partners worldwide, consolidation and integration (both vertically and horizontally) of the IAM have recently been taking place and several large IAM players have emerged in the region (also in response to OEM actors diversifying their portfolio and entering the aftermarket as a crisis management strategy creating additional competition particularly threatening to SMEs) (see also Berger, 2020).

[8] Interviewees state that local, national or regional policies have no direct role in the development of the automotive aftermarket in their countries (e.g. The Croatian Chamber of Economy— Croatian Association of Car Importers and Distributors, personal communication, August 12, 2022).

[9] The latter exceeds 19 years in Kosovo, North Macedonia and Albania, and 14 years in Croatia, Montenegro and Serbia. It is the 'lowest' in Slovenia at 12 years of age (ACEA, 2022; FIPA, personal communication, August 10, 2022; Jovanović, 2018; Kolekevski, 2022; N1 Belgrade, 2019; The Croatian Chamber of Economy— Croatian Association of Car Importers and Distributors, personal communication, August 12, 2022; The Investment Promotion Section at the Montenegrin Investment Agency, personal communication, August 19, 2022; World Bank Group, 2019a). The average car age in Europe was 9.7 in 2021. This increased from 8.4 in a decade (Market Data Forecast, 2022).

of Car Importers and Distributors, personal communication, August 12, 2022; The Investment Promotion Section at the Montenegrin Investment Agency, personal communication, August 19, 2022; World Bank Group, 2019a) as well as upgrades reducing the environmental footprint of these vehicles (OECD, 2008; Šulin Košar, 2017). The low purchasing power of consumers in addition favours IAM actors in this respect.[10] According to the Croatian Association of Car Importers and Distributors at the Croatian Chamber of Economy (personal communication, August 12, 2022), this trend will become even more pronounced with inflation, which is also likely to contribute to the expansion of the grey market and change consumer habits. Used car sales have already increased with the Covid-19 pandemic and the war in Ukraine (due to difficulties and delays in the delivery of new cars and their increasing prices) (Rosenbaum, 2020) and have additionally contributed to automotive aftermarket growth and growth potential, especially in the replacement and repair services.

The developments in e-commerce in particular are also relevant for the development of IAM in the region (The Croatian Chamber of Economy— Croatian Association of Car Importers and Distributors, personal communication, August 12, 2022). Online sales and digitalisation, the two important drivers of market growth, are in their infancy in most of the region, however (see e.g. OECD, 2021). According to Jovanović and Vujanović (2023), the region's automotive industry has the greatest potential for catch-up and technological upgrading (including digitalisation)—also due to the strong inflows of FDI and a pronounced scope for linkages to other technology-intensive industries. A looming recession and the rising energy costs related to the current global energy crisis that began in 2021 with the Covid-19 pandemic could create some additional challenges for the regional automotive aftermarket sector by increasing operational costs. They may also introduce opportunities for the restructuring of the sector, aftermarket actors taking over business from competitors not able to supply the markets with new vehicles and mastering the transition to (servicing) electric vehicles (also in response to changed consumption patterns and green policies).

This chapter summarises the strengths and weaknesses of the region for the automotive aftermarket development, outlines the specifics of individual Western Balkan countries in this respect and provides policy recommendations based on an extensive desktop analysis and anecdotal evidence from our interviews with business, cluster and support environment representatives in the region.

2 Strengths of Western Balkans for the Development of the Automotive Aftermarket

The region is characterised by several strengths in terms of the automotive aftermarket development. Access to a sizeable duty-free market due to multilateral (e.g. the Stabilisation and Association Agreement with the EU, The European Free Trade

[10] IAM gained 2% points of market share in 2020 against the authorized channel— despite the negative effect of the pandemic on the economy (Waas et al., 2021).

Association with Switzerland, Norway, Iceland and Liechtenstein and The Central European Free Trade Agreement within the Western Balkans) and bilateral *trade agreements* (Several countries have free trade agreements with Turkey and Ukraine, whereas Serbia is the only country in the Western Balkans that also has a free trade agreement with Russia. The region has also been recording intensified regional trade relations with China— especially in terms of imports— due to the 16 + 1 initiative)[11] combined with (a promise of) EU membership[12] attracts foreign investors to the region. This especially holds true for investors from partner countries, but it also has spillover effects on other investors as it signals economic security for businesses overall. Foreign investors thereby often work to upgrade technical and engineering skills in the host countries— also in collaboration with higher education institutions (e.g. by forming joint ventures) (International Bank for Reconstruction and Development/The World Bank, 2017).[13] This in turn not only boosts (domestic and international) economic activity but also leads to local businesses moving up the value chain (through development and provision of value added services and innovation) (see also OECD, 2008). National governments are thus committed to the attraction of foreign direct investments (FDI), which is reflected in regulatory, fiscal and tax reforms as well as investor incentives (Agency for Foreign Investments and Export Promotion, Invest North Macedonia, Investor servicing Unit, Investor servicing and aftercare department, personal communication, August 24, 2022; FIPA, personal communication, August 10, 2022; Redžepagić & Richet, 2008).

The region's *strategic location and geographical proximity* to the European Union also enables firms in the region to ensure just-in-time delivery. Besides time to market, just-in-time delivery is a fundamental requirement in the automotive supply chain (OECD, 2008). For this reason, Western Balkans is becoming an even more important player in the European regional value chain— also through reshoring and nearshoring. Following the Covid-19 pandemic[14] and the Russian invasion of Ukraine the European regional value chain has been experiencing issues in supply chains (both in terms of delivery times and prices) and a greater need for bringing manufacturing and other business operations closer to their home markets. The Western Balkans is thereby at a comparative advantage relative to the Central and Eastern Europe, since the region's automotive (labour) markets are not yet saturated, which makes finding suitable labour easier (Jovanović et al., 2021).

[11] In 2013, Serbia also signed a strategic partnership agreement that further deepened the already strong economic and political cooperation with Russia (OECD, 2008; Savic, 2013).

[12] Alignment with the EU reduces the 'sovereign country risk' (i.e. the likelihood of a country defaulting on its financial agreements due to political and economic uncertainties)— especially for EU-based companies (Frink, 2008).

[13] One such example is Bosch, which has developed a strong regional presence over the past decade (industry expert, personal communication, 2023a).

[14] The automotive sector is the second most affected sector by issues in supply chains following the Covid-19 pandemic, as well as one of the sectors most likely to introduce changes to its supply chains (Jovanović et al., 2021).

A sufficient pool of cost competitive and quality labour is another advantage of the Western Balkans. This stems from a long legacy of automotive components manufacturing in the region that has generated extensive market knowledge. The educational system of former Yugoslavia also provided a high level of automotive specific training and a strong skills base (especially in comparison with the situation in other emerging markets) (Jovanović et al., 2021; OECD, 2008). Foreign investors further promote knowledge upgrading and in some cases even participate in the modernisation of the scientific higher education system (Jovanović et al., 2021; OECD, 2009).

3 Weaknesses of Western Balkans for the Development of the Automotive Aftermarket

Political unrest and economic uncertainty across Western Balkans have culminated into several regional weaknesses that hinder the development of the automotive industry and its aftermarket sector. *Limited resources* at local, national and regional levels, resulting from persistent conflicts, for example, hinder infrastructure building, institutional development and the overall quality of business environment. The region was long known for lengthy and complex procedures related to starting a business (OECD, 2008). While these procedures have simplified and significantly shortened in several Western Balkan countries (industry expert, personal communication, 2023b), an image of bureaucratically burdensome business environment remains. This in turn deters FDI[15] and the related capital inflows: including new technologies and knowledge (Demekas et al., 2005; OECD, 2008).[16] As a result, the region is characterised by slow technological development and innovation progress as well as low R&D investment and labour productivity (with the exceptions of Croatia and Slovenia), which is particularly pressing for local suppliers (Krasniqi et al., 2022; OECD, 2021), unable to move up the value chain and become developmental partners to their foreign counterparts. This issue is additionally deepened by business undercapitalisation, limited financing and a lack of strategies for reinvesting in technology and human capital (Jovanović et al., 2021; OECD, 2008).

The region's automotive industry and aftermarket development is also affected by the *skill gaps* encountered by businesses operating in the Western Balkans. Forty-four percent of automotive component companies classify the unavailability of skills as a top challenge they face in expanding their operations in the region. Skill gaps are especially pressing in the areas of design, supply chain management and quality assurance, and electrification (FIPA, personal communication, August

[15] FDI in the Western Balkans' automotive industry and aftermarket is unequally distributed across the region (The Investment Promotion Section at the Montenegrin Investment Agency, personal communication, August 19, 2022).

[16] Some Western Balkan countries are facing a challenge of overreliance on extant partnerships, as new investments are rare (FIPA, personal communication, August 10, 2022).

10, 2022; OECD, 2008).[17] These gaps are additionally deepened by notable emigration flows from the region that is shrinking the available pool of (skilled) employees (see e.g. Koleša, 2019). Another factor that increases the impact of skill gaps on the industry is the lack of mechanisms for institutionalisation and dialogue between ministries of economy, businesses and education, and limited partnerships between universities and business associations. This in turn reduces the possibilities to match skilled employee supply with demand for it at the structural level (OECD, 2008) as well as immediate knowledge transfers from research institutes to business (and vice versa).

In addition, there are *limited linkages between domestic suppliers and foreign investors*. Many manufacturers from the region (skills and modern technologies allowing them to participate in the innovation processes) act as non-developmental suppliers rather than developmental partners to their counterparts. The developmental potential of local suppliers is further limited by a lack of collaborative mechanisms enabling innovation to flow between research centres, academia and foreign and local companies. Nonetheless, there are a few individual cases that manage to attract foreign capital to the region and get involved in modern manufacturing models and programmes, such as programmes of lifecycle solutions aimed at extending the use of old cars. Hella Saturnus Slovenia in Ljubljana, which develops, produces and markets products for the independent parts trade and for workshops, is one such example (see e.g. Tasevski, 2019).[18] Regional and local offices of Renault and Volkswagen have also intensified their purchasing of used vehicles and established units for recycling them (see also Prebil, 2022).

According to PwC (2019), automotive (parts and components) suppliers' value chains in the Western Balkan countries are rather small and synergies between them weak. Regional products are rarely used as inputs by other regional producers in the sector, which suggests that cooperation at the regional level is also limited. This is

---

[17] While the Western Balkans was also marked with a shortage of digital and software engineering skills only a few years ago, the information and communication technology (ICT) industry has quickly developed into one of the largest in the region (industry expert, personal communication, 2023b). The ICT industry accounts for almost 3% of value added in Serbia, almost 2% in Bosnia and Herzegovina and North Macedonia, and around 1% in Albania, Kosovo and Montenegro. With the exception of the latter, these percentages are increasing in all listed countries (Jovanović & Vujanović, 2023). Corresponding with these developments and the related growing demand for ICT experts, about 10% of graduates from the six non-EU Western Balkan countries come from the ICT sector— with this share increasing (Stojkovski, 2023). There are nonetheless differences across (and within) countries in the uptake of digital skills and technologies. Both Croatia and Serbia act as drivers of growth to the region's ICT industry (Stojkovski, 2023). Serbia, in particular, has noted prompt growth in ICT (industry expert, personal communication, 2023b). Ten percent of Serbia's GDP is generated by the ICT sector, which is also among the country's top four export sectors (International Trade Administration, U.S. Department of Commerce, 2022; Oxford Analytica, 2022). Export in ICT in Serbia amounted to 2.7 billion EUR in 2022— a 45% growth on the previous year (Agenzia Nova, 2023; industry expert, personal communication, 2023b).

[18] With its Business Group Lifecycle Solutions, Hella aims to build a holistic ecosystem that covers the entire process chain in the aftermarket from diagnostics to distribution of spare parts to services from a single source in order to create added value for all parties involved and contribute to the responsible use of resources and materials (aftermarketNews Staff, 2022).

also because the quality of products by local producers of raw materials for the most part does not meet the requirements of the automotive industry. Another reason for the lack of regional cooperation includes limited presence of vehicle producers (or at least Tier 1 suppliers) in the region who could act as connectors of regional suppliers. As a result, the regional automotive industry is *fully dependent on foreign, mainly European, vehicle producers* and Tier 1 suppliers. Regional suppliers of automotive parts and components to the European and global value chains can thereby be classified into: (1) subsidiaries of renowned European producers and (2) domestic producers with strong partnership relations with their buyers, who can be either intermediaries or vehicle producers. The former provide processing of materials in line with the specifications of their parent companies (who are usually their key raw material suppliers, buyers and owners at the same time). The latter, on the other hand, also act as developmental partners for multiple buyers. The first group is prevalent in the region, however (PwC, 2019).

Dependence on the European business partners is also reflected in the impact that the policy shifts in EU have on the region. For example, the recent European Green Deal (European Commission, 2019) has affected competitiveness of the European automotive sector by imposing additional environmental constraints and expectations of 'clean innovation' on businesses. This has in turn had an indirect effect on the Western Balkans due to the Western Balkan businesses' ties with European business partners whose expectations regarding product, service and process specifications have changed. The policy shift has also resulted in changed investment behaviour, whereby ESG (environmental, social and governance investing) requirements have increased. This has been both a challenge and an opportunity for the automotive aftermarket players in the region to introduce innovative processes as well as develop products and services not only for the newly produced vehicles but also for the existing (i.e. used) cars in the market (e.g. innovative upgrading hardware or software decreasing the environmental footprint of the car or local upgrading garages and workshops).

The slow progress on the supply side is accompanied by peculiarities on the demand side as well. Low economic growth has limited the purchase of new cars and reduced the number of premium customers. These circumstances affect both the volume of repair services (consumers become more motivated to repair, refurbish and maintain their vehicles, as they cannot afford new ones) and the structure of the market for automotive parts. Anecdotal evidence indicates that both customers and retailers in the service market prefer non-certified parts over certified or original parts due to the lower purchasing power in the region. Consumer loyalty to original equipment can also be lower as a result. In addition, service through the 'do-it-yourself' channel seems to be comparatively more attractive and is expected to continue to grow. While the original parts segment dominates the Western markets, demand for these parts is likely to grow slower in the Western Balkan region, because original parts are often expensive and only available through select retailers. In addition, parts for fossil fuel rather than hybrid and electric cars prevail due to the older structure of the carpark in the region— at least in terms of servicing domestic markets (this also holds true for the more developed markets in the region, such as Slovenia (see Šulin Košar, 2017)).

All these challenges slow down the adaptation of local actors to global trends (businesses and governments alike). Inequalities between countries, regions and cities (also in terms of infrastructure (FIPA, personal communication, August 10, 2022)) thereby additionally hinder the development of the automotive industry and its aftermarket further. We discuss the regional diversity in the next section of the chapter.

4 Country Specificities in Terms of the Automotive Industry and Aftermarket

An additional challenge for policy-makers and businesses in the Western Balkans is the *regional diversity* that requires these actors to prepare and implement strategies tailored to each market. "That is the joy of the region, where nothing is the same" (Lacey in Frink, 2008). This diversity is also evident in the automotive industry and its aftermarket. Western Balkan countries vary greatly in terms of both their automotive industries' level of development and characteristics. Croatia, Slovenia, Bosnia and Herzegovina, North Macedonia and Serbia in particular have established a considerable presence in the industry— with Bosnia and Herzegovina boasting a well-organised automotive aftermarket; while Albania, Kosovo and Montenegro[19] are lagging behind (Agency for Foreign Investments and Export Promotion, Invest North Macedonia, Investor servicing Unit, Investor servicing and aftercare department, personal communication, August 24, 2022; bne IntelliNews, 2022; FIPA, personal communication, August 10, 2022; The Croatian Chamber of Economy— Croatian Association of Car Importers and Distributors, personal communication, August 12, 2022; The Investment Promotion Section at the Montenegrin Investment Agency, personal communication, August 19, 2022).

4.1 Croatia: A Small, Yet Highly Innovative and Growing Automotive Industry

The Croatian automotive industry mainly consists of companies manufacturing parts and components for conventional and electric vehicles and businesses engaging in new trends in the industry. Both these groups of firms base their operations on a long-term tradition of metal and plastics processing and engineering (PwC, 2019). However, many of the newly established firms fall under the second category and operate in electro-mobility, engineering and information technologies (IT) in the

[19] The automotive sector is one of the least developed sectors in Montenegro: the country does not have its own car production and only has a few car components producers. The automotive aftermarket does not include manufacturing or remanufacturing but is mostly involved in distribution and retailing of imported auto components. Montenegro (like Kosovo) is also not a large recipient of inward FDI aimed at the automotive industry (The Investment Promotion Section at the Montenegrin Investment Agency, personal communication, August 19, 2022).

automotive sector. Several are also active in the aftermarket. One such example is Amodo, which is offering insurance models like Pay How You Drive, Pay As You Drive and Manage How You Drive (Republic of Croatia Trade and Investment Promotion Agency, n.d.). Although the automotive industry in Croatia is rather small (it accounts for 1.1% of total production and 4.7% of exports in the country,[20] includes 144 firms[21] and employs less than 3.000 people), it is highly innovative (Republic of Croatia Ministry of Economy and Sustainable Development, 2022; Simmonds, 2020). Croatian automotive parts producers, for example, deliver more complex, high quality and high value added products that require higher technical skills, innovation, flexibility and design. They also deliver these products to some of the upper-market car producers such as Bentley, Ferrari, Mercedes, Alfa Romeo, McLaren, Lamborghini and Aston Martin (Republic of Croatia Trade and Investment Promotion Agency, n.d.).

The country is thereby attracting additional investments in the automotive industry.[22] Especially investments in R&D are recording the greatest growth. For example, dSPACE, AVL, Atron and Porsche all have R&D centres for the automotive industry in Croatia. Foreign investors thereby engage in the development of autonomous vehicles and robots (e.g. DOK-ING, Gideon Brothers and Inetec), as well as cyber security (e.g. Reversing Labs), navigation (e.g. Mireo and OptimoRoute) and hardware for them (e.g. Xylon) (Simmonds, 2020).

Companies registered in Croatia can thereby also benefit from EU funds for R&D. One such example has been Rimac Automobili— Croatia's unicorn company that produces its own high-performance electric cars as well as develops and manufactures drive and battery systems for other manufacturers. Its solutions also help other companies reduce fossil fuel consumption and harmful emissions of industrial CO2 in vehicles (HAMAG-BICRO, 2020).[23] In 2018, the European Investment Bank (EIB) and Rimac Automobili signed a 30 million EUR loan for investment in research and development of electric vehicles and related components guaranteed by the European Fund for Strategic Investments (EFSI) (European Investment Bank, 2018). Less than 3 years later, the company was awarded a 200 million EUR grant from the EU's 6.3 billion EUR aid package in support of Croatia's Covid-19

[20] Companies from the industry thereby export more than 90% of their R&D, software (e.g. simulation software and automation software for test beds) or goods— mostly to EU markets (Republic of Croatia Trade and Investment Promotion Agency, n.d.).

[21] SMEs dominate the industry: their small size thereby gives them greater flexibility, increases their responsiveness and adaptability to new products and technologies, as well as provides them with an opportunity to enter market niches (Republic of Croatia Trade and Investment Promotion Agency, n.d.). This in turn supports their role as a developmental partner to other businesses.

[22] Croatia offers various investment incentives for investment, R&D and employment, such as 0% corporate income tax up to 10 years, up to one million EUR cash grants for capital investments, up to 0.5 million EUR cash grants for R&D investments and up to 18.000 EUR cash grants for individual employees (Republic of Croatia Trade and Investment Promotion Agency, n.d.). These are not exclusive to automotive industry.

[23] The Rimac Group is an example of a local company moving up the automotive value chain through outward FDI.

recovery plan for developing a driverless passenger vehicle (Tsoneva, 2021). The success of Rimac Automobili resulted in both investors and talents from the industry migrating to the country.[24] It also sparked growth and innovation in related sectors— including the automotive aftermarket. The company is additionally engaged in shaping public policy (domestically and at the level of the EU), specifically in the fields of the battery economy, smart transport, e-mobility and autonomous driving (Euronavigator, 2022) and in developing study programmes for automotive professions (together with some other businesses from the industry and universities) (Republic of Croatia Trade and Investment Promotion Agency, n.d.). Rimac Automobili has also announced the establishment of the Association for the Promotion and Regulation of Autonomous Vehicles.[25] This will connect and support the work of companies, regulators, cities and other stakeholders interested in participating in the future autonomous vehicle ecosystem (combining electrification, vehicle networking and autonomous driving) (Simmonds, 2020).

The rise of Rimac Automobili and the enthusiasm for car racing in Croatia have recently also fuelled the demand for racing and other accessories as well as the installation of digital equipment in older vehicles. This has in turn resulted in rapid development of IAM in the country. Tokić Ltd., for example, developed a car parts shop in Zabok, which offers 300 brands and about 300.000 different items for over 200 vehicles and 100 branches, into a kind of 'rehabilitation centre' for cars and drivers in Croatia. With ten employees in this shop, most of the activity takes place behind the counter. In addition, out of the total 1000 Tokić Ltd. employees, more than 40 work on technology-related processes and support the network of rehabilitation centres around the country. A team working on artificial intelligence, object recognition, machine learning and robotic process automation completed over five million tasks in 2022. Together with Bartog from Slovenia and franchising partners, Tokić is the largest retail chain for automotive spare parts in the region. It has more than 140 shops in Croatia and Slovenia. Similar suppliers are gaining recognition not only in Croatia but also in the overall region (industry expert, personal communication, 2023a).

Some of the country's advantages for the development of the automotive industry and aftermarket include good road infrastructure (its quality ranks 13th among 141 countries according to Schwab, 2019) and cost competitiveness with respect to investment, energy prices, education and salaries. Investment costs are 34% below the EU average, construction costs 50%, and energy and gas prices 36% below the EU average (Republic of Croatia Ministry of Economy and Sustainable Development, 2020). Estimated hourly labour costs are also at 40% of the EU-27 average (data for

[24] While this development may be advantageous for the Croatian economy, it also presents a danger of the single firm exhausting the regional skilled labour force pool.

[25] According to Mate Rimac, founder and CEO of Rimac Automobili (in Simmonds, 2020), vehicles will be 80 percent autonomous by 2030. This means they will also be networked and shared, which will increase their time in operation from 4 to 70% (covering more than 250,000 miles a year). This implies disruption of current business models in the industry as well as significant changes in the regulatory framework (Simmonds, 2020). It also indicates tremendous potential for the automotive aftermarket: especially in terms of in-car data-based service provision.

2021). (Eurostat, 2022). In addition, Croatia has 200 fully equipped business zones prepared for investments, whereby land prices start at 1 EUR/m2 in some locations. Investors registered in the country are also eligible for attractive incentives for investment, R&D and employment, as well as EU grants (Republic of Croatia Ministry of Economy and Sustainable Development, 2020). The country also records relatively high levels of labour productivity and has good know-how of the automotive components industry. High precision manufacturing with zero-defect tolerance, high level of quality standards (often supported by ISO and industry-specific certification) and the businesses' capability to establish developmental collaboration with customers along with the strong relationships between local suppliers and international customers are also recognised as the country's advantages (OECD, 2008).

Croatia is also actively developing talent for the automotive industry (and the related sectors with high value added products and services). In the 2018/2019 school year, about 15% of students graduated in the field of engineering. There are also several faculties specialising in electrical and mechanical engineering, as well as computing. An Automotive Computing and Communications— AUTOCOM programme co-developed by companies from the automotive industry has been offered since 2017 at the Osijek Faculty of Electrical Engineering, Computing and IT (available in English from 2019) (Republic of Croatia Trade and Investment Promotion Agency, n.d.). These institutions are developing talents with relevant knowledge and skills for technical occupations (PwC, 2019). Croatia also records high levels of foreign language knowledge. 95% of the population aged 16–55 has at least a basic proficiency of English, 49% of German, 22% of Italian, 11% of Spanish and 6% of French (Republic of Croatia Trade and Investment Promotion Agency, n.d.). The country's entrepreneurial culture and marketing mindset as well as the availability of locally produced management talent and the underlying competitiveness of cost/quality/engineering skills combination in certain niches of the automotive components industry are additional advantages of Croatia's labour market. For example, the use of AutoCAD and other software for the design and development of components is relatively widespread compared with other economies in the Western Balkan region (OECD, 2008; South-East European INDUSTRIAL Market, 2016a). Availability of complementary skills and resources relevant for automotive industry and its aftermarket presents excellent potential for the future development of the sector.

4.2 Slovenia: Investments in R&D and Innovation Supporting Automotive Industry Growth

The automotive industry is the main driver of Slovenia's economy (Slovenia Business, 2022a). It makes up 20% of total Slovenian exports, contributes 10% to the country's GDP (the European average is approximately 7% (Mesojedec, 2022)) and is highly innovative (Slovenia Business, 2022b). In fact, the country's competitiveness is based on strong innovation processes and R&D. These are additionally incentivised by the government (e.g. through financial incentives and tax reductions) (Slovenia Business, 2022b). Many companies already have modern processes and technologies that follow industry 4.0 principles in place (Slovenia Business, 2022c). Slovenia also records the highest innovation activity in CEE (European Patent Office, Annual Report 2021 in Slovenia Business, 2022b). Between 2016 and 2018, more than 50% of Slovenian companies developed or introduced product, process or service innovations (SURS, 2020 in Slovenia Business, 2022b). This results in high value added in manufacturing: Slovenia ranks eighth among 200 countries in this aspect (Bloomberg Innovation Index 2021 in Slovenia Business, 2022b).

Automotive companies are among the most innovative in the country (Slovenia Business, 2022d). In 2021, the Slovenian automotive industry comprised of 253 companies with 15,300 employees and 3.8 billion EUR in revenues. It includes manufacturers of components and parts, end-product manufacturers, as well as highly specialised niche businesses (e.g. developers of high-torque in-wheel electric drive systems, providers of sensor systems and measurement solutions and firms offering AI-based real traffic simulation solutions), supplying both popular and prestigious automotive brands (Slovenia Business, 2022b, d, e). The mechatronic components, electromotors and their parts are some of Slovenian automotive industry's strengths (Slovenia Business, 2022d). Slovenian firms have been engaged in both the development and supply of electrical components and software for smart, autonomous and electric vehicles. Dewesoft is one such example of a highly innovative company that develops and produces high-quality measuring devices and instrumentation for data acquisition, testing and measurement of all kinds of physical phenomena. The company's solutions are also used for the development of new safer vehicles (e.g. by General Motors, Porsche, BMW, Audi and Volvo) (Slovenia Business, 2022e).

Slovenia has a dense and high-quality research network that can be facilitated for business R&D, efficient network of excellence and competence centres that ensure a systematic flow of knowledge to businesses, high-quality educational system that produces highly qualified multilingual workers with a high level of technical and engineering skills[26] and well-established cooperation between business and academia (ACSEE in Slovenia Business, 2022a; Jaklič & Koleša, 2020, 2022; Slovenia Business, 2022b, d). The availability of local partners from a wide array of supporting industries, including tooling, robotics,[27] the measurement devices industry, composite material development, coating, and software and AI development, is also an important factor in the automotive industry development (Slovenia Business,

[26] There is a growing number of engineering students in key areas such as machinery, mechatronics, chemistry and IT (Slovenia Business, 2022d).

[27] Slovenia is among top 20 countries with the highest robot density in the world (Slovenia Business, 2022c, f). It has more robots per 10,000 workers than Switzerland or the UK. Attracting the Yaskawa robot production facility with its European robotics development centre to Kočevje in Slovenia servicing OEM and IAM alike is likely to further promote robotisation in Slovenia (Slovenia Business, 2022f).

2022b, d, f). Slovenia's automotive industry also benefits from FDI. Many international automotive companies locate their regional headquarters in Slovenia due to its proximity to the emerging regional markets and strong ties of local managers, engineers and other professionals with their colleagues in the Western Balkans (Slovenia Times, 2011). Due to its small size, yet relatively high level of development, especially in terms of the automotive industry; openness to innovations and regional connections (spanning both the EU and the Western Balkans), Slovenia has also been recognised as a good testing market for new operating vehicle systems (Slovenia Business, 2022d).

Like most countries in the region, Slovenia does not have a special database of businesses active in the automotive aftermarket— nor a strategy targeting this sector. While the country has its own vehicle production facilities, most Slovenian companies in the automotive industry are Tier 2 suppliers, while some have managed to progress to systemic Tier 1 suppliers that supply their products and services directly to OEM. Although the majority of these companies still produce car parts and components based on specifications by their OEM customers, several have also become developmental partners to the latter. Operations in the automotive aftermarket are thereby part of the arrangement with the OEM customer and as such tied to servicing the vehicle for which the part is produced for the duration of the car lifecycle as determined by OEM (SRIP ACS GIZ, personal communication, April 19, 2022).

The most impactful global trends on Slovenian automotive industry include electrification, development of autonomous vehicles, new (data-driven) business models, digitalisation and transition towards green mobility— also in relation to delivery models; increased prices of raw materials, uncertain delivery times, seasonality and limited size of orders, and financial issues among distributors resulting from the Covid-19 pandemic, inflation and changed consumption patterns, such as increased purchasing of used cars (SRIP ACS GIZ, personal communication, April 19, 2022, SRIP ACS GIZ, 2022a; survey among SRIP ACS GIZ members, 2022). Firms thereby stress that investing in employee development, automation and digitalisation will be the most important ways of preparing for changes in the automotive aftermarket. They also believe strengthening the sales network and strategic partnerships as well as developing new business models and changing their product portfolio to be somewhat important (survey among SRIP ACS GIZ members, 2022).

In 2020, the Strategic Research and Innovation Partnership in Mobility (SRIP ACS GIZ) prepared a projection of the growth potential of the automotive industry in Slovenia following the pandemic. The calculations indicated that the industry could increase its revenues from 3.5 billion EUR to 5 billion EUR by 2030 due to electrification and Industry 4.0 if the level of investments in R&D was maintained at 500 million EUR annually. However, it was estimated that additional 600 million EUR was needed for the industry's transformation. The GREMO (i.e. GREen MObility) mission— strategic partnership among the largest automotive industry suppliers in the country[28]— was formed with an aim to transform the industry and

[28] These are: Domel, Hidria, Iskra Mehanizmi, Kolektor, LTH Castings, MAHLE, SIJ Acroni, TAB, Talum, TPV Automotive and Unior (SRIP ACS GIZ, 2022b).

transition to digital and green mobility (e.g. through electrification). Two thirds of the sum for this transition was secured by the GREMO partners, whereas one third is being funded by the government (SRIP ACS GIZ, personal communication, April 19, 2022).

4.3 Bosnia and Herzegovina: Prepared to Adapt to Regional and Global Trends But Slow in Doing So

Bosnia and Herzegovina's production of automotive parts and components dates back to the 1950s. Many companies ceased to exist during the 1990s Balkans Wars. Only a limited number have undergone privatisation and managed to survive. After 2000, new enterprises started emerging in the sector, which now includes about 30 medium-sized and large enterprises along with approximately 40 small automotive companies that altogether employ around 11,000 persons. They are for the most part almost exclusively export-oriented (with constant export growth)[29] and are some of the fastest growing and best developed businesses in terms of technology, management and organisation in the country. The sector is thus becoming the leading branch of the country's metal processing industry (FIPA, 2012; PwC, 2019; South-East European INDUSTRIAL Market, 2021).

With more than 95% of automotive companies specialising in automotive components, the local industry is highly dependent on the developments in the European and global automotive markets.[30] The prevalence of SMEs in the sector is the country's advantage in this respect, as these firms can be more responsive to the changing customer needs (if they are ensured support for R&D and technology investments (South-East European INDUSTRIAL Market, 2021) or share their R&D budgets (PwC, 2019) as their individual finances often do not suffice). According to PwC (2019), automotive companies in the country are well prepared for the changes in the industry. This includes the larger firms that constantly monitor developments in the automotive suppliers' components markets and obtain information from their parent companies, buyers and suppliers, as well as at trade fairs. The newly founded firms in the sector are similarly responsive to the latest trends and already mainly planning to pursue automotive design, development of electric vehicles and their components and application of software solutions in the automotive sector (PwC, 2019; South-East European INDUSTRIAL Market, 2021).

Some of the country's advantages include a long and extensive tradition in automotive components supply and complementary metal processing sector, cost

[29] Automotive industry presents 16.6% of total exports in Bosnia and Herzegovina (Sabha et al., 2020).

[30] The automotive sector companies in the country can be divided into: (1) firms whose production programmes are not directly threatened by new trends (around 31 enterprises); (2) firms that are, to some extent, threatened by new industry developments (also around 30); and (3) firms whose production programmes could be exposed to the anticipated changes (around six companies) (South-East European INDUSTRIAL Market, 2021). For a list of firm-level actions recommended depending on the firms' readiness for new trends, see PwC (2019).

competitive, highly qualified and educated labour, availability of R&D institutes (there are currently six faculties of mechanics in Bosnia and Herzegovina), wide range of production facilities and expertise, responsive, flexible, prompt and quality production (with many firms having ISO and industry-specific certificates),[31] reliable delivery and honouring of contracts and agreements (FIPA, 2012; PwC, 2019). However, firms operating in the automotive industry in Bosnia and Herzegovina also note they have difficulties in obtaining skilled staff due to its low supply, which is further reduced by migration, lack of quality standards, weaker technology compared to competitors,[32] poor infrastructure, persisting political and economic instability in the country, poor country image and low investment and development budgets (PwC, 2019). Companies in the sector also lack their own R&D facilities, are at a low level of digitalisation and automation and have modest investments in marketing activities. Despite subsidies for FDI, the extent of foreign investments in the automotive sector is far below its potential. Serbia and North Macedonia, for example, offer far more benefits for investors in the automotive industry (South-East European INDUSTRIAL Market, 2021).

Bosnia and Herzegovina describes the European Union, its policy framework and strategic orientation (especially that related to sustainability and green innovations) as an important contributor to the automotive industry (including its aftermarket) development and growth. The country has failed to develop complementary automotive aftermarket-specific sustainability supporting tools or policies mirroring the trends in the European OEMs, however (FIPA, personal communication, August 10, 2022). As a result, the uptake of novelties (regardless of individual firms' readiness for it) is still slow. With an exception of the growing auto components manufacturers described above and whose growth and modernisation can largely be contributed to their cooperation with the big European manufacturers, most segments of the automotive industry in Bosnia and Herzegovina (including the aftermarket) are stagnant (FIPA, 2015).

4.4 North Macedonia: Strategically Promoting FDI in Automotive Components Manufacturing

North Macedonian automotive industrial facilities were completely destroyed in the 2000s. However, the country has managed to develop into an attractive investment location (PwC, 2019) and has recently been recognised as a hub for automotive parts and bus manufacturing within the European periphery automotive regional value chain. This is because of the proactive efforts to attract FDI in higher value

[31] Thirty percent of the 67 companies are IATF-certified (Automotive Quality Management System) and 76% have introduced ISO 9001 quality standard (PwC, 2019; South-East European INDUSTRIAL Market, 2021).

[32] Fifty-nine percent of automotive companies from Bosnia and Herzegovina have a below-average, 28% average and 13% above-average technological competence index (PwC, 2019; South-East European INDUSTRIAL Market, 2021).

added manufacturing and an effective Special Economic Zones incentives regime,[33] which has been particularly successful in the automotive sector (Macedonia and the European Periphery: Automotive Regional Value Chain, 2015; World Bank Group, 2019b). In the latter, North Macedonia focused on promoting itself as a low-cost medium-technology location for components production since 2006 (Macedonia and the European Periphery: Automotive Regional Value Chain, 2015). In 2017, the country's automotive industry consisted of 50 companies that employ 20,000 people (i.e. almost 1% of the population). Automotive components and parts manufacturing is the prevalent sector in the industry (PwC, 2019)[34] and is mentioned among the sectors with the biggest potential for investments according to the Programme for Stimulating Investment in the Republic of Macedonia (2007–2010) as well as the Industrial Policy of the Republic of Macedonia (2009–2020) (OECD, 2021).

Subsidies and a strategy aimed at leveraging legacy capabilities in the manufacturing sector have been particularly attractive for automotive FDI from OECD countries. This has had a significant impact on the country's export profile and labour market, causing a 7 percent increase in total merchandise exports from 2011 to 2016 and a huge increase in the share of export by the top ten automotive exporters in the country, which now account for 50% of total exports (compared to 26% in 2013).[35] The share of medium-tech exports (as a percentage of manufactured exports) also increased tremendously since 2009 and thousands of jobs, particularly in the automotive components sector, have been created (World Bank Group, 2019b).

The country aims to continue this trend through different incentives. The Law for Financial support from 2018, for example, grants businesses financial aid for generating new employments, establishing and promoting cooperation with Macedonian suppliers, investments in technological development and R&D, investment projects of significant economic interest to North Macedonia, capital investments, acquisition of companies that are facing difficulties, increasing trade, competitiveness in the market and promoting balanced regional development (Agency for Foreign Investments and Export Promotion, Invest North Macedonia, Investor servicing Unit, Investor servicing and aftercare department, personal communication, August 24, 2022). According to Adela (2021), the establishment of the automotive components sector has been responsible for the Macedonian economy's integration in the European supply chains.

North Macedonia is nonetheless still experiencing several *challenges* in terms of growing and furthering its automotive industry and the related aftermarket sector. For example, there is limited presence of supporting industries relevant for the automotive sector. Automotive industry companies in the country also manufacture products for other industries. One of the most pressing issues includes the scarce

[33] North Macedonia has launched its first high-tech free zone in 2022 (Leiva, 2022), which indicates the country's progression towards higher value added products and services (including those in the automotive industry or supporting it).

[34] In 2018, the automotive parts accounted for almost 50% of exports from North Macedonia (PwC, 2019).

[35] These did not exist before 2009 (World Bank Group, 2019b).

linkages between FDI and local suppliers that typically take a long time to develop. The limited degree of co-location between local firms and Tier 1 FDI additionally contributes to the minimal clustering effects and backward linkages between domestic and foreign companies. To address this issue, creating backward linkages has become a top policy priority of InvestMacedonia and TIDZ (Macedonia and the European Periphery: Automotive Regional Value Chain, 2015). In 2007, a Group for Automotive Industry in Macedonia was established to develop an automotive cluster with the support of the government, and by establishing links with other relevant stakeholders such as academia and research institutes (Groboljsek, 2012). The World Bank (through the CIIP MultiDonor Trust Fund— MDTF) and the International Finance Corporation (IFC) have also implemented the automotive backward links pilot programme in 2015–2016. This was aimed at both relationship and expertise building. The Government's 2018 Plan for Economic Growth introduced subsidies for domestic purchases. However, it did not address the needs of upgrades on the supply side. In 2015, IFC invested eight million EUR in a greenfield automotive FDI in an underdeveloped region of the country to provide long-term funding, support quality job creation and development of a skilled labour force, transfer state-of-the-art quality management, technologies and environmental standards, as well as signal the investment potential of the region to investors. This has had a notable demonstration effect and transformed the country's export profile from low- to medium-tech merchandise. Spillovers to the local economy were limited due to the limited upstream links to local suppliers.[36] Plans for a South East Europe (SEE) Light Manufacturing Advisory Services Program, that was meant to address these issues, were delayed due to a political crisis (World Bank Group, 2019b).

North Macedonia is also experiencing significant structural weaknesses that undermine competitiveness. These include weak rule of law, governance and institutions, as well as large implementation gaps relative to reforms and regulations. In addition, underdeveloped logistics, transportation and ICT infrastructure have a similar effect (OECD, 2021; World Bank Group, 2019b). The World Bank Group (2019b) has recorded limited ability of local firms to join GVCs as the country's weakness in terms of furthering the local automotive industry as well. OECD (2021) attributes this to the lack of investment in R&D and innovation, technology uptake, meeting quality standards and related employee development— also due to difficulties in accessing financial resources needed for this (Council of the EU, 2022).

Employers report that skill shortages require significant on-the-job training. This includes technical skills, as well as cognitive and soft skills, such as communication, management and interpersonal skills (World Bank, 2017). Shortcomings in the quality and relevance of the education system, along with a lack of adequate labour market information systems and active labour market policies are identified as

[36] Foreign investors rely on imports for most of their inputs other than low-cost local labour. This share is higher than in other regional and global peers. Almost 80% of intermediary goods for automotive exports are imported. At the same time, non-exported goods include 55% of imported content (OECD, 2021; World Bank Group, 2019).

important issues for further economic development— particularly in the automotive industry (OECD, 2021). Unlike the majority of firms, ICT and automotive sectors do tend to communicate regularly with educational institutions, though, which gives them an opportunity to influence the curricula (World Bank, 2020). There has nonetheless been only slow progress in the technological and educational development of the human resources necessary for the advancement of the automotive industry and its components manufacturing sector in particular (Adela, 2021).

Finally, the economy in North Macedonia has been strongly affected by the Covid-19 pandemic: the slowing economic activity in key trading partners and related disruptions in the automotive value chain in particular notably affected external demand, resulting in a first decline in the automotive components sector in 2020— following a continuous growth since 2015 (Adela, 2021; OECD, 2021). While the pandemic has influenced exports in the sector (reducing them by 13.9% in 2020 compared to 2019), employment in the sector grew by 0.1% during the pandemic. It was primarily due to the productivity of the automotive sector that overall exports in North Macedonia have experienced positive developments in the first quarter of 2021 already (Adela, 2021).

4.5 Serbia: Skill Development and Networking for Attracting FDI

The automotive industry in Serbia is the most important industrial sector in the country: it employs 45,000 persons, accounts for 14% of all FDI, presents 10% of total exports, and generates a total income of 3.3 billion EUR (PwC, 2019). Serbian automotive industry has a long tradition;[37] starting with the first assembly line for truck manufacturing in 1939 in Kraguljevac that later transformed into a state-owned factory producing motor vehicles under the license and quality standards of FIAT (Zastava Automobili).[38] Development of a growing new industry in the region also resulted in a flourishing automotive component manufacturing sector,[39] which had limited exporting capabilities and was highly dependent on Zastava, though (this is despite sourcing of components by several foreign firms following WWII). After peaking in 1989, the breakup of former Yugoslavia and the economic sanctions imposed on Serbia during the 1990s halted the development of the sector. Being confined to more or less serving only the Serbian market reduced companies' profits and hindered their investment in R&D and new technologies (South-East European INDUSTRIAL Market, 2016b).

[37] This is particularly true for metal and plastics processing and engineering (PwC, 2019).

[38] Zastava Automobili has an annual capacity to produce 300,000 vehicles and is one of the most cutting-edge production facilities of FIAT Chrysler Automobiles Group. The firm is currently preparing for production of hybrid and electric vehicle variants (PwC, 2019).

[39] This was also facilitated by state subsidies that resulted in significant brownfield and greenfield investments in the sector (PwC, 2019).

Privatisation, FDI[40] (promoted by both cost competitiveness and quality levels that meet international standards (see e.g. OECD, 2008)) and government support have contributed to the automotive industry's recovery after 2000. By 2016, approximately 100 companies employing 25,000 people were producing parts and systems for motor vehicles in Serbia. Engine and engine component manufacturing for foreign OEMs rather than the spare parts market is prevalent. This is followed by vehicle plastic and rubber parts, which are mostly directed at the domestic market (also due to the lack of FDI in the sector). Electrical, electronic and power supply production is also significant and has attracted foreign investors (particularly in wire harness production). The latter also invest in the production of brake discs and drums. Foreign-owned firms record greater turnover and exports growth. More than 90% of foreign companies' production is exported (predominantly to EU and former Yugoslav markets). In comparison, domestic companies export only around 45% and are more focused on local demand. Italian and German investors have a notable presence in the market (also locating their research and development as well as logistics activities rather than solely manufacturing in Serbia) (industry expert, personal communication, 2023b). There are also considerable South Korean, French and Chinese investments in Serbian steel, rubber and tire industry (industry expert, personal communication, 2023b; Shehadi, 2020; Vladisavljev, 2023).

Besides the long tradition, high quality manufacturing, well-established industry networks[41] and resilience of the sector, Serbia's strength in terms of automotive industry development includes its network of technical secondary schools and universities[42] with five faculties of mechanical engineering specialised in automotive mechanics and three institutes specialised in design, testing, homologation and certification for the automotive industry. Collaboration between these academic and research bodies and firms has been established within the Serbian Automotive Cluster (OECD, 2008; South-East European INDUSTRIAL Market, 2016b). In 2018, Serbia introduced a dual education system based on the German model in order to boost the effectiveness of technical schools. This allows secondary technical school students to spend 3 days per week in class and 2 days per week in a factory. A vast majority of job applicants also speak foreign languages (86% English) (RAS, 2021). Development of skills in key business functions such as purchasing, supply chain management, product engineering and manufacturing engineering needed for the realisation of the government's aim to produce high end products with high value added remains a challenge, however (OECD, 2008; RAS, 2021).

[40] Financial Times ranked Serbia as number one in the world in Greenfield FDI performance for 2019 based on the number and size of greenfield investment projects relative to the country's GDP. While general FDI incentive schemes are in place, the country lacks sector-specific linkage programmes with foreign investors that would further promote FDI, technology and specialised knowledge inflows to the country. The automotive sector nonetheless ranks first in terms of the share of all FDI projects in the country (with 18.5%). A sector-specific supplier database allowing for a more targeted approach to further activities in developing the sector has been established (OECD, 2008; RAS, 2021).

[41] Serbian automotive cluster was created in response to the crisis experienced by the Serbian automotive industry in the wake of the dissolution of the former Yugoslavia (OECD, 2008).

[42] The first technical higher education school in the country was established in 1963 (PwC, 2019).

Some of Serbia's advantages in the automotive industry include the low construction, labour and energy costs compared to other countries in Europe, second best cost-to-quality ratio in CEE (together with Hungary this is only surpassed by Romania according to fDi Benchmark in Shehadi, 2020), educated and skilled labour (Chamber of Commerce and Industry of Serbia, 2016), and governmental efforts aimed at incentivising FDI in manufacturing, such as free economic zones and subsidies, that are particularly generous towards investments that are either capital heavy, located outside the capital Belgrade or create a large number of full-time jobs. Serbia also benefits from numerous free-trade agreements (e.g. the Central European Free Trade Agreement, with the European Free Trade Association, Russia and Turkey, and a Generalised System of Preferences with the US, Australia and Japan). The ability to develop products based on customer specifications and flexibility in adapting to potential customers' new needs and requirements are two additional differentiators of Serbian firms (OECD, 2008). The increasing number of companies complying with ISO and industry-specific standards and obtaining the related certificates (industry expert, personal communication, 2023b) is promoting these firms' international collaboration and a more developmental role in the value chains they are integrated into. Finally, the developments in the ICT industry (see e.g. Stojkovski, 2023) present potential for synergies in the automotive industry and its aftermarket— particularly in terms of additional or upgraded digital in-vehicle offerings for new and used vehicles and online consolidation of automotive aftermarket product and service supply (not solely at the country but also at the regional level).

The country's automotive industry is highly integrated into the European market with EU being Serbia's main trading market. This makes it susceptible to crises in the EU. In addition, the country's legal framework is changing, as it is in the process of adjustment with that of the EU, which suggests uncertainties in this respect as well (Chamber of Commerce and Industry of Serbia, 2016). Investments in improving Serbia's logistics are also needed: the country ranks third worst in the Western Balkans in terms of road quality— with only North Macedonia and Bosnia and Herzegovina performing worse in this aspect (Schwab, 2019). Larger foreign investors may be able to contribute to resolving this issue. Locating large parts of their logistics teams in Western Balkan countries implies that some are already active in this area. For example, the European Bosch LOT team (i.e. logistic, operations, sourcing and transport) has about 25% of its members located in Serbia (industry expert, personal communication, 2023b).

4.6 Montenegro, Kosovo and Albania: Lagging Behind in Manufacturing, Having Potential in Service Development and Innovation

In Montenegro, Kosovo and Albania, automotive manufacturing and remanufacturing are not particularly developed, which is why the high demand in the aftermarket of these countries has had to be met by import. Montenegrin businesses in the automotive aftermarket have, for example, mainly been engaged in distribution and retail of these imported goods (The Investment Promotion Section at the Montenegrin Investment Agency, personal communication, August 19, 2022). The main challenges faced by Montenegro's automotive (components) sector include the cost structure and further investment generation (OECD, 2008). According to PwC (2019), Montenegrin aluminium industry nonetheless has some potential to become an automotive industry supplier.

Automotive suppliers in Kosovo, on the other hand, suffer from a lack of reliability both in terms of the volumes required to fulfil their delivery commitments and the quality of their products. They also face issues due to skilled labour deficiencies, lack of international quality certificates, poor IPR protection, outdated equipment and machinery (due to poor maintenance) and financial mismanagement. According to OECD (2008), Kosovo could be a good location for manufacturing of components with high labour content and a low level of technological sophistication, but only if its firms manage to achieve high enough levels of quality.

Albania's auto-parts sector can be described as nascent. The country is a latecomer to the automotive industry and lags behind the remainder of the region. There has been more FDI in the automotive parts and components sector recorded over the past decade.[43] This is because of cost competitive labour, educated and young employee base with good knowledge of foreign languages (especially English, Italian, French, German and Spanish), affordable electricity and sustainable energy,[44] economic stabilisation and an investor-friendly legislation and incentives for FDI targeted specifically at promoting automotive manufacturing investments (Albaniatech, 2021; bne IntelliNews, 2022; Invest in Albania, 2020; MarketResearch, 2019).[45] However, stringent environmental regulations on pollution and carbon emissions require heavy investments (MarketResearch, 2019). The country also lags behind other countries in the region in terms of infrastructure development. This includes ICT infrastructure (Schwab, 2019). Companies operating in the automotive industry in Albania are nonetheless starting to focus on connected and autonomous driving vehicles (MarketResearch, 2019).

In terms of the automotive aftermarket, this has potential especially in terms of service development and innovation in Montenegro, Kosovo and Albania, since local manufacturing is very limited.

[43] There are currently six foreign companies present in the Albanian market specialising in niche components (bne IntelliNews, 2022).

[44] Ninety-five percent of energy is generated from renewable sources (Albaniatech, 2021).

[45] Companies can sign a symbolic lease for 1 EUR per year up to 30 years when renting government property (the so-called 1 euro contract), are exempt from VAT on imported machineries and equipment, eligible for different financing incentives for professional on-the-job training and can benefit from eased customs procedures (AIDA, n.d.; Albaniatech, 2021).

5 Conclusion: Strengths, Weaknesses, Opportunities and Threats for the Automotive Aftermarket in the Region and Policy Recommendations

A competitive and sustainable automotive industry and the related aftermarket sector in the Western Balkans is not only in the best interest of the region but also in the strategic interest of the European Union (OECD, 2009). This is because the Western Balkans' automotive industry and aftermarket are highly integrated in the European regional value chain (bne IntelliNews, 2022). The EU member states are the most important trading partners and investors in the region and as such contribute to the development of well-functioning market economies and political stabilisation. Despite significant economic connections between the two regions, the EU accession process for most Western Balkan countries remains at a standstill. This has opened the Western Balkans to other countries looking to gain an economic foothold in the region— including China, Turkey and Russia (particularly influential in Serbia) (see Pizzolo, 2023). China in particular has been pursuing a targeted strategy to integrate the infrastructure of the Western Balkans into its Eurasian trade network, the 'New Silk Road'. FDI, large (international) infrastructure projects (e.g. the Pelješac Bridge), political influence and migration from these powers have impacted the development of the automotive industry and aftermarket in the Western Balkans (see e.g. Gruebler, 2021; Holzner, 2022; Holzner & Schwarzhappel, 2018). Paradoxically, the strengthened connections with additional emerging markets, such as China and Turkey, have reignited the interest of European investors in the region (see also Lachert, 2018), which acts as a bridge between the developed and emerging economies— especially in the automotive industry (automotive components manufacturing and aftermarket in particular) in which it has a long history, established partnerships and built capacities.

Croatia and Slovenia have thereby benefited from the policy framework, strategic collaboration, supporting tools aimed at green, digital and sustainable transformation of the industry, as well as the (perceived) stability related to their European Union membership (see also Jaklič & Koleša, 2020, 2022). This has not only been an important factor in investor attraction but has also contributed to the modernisation of the industry more directly (e.g. through various funding mechanisms).[46] There are also specific mechanisms and instruments developed for the Western Balkans in the EU. One such example is the Economic and Investment Plan for the Western Balkans adopted by the European Commission in 2020. This is expected to increase the competitiveness of the region through a substantial investment package for key productive investments, human capital development and infrastructure

[46] Hidria, a Slovenian company, which controls 30% of the European market in core solutions for hybrid and electrical powertrains, for example, established an international innovation centre for mobility research located in Slovenia with the support of EU funds (Hidria, 2019; Slovenia Business, 2022c). Rimac Automobili from Croatia has also successfully been using EU funds for its R&D activities (HAMAG-BICRO, 2020).

promoting a green and digital transition in the region (Council of the EU, 2022; WeBalkans, 2022).

In the absence of data specific to automotive aftermarket in the Western Balkan countries, our analysis outlines the broader automotive industry trends and developments in the region that also have implications for the automotive aftermarket. The main strengths and opportunities for its growth as well as weaknesses and threats that need to be addressed for the sector to also progress to higher value added products and services provided for domestic, regional and international markets are outlined in Table 1.

Some of the greatest strengths of the region include its location, human capital, international trade agreements and the SME structure of the automotive industry. The increasing industry-specific certification also presents a strength as it is promoting international business partnerships and the internationalisation of the industry. Employees' and consumers' passion for automotive innovation and car racing, on the other hand, further promotes the development of innovative products and services tailored to the region (also in the automotive aftermarket as they allow for upgrades of used vehicles consistent with consumer wants— within their financial capabilities). The weaknesses include limited resources for R&D and innovation, low purchasing power that can result in customers turning to the grey market and a 'do-it-yourself' approach, limited backwards linkages between domestic suppliers and foreign investors that could result in knowledge and technology spillovers, lack of collaborative mechanisms— especially with respect to innovation (this differs greatly across countries, with Croatia and Slovenia having relatively well-established networks between businesses, academia and research centres) and underdeveloped infrastructure (having a negative impact on both international trade and innovation).

Further integration into European and global automotive value chains (especially with high added value products and services) following the Covid-19 pandemic due to business tendencies for nearshoring and reshoring, servitisation, trends related to the transition to digital and green mobility, electrification and use of AI and the related product, process and service innovations, as well as the introduction of new business models, regional collaboration and smart specialisation, servitisation and utilisation of e-commerce present opportunities for further development of the automotive aftermarket. Consumers in the region are also enthusiastically adopting digital in-vehicle offerings, such as the hands-free power tailgate system or the car parking sensor system, as well as expressing demand for car racing and other accessories (industry expert, personal communication, 2023a). Changed consumer habits and consumption patterns due to policy changes (e.g. promotion of sustainability), economic and other factors (e.g. increased prices of new cars) can be utilised in favour of both the traditional and more innovative automotive aftermarket products and services.

Moreover, integration into the EU is expected to provide the Western Balkans with additional access to European markets and partners as well as EU funds and a more unified as well as stable legislative framework. Currently, one of the important weaknesses to the development of the automotive aftermarket (and business in

Table 1 SWOT analysis of the Western Balkans' automotive industry and its aftermarket

Strengths	Weaknesses
• Strategic location • Access to a sizeable duty-free market due to multilateral and bilateral agreements (time to market, delivery time) • Tradition of automotive car components industry and new innovative players • Flexibility, adaptability and responsiveness due to the predominantly SME structure of the sector • Cost competitive and skilled labour, relative abundance of technical and engineering knowledge • Increasing industry-specific certification • Passion for automotive innovation and car racing (among both employees and consumers)	• Unfinished structural reforms • Limited financial resources for (R&D) investment (components manufacturers and repair shops prevail) • Low purchasing power (lower use of genuine parts compared to (un)certified parts) • Limited backward linkages between domestic suppliers and foreign investors • Lack of collaborative mechanisms in innovation • Underdeveloped infrastructure
Opportunities	**Threats**
• Further integration into the European and global automotive value chains after the Covid-19 pandemic (reshoring, nearshoring) • Transition to digital and green mobility (promoted at policy and business levels) • Electrification[a] and use of AI • Regional collaboration and smart specialisation • Servitisation • E-commerce (internet component sales and digitally driven product and services) • Changed consumption patterns (high average car age, increased focus on used cars and car maintenance after the Covid-19 pandemic and war in Ukraine) • Integration into the EU • Rising demand for racing and other vehicle accessories as well as digital in-vehicle offerings	• Regional conflicts (influencing demand, delivery times and prices) • Slow EU accession process • Turbulence in European (and global) markets and the related disruptions in value chains • Recession and inflation • Emigration of skilled labour

[a]Electrification and the related transition to new products, services and processes provide an opportunity especially for the more flexible SMEs in the industry. For them to be able to engage in R&D and innovations needed for such transition, the education system needs to support businesses in skill development. Partnerships between academia and business in particular present an opportunity for resource sharing and co-creation of both new knowledge and products or services. Since the aforementioned transformation will also require significant funds, business partnerships combining resources and public funding (at either national or EU levels) should also be utilized. Several companies in the region have already successfully used these options and some have also developed their own support systems

Source: Own analysis

general) is the unfinished structural reforms and the related weak institutions in most parts of the region. This (combined with regional conflicts) discourages foreign investments and international business partnership formulation as well as hinders international trade.

Threats to the sector and industry as a whole include emigration of skilled labour from Western Balkan countries (also to individual countries within the region, such as in the case of Rimac Automobili, which is attracting regional and international talents to Croatia). High dependence on European and global markets can also be a threat, as recession and other negative trends in these markets are reflected in the Western Balkan region as well. This necessitates constant monitoring and great responsiveness of regional business operations. Persisting regional conflicts are also hindering business in the region in general (including the automotive industry and its aftermarket).

Based on the SWOT analysis, we provide the following *recommendations for policy measures* to promote the automotive aftermarket development (that could also contribute to greater sustainability) in the region:

- Investing in transport and ICT infrastructure;
- Attracting industry-specific FDI (also by speeding up the EU accession process) for technology and knowledge spillovers as well as greater global value chains integration;[47]
- Supporting R&D and innovation (also through stimulating collaborative mechanisms for innovation);
- Facilitating digitalisation and automation;
- Investing in upgrades of the education system and enhancing the private sector's engagement and contribution to skills development (i.e. ensuring continuous and lifelong employee development responsive to business trends and market needs). Specific actions should target both formal education and on-the-job training;[48]
- Promoting industry-specific interorganisational networking (including networking between businesses, between businesses and academia, and businesses and research institutions— also across supporting sectors) and eased formulation of partnerships (also through co-funding attendance at international trade fairs) with a particular focus on interpersonal relations that can drive collaboration (i.e. at the level of management and professionals alike);
- Promoting (developmental) partnerships with large and innovative international businesses and hubs (also through awareness raising about the local businesses' strengths and capabilities among potential international partners, as well as

[47] FDI attraction policies should thereby progress to also promoting support linkages with domestic firms. For example, North Macedonia's Economic Reform Programme 2019–2021 includes measures and funds to encourage foreign-owned companies to develop backward linkages with local enterprises (MoF, 2019). This policy could be scaled up to a regional (i.e. Western Balkans) level (Krasniqi et al., 2022).

[48] Investments in skills development are all the more relevant, as cost competitiveness is not sustainable as a source of differentiation (OECD, 2008). According to Jovanović et al. (2021) availability of skilled labour is the most important factor for investors when deciding where to invest. Despite their high levels of technical and engineering skills, upskilling in some of the Western Balkan countries as well as continuous updates of employees' skills in general are still needed due to the constant changes in the industry. Research shows that companies are more willing to invest in such upskilling and continuous training in the region because of the generally low labour costs in the Western Balkans. However, they are also discouraged from training the workers, because many of them tend to leave the region (Jovanović et al., 2021)— especially when equipped with the skills and resources needed to do so. Limited access to finance and a lack of strategies for reinvesting in technology and human capital have been identified as a risk for the future.

through quality certification and investments in technology and equipment upgrading)— within the region and broader;
- Engaging in legislative and administrative simplification;
- Improving customs and trade regulations as well as the import regime for capital goods that hamper the ability of suppliers in the Western Balkans to engage in international trade and create discrimination among domestic and foreign-owned companies when operating in Western Balkan markets (often at the disadvantage of the former);
- Taking advantage of the diversity in the region by following the smart specialisation approach and capitalising on complementarities between specialist (often niche) automotive companies from different Western Balkan countries as well as creating spillover effects through regional collaboration;[49]
- Developing (a) regional ecosystem(s) for lifecycle solutions;
- Improving cost of and access to finance and taking advantage of EU funds for R&D as well as other available financial sources for development[50] (see also Jovanović et al., 2021; Krasniqi et al., 2022; OECD, 2008, 2009; Sabha et al., 2020; survey among SRIP ACS GIZ members, 2022).

A combination of policies is essential. The region would also benefit from more general structural reforms aimed at strengthening governance, boosting productivity, competitiveness and resilience of individual countries as well as the region, and institutional development as crucial components of a quality business environment (OECD, 2008). This includes strengthened private sector development policies at country and regional levels (OECD, 2009). A policy focus on the automotive aftermarket seems unrealistic, yet a more holistic and regionally complementary strategic approach to the overall automotive industry could generate spillover effects to this sector as well.

Although the Western Balkan region does not strategically focus on the automotive aftermarket, its long tradition in the automotive industry also supports the development of this sector. There are still significant differences between countries in the region, however. In countries where (either domestic or foreign-owned) car production facilities are located, their operations act as catalysts for the development of support sectors (including the aftermarket). The latter can thereby be more traditional or innovative— depending on the level of innovativeness at the level of OEM (domestically and internationally) and the needs of end consumers. For progression towards the innovative aftermarket products and services integration into regional and global value chains, investments in education (especially in terms of

[49] Previous industry studies (e.g. OECD, 2008) suggest a similar approach to FDI: i.e. an integrated investment promotion strategy in the form of a cross-border clustering approach, where each country could target specific categories of potential investors and customers based on their position along the automotive value chain.

[50] Credit constraints present a major barrier to the growth and sustainability of businesses in the automotive industry in the Western Balkans (especially in Bosnia and Herzegovina, North Macedonia and Serbia) (OECD, 2008).

technical talent development) and R&D, as well as FDI are crucial. In markets, where these are not as pronounced, the aftermarket remains limited to basic services of vehicle maintenance for local customers. Regional collaboration could support these markets to evolve further.

References

ACEA. (2022, April 2). *Average age of the EU vehicle fleet, by country*. https://www.acea.auto/figure/average-age-of-eu-vehicle-fleet-by-country/

Adela, G. M. (2021). North Macedonia economy briefing: Automotive components remain a key driver of Macedonian exports in spite of the pandemic. *China-CEE Institute, 40*(2), 1–5. https://china-cee.eu/wp-content/uploads/2021/05/2021e05_North-Macedonia.pdf

aftermarketNews Staff. (2022, May 3). *HELLA establishes business group lifecycle solutions*. https://www.aftermarketnews.com/hella-lifecycle-solutions-business-group/

Agenzia Nova. (2023, March 29). *Serbia, information technologies: exports grow more than seven times in ten years*. https://www.agenzianova.com/en/news/Serbia%27s-information-technology-exports-grow-more-than-sevenfold-in-ten-years/

AIDA. (n.d.). *Albania: Automotive parts manufacturing: Explore new possibilities*. https://aida.gov.al/images/PDF/Publikime/Automotive-Parts-Manufacturing.pdf

Albaniatech. (2021). *Think Albania: Automotive manufacturing in Albania*. https://albaniatech.org/wp-content/uploads/2021/06/Automotive-Manufacturing-in-Albania.pdf

Berger, R. (2020, April 6). *COVID-19— the automotive independent aftermarket (IAM) opportunity: European perspective*. https://www.vft.at/fileadmin/user_upload/pdf/20200406_Roland_Berger_AUTOMOTIVE_IAM_on_the_other_side_of_COVID-19_vF2.pdf

bne IntelliNews. (2022, June 17). *Albania trails neighbours in auto-parts production but economists see a chance to catch up*. https://www.intellinews.com/albania-trails-neighbours-in-auto-parts-production-but-economists-see-a-chance-to-catch-up-247830/

Chamber of Commerce and Industry of Serbia. (2016). *Automotive industry in Serbia*. https://aplikacije.pks.rs/Sadrzaj/Files/Centar%20za%20podr%C5%A1ku%20investicijama%20i%20JPP/Automotive%20brochure,%20decembar%202016.pdf

Council of the EU. (2022, April 25). *Commission staff working document - Economic reform programme of North Macedonia (2022-2024) - Commission assessment* (Doc. No. SWD(2022) 123 final). https://data.consilium.europa.eu/doc/document/ST-8299-2022-INIT/en/pdf

Demekas, D. G., Horváth, B., Ribakova, E., & Wu, Y. (2005, June 1). *Foreign direct investment in Southeastern Europe: How (and how much) can policies help?* (IMF Working Paper No. 2005/110). Washington: IMF. https://www.imf.org/-/media/Websites/IMF/imported-full-text-pdf/external/pubs/ft/wp/2005/_wp05110.ashx

Euronavigator. (2022). *The new automotive industry in Croatia*. https://euronavigator.be/project/the-new-automotive-industry-in-croatia/

European Commission. (2019). *A European Green Deal*. https://commission.europa.eu/strategy-and-policy/priorities-2019-2024/european-green-deal_en

European Commission. (2020, October 6). *Commission communication on an economic and investment plan for the Western Balkans* (COM(2020) 641 final). https://neighbourhood-enlargement.ec.europa.eu/system/files/2020-10/communication_on_wb_economic_and_investment_plan_october_2020_en.pdf

European Investment Bank. (2018, December 7). *Croatia: Investment Plan for Europe - EIB supports Croatian Rimac Automobili in development of high performance electric vehicles and components*. https://www.eib.org/en/press/all/2018-330-investment-plan-for-europe-eib-supports-croatian-rimac-automobili-in-development-of-high-performance-electric-vehicles-and-components

Eurostat. (2022, March 28). *Hourly labour costs ranged from €7 to €47 in the EU.* https://ec.europa.eu/eurostat/web/products-eurostat-news/-/ddn-20220328-1

FIPA. (2012). *Automotive industry.* http://www.fipa.gov.ba/atraktivni_sektori/automobilska_industrija/default.aspx?id=42&langTag=en-US

FIPA. (2015). *Bosnia and Herzegovina : Automotive Industry.* FIPA. http://www.fipa.gov.ba/publikacije_materijali/brosure/Automotive%20Industry.13.05.2015.pdf

Frink, L. (2008, March 31). *Auto industry revives in western Balkan markets.* Automotive News Europe. https://europe.autonews.com/article/20080331/ANE09/592643359/auto-industry-revives-in-western-balkan-markets

Groboljsek, M. (2012, June 20). *Automotive cluster of Macedonia.* Western Balkans Info Hub. https://wbc-rti.info/object/organisation/10223

Gruebler, J. (2021). China connecting Europe? *Asia Europe Journal, 19*(1), 77–101. https://doi.org/10.1007/s10308-021-00616-4

HAMAG-BICRO. (2020). *Croatia on the map of the future's automotive industry.* https://en.hamagbicro.hr/croatia-on-the-map-of-the-futures-automotive-industry/

Hidria. (2019, November 20). *The EDISON WINCI project as a transition to new technologies of electromobility.* https://www.hidria.com/int/en/news/the-edison-winci-project-as-a-transition-to-new-technologies-of-electromobility/

Holzner, M. (2022, September 16). *Serbia, China's gateway to Europe.* The Vienna Institute for International Economic Studies. https://wiiw.ac.at/serbia-china-s-gateway-to-europe-n-566.html

Holzner, M., & Schwarzhappel, M. (2018). *Infrastructure investment in the Western Balkans: A first analysis (wiiw Research Report, No. 432).* The Vienna Institute for International Economic Studies. https://www.econstor.eu/bitstream/10419/204204/1/wiiw-research-rep-432.pdf

International Bank for Reconstruction and Development/The World Bank. (2017). *The Western Balkans: Revving up the engines of growth and prosperity (Report No. ACS22690).* The World Bank Group. https://openknowledge.worldbank.org/bitstream/handle/10986/28894/ACS22690.pdf?sequence=4&isAllowed=y

International Trade Administration, U.S. Department of Commerce. (2022, 5 August). *Serbia country commercial guide: Information and communications technology market.* https://www.trade.gov/country-commercial-guides/serbia-information-and-communications-technology-market

Invest in Albania. (2020, February 25). *Albania to jump-start automotive industry with tax incentives.* https://invest-in-albania.org/albania-to-jump-start-automotive-industry-with-tax-incentives/

Jaklič, A., & Koleša, I. (2020). *Tuji investitorji o slovenskem poslovnem okolju 2020: rezultati raziskave med podjetji s tujim kapitalom v Sloveniji 2020 [Foreign investors about Slovenian business environment 2020: results from a study among businesses with foreign capital in Slovenia 2020].* Center za mednarodne odnose [Centre for international relations].

Jaklič, A., & Koleša, I. (2022). *Tuji investitorji o slovenskem poslovnem okolju 2022: rezultati raziskave med podjetji s tujim kapitalom v Sloveniji 2022 [Foreign investors about Slovenian business environment 2020: results from a study among businesses with foreign capital in Slovenia 2020].* Center za mednarodne odnose [Centre for international relations].

Jovanović, S. (2018, November 5). *Albania bans imports of cars over 10 years old, not meeting Euro 5 emission standard from 2019.* Balkan Green Energy News. https://balkangreenenergynews.com/albania-bans-imports-of-cars-over-10-years-old-not-meeting-euro-5-emission-standard-from-2019/

Jovanović, B., & Vujanović, N. (2023). *Towards effective industrial policy in the Western Balkans (Policy Notes and Reports 66, February).* The Vienna Institute for International Economic Studies. https://wiiw.ac.at/towards-effective-industrial-policy-in-the-western-balkans-dlp-6493.pdf

Jovanović, B., Ghodsi, M., van Zijverden, O., Kluge, S., Gaber, M., Mima, R., et al. (2021). *Getting stronger after Covid-19: Nearshoring potential in the Western Balkans (Research Report 453).* The Vienna Institute for International Economic Studies. https://img.innoloft.com/platforms/8197452/getting-stronger-after-covid-19-nearshoring-potential-in-the-western-balkans-dlp-5814.pdf

Kolekevski, I. (2022, August, 5). *Vehicle usage and average vehicle age in North Macedonia rising.* MIA. https://shqen.mia.mk/vehicle-usage-and-average-vehicle-age-in-north-macedonia-rising/

Koleša, I. (2019). Emigrants from the Western Balkans: The region's money sacks? In M. Elo & I. Minto-Coy (Eds.), *Diaspora networks in international business: Perspectives for understanding and managing diaspora business and resources* (pp. 557–590). Springer. https://link.springer.com/book/10.1007/978-3-319-91095-6

Krasniqi, B., Ahmetbasić, J., & Bartlett, W. (2022). Foreign direct investment and backward spillovers in the Western Balkans: The context, opportunities and barriers to the development of regional supply chains. *Southeastern Europe, 46*(1), 1–22. https://doi.org/10.30965/18763332-46010001

Lachert, J. (2018, November 25). *Western Balkans in the context of competition between China, Russia, the European Union, and Turkey.* Warsaw Institute. https://warsawinstitute.org/western-balkans-context-competition-china-russia-european-union-turkey/

Leiva, M. (2022, August 3). *North Macedonia launches its first high-tech free zone.* Investment Monitor. https://www.investmentmonitor.ai/features/north-macedonia-launches-first-high-tech-free-zone/

Macedonia and the European Periphery: Automotive Regional Value Chain. (2015). https://docslib.org/doc/7200853/macedonia-and-the-european-periphery-automotive-regional-value-chain

Market Data Forecast. (2022). *Europe automotive aftermarket market.* https://www.marketdataforecast.com/market-reports/europe-automotive-aftermarket-market

MarketResearch. (2019). *2019 Future of Albania automobile market— trends, outlook and growth opportunities.* https://www.marketresearch.com/OG-Analysis-v3922/Future-Albania-Automobile-Trends-Outlook-12182877/

Mesojedec, M. (2022, July 28). *Slovenska avtomobilska industrija na prelomni točki [Slovenian automotive industry at a breaking point].* Družina. https://www.druzina.si/clanek/slovenska-avtomobilska-industrija-na-prelomni-tocki

MoF. (2019). *Republic of Macedonia : Economic reform programme 2019–2021.* Ministry of Finance. https://finance.gov.mk/wp-content/uploads/2018/11/Economic-Reform-Program-2019-EN-Nov-2018.pdf

N1 Belgrade. (2019, April 23). *Average age of cars over 17 years in Serbia.* N1. https://rs.n1info.com/english/news/a478500-average-age-of-cars-over-17-years-in-serbia/

OECD. (2008). *Defining and strengthening sector specific sources of competitiveness in the Western Balkans: Recommendation for a regional investment strategy* (Advanced copy). https://www.rcc.int/download/docs/1-229WBalkans.pdf/c531fbe588665815164709832be6d9c

OECD. (2009). *Sector specific sources of competitiveness in the Western Balkans: Key conclusions and next steps.* https://www.oecd.org/global-relations/45375074.pdf

OECD. (2021). *Competitiveness in South East Europe 2021: A policy outlook. Competitiveness and Private Sector Development.* OECD Publishing. https://doi.org/10.1787/dcbc2ea9-en

Oxford Analytica. (2022, February 28). *Balkans tech ecosystem is poised for robust growth* (Expert Briefings). https://doi.org/10.1108/OXAN-DB267596.

Pizzolo, P. (2023, January 13). *A triangle of interests: Russian, Chinese, and Turkish presence in the Balkans.* OSMED. https://www.osmed.it/2023/01/17/a-triangle-of-interests-russian-chinese-and-turkish-presence-in-the-balkans/

Prebil, G. (2022, November 17). *Renault objavil veliki načrt, to pravijo v Revozu* [Renault published its great plan, this is what they say in Revoz]. Žurnal 24. https://www.zurnal24.si/avto/renault-objavil-veliki-nacrt-to-pravijo-v-revozu-396577

PwC. (2019). *BIH automotive industry assessment: Trends, challenges and opportunities* (Final report). https://www.undp.org/sites/g/files/zskgke326/files/migration/ba/EGG_Automotive_ENG_FIN.pdf

RAS. (2021). *Invest in Serbia: Automotive industry Serbia 20.20: Time to advance.* Development Agency of Serbia. http://www.berlin.mfa.gov.rs/odrzavanje/uploads/RAS_Automotive-_presentation_compressed.pdf

Redžepagić, S., & Richet, X. (2008). The attractiveness of the Western Balkans for the FDI. *Economic Analysis, Institute of Economic Sciences, 41*(1–2), 48–59. https://core.ac.uk/download/pdf/33812183.pdf

Republic of Croatia Ministry of Economy and Sustainable Development. (2020). *Automotive industry in Croatia*. https://investcroatia.gov.hr/wp-content/uploads/2018/11/Automotive-Industry-in-Croatia-web-1.pdf

Republic of Croatia Ministry of Economy and Sustainable Development. (2022). *Automotive industry*. https://investcroatia.gov.hr/en/automotive industry/

Republic of Croatia Trade and Investment Promotion Agency. (n.d.). Automotive industry. http://www.ukrexport.gov.ua/i/imgsupload/Croatian%20Automotive%20Industry.pdf

Rosenbaum, E. (2020, October 15). *The used car boom is one of the hottest, and trickiest, coronavirus markets for consumers*. CNBC. https://www.cnbc.com/2020/10/15/used-car-boom-is-one-of-hottest-coronavirus-markets-for-consumers.html

Sabha, Y., Liu, Y., & Jedlicka, H. (2020, April 23). *The Western Balkans should leverage foreign direct investment to integrate in global value chains*. https://blogs.worldbank.org/psd/western-balkans-should-leverage-foreign-direct-investment-integrate-global-value-chains

Savic, M. (2013, May 16). Putin and Nikolic to sign Russia-Serbia partnership agreement. *Bloomberg*. https://www.bloomberg.com/news/articles/2013-05-16/putin-and-nikolic-to-sign-russia-serbia-partnership-agreement?leadSource=uverify%20wall

Schwab, K. (Ed.). (2019). *The global competitiveness report*. World Economic Forum. https://www3.weforum.org/docs/WEF_TheGlobalCompetitivenessReport2019.pdf

Shehadi, S. (2020, March 9). *Serbia's automotive companies drive inward investment*. fDi Intelligence. https://www.fdiintelligence.com/content/feature/serbias-automotive-companies-drive-inward-investment-76803

Simmonds, L. (2020, November 11). *Does chance for Croatian automotive industry lie with electric cars?* Total Croatia News. https://www.total-croatia-news.com/business/48078-croatian-automotive-industry

Slovenia Business. (2022a). *Automotive industry— number one national industry*. https://www.sloveniabusiness.eu/hot-topics/automotive-industry-number-one-national-industry

Slovenia Business. (2022b). *Research and development in Slovenia: Top local talent for R&D in Slovenia*. https://www.sloveniabusiness.eu/business-environment/research-development-and-innovation

Slovenia Business. (2022c). *Towards Industry 4.0*. https://www.sloveniabusiness.eu/hot-topics/towards-industry-40

Slovenia Business. (2022d). *Automotive industry: A trusted partner in the European value chain*. https://www.sloveniabusiness.eu/industries-and-technologies/automotive-industry

Slovenia Business. (2022e). *The company Dewesoft is the ITM exporter of the year 2022*. https://www.sloveniabusiness.eu/hot-topics/the-company-dewesoft-is-the-itm-exporter-of-the-year-2022

Slovenia Business. (2022f). *Hot topic: The rise of the robots*. https://www.sloveniabusiness.eu/hot-topics/the-rise-of-the-robots

Slovenia Times. (2011, November 14). *Made in Slovenia: Automotive industry*. https://slovenia-times.com/made-in-slovenia-automotive-industry/

South-East European INDUSTRIAL Market. (2016a). *Automotive industry in Croatia*. https://www.see-industry.com/en/automotive-industry-in-croatia/2/1321/

South-East European INDUSTRIAL Market. (2016b). *Automotive industry in Serbia*. https://www.see-industry.com/en/automotive-industry-in-serbia/2/1325/

South-East European INDUSTRIAL Market. (2021, November 3). *Automotive industry in Bosnia and Herzegovina*. https://www.see-industry.com/en/automotive-industry-in-bosnia-herzegovina/2/2050/

SRIP ACS GIZ. (2022a). *Vpliv krize Covid-19 na avtomobilsko industrijo* [Impact of the Covid-19 crisis on the automotive industry]. https://www.acs-giz.si/uploads/vpliv_covid-19_na_avtomobilsko_industrijo_april_2020_koncno.pdf

SRIP ACS GIZ. (2022b). *Vlada na srečanju s predstavniki slovenske avtomobilske industrije o misiji GREMO (Green Mobility)* [The Government at a meeting with the Slovenian automotive industry representatives about the GREMO mission]. https://www.acs-giz.si/novice/vlada-na-srecanju-s-predstavniki-slovenske-avtomobilske-dobaviteljske-industrije-o-misiji-gremo-green-mobility-2022-09-07

Stojkovski, B. (2023, February 28). A bright future for the Western Balkan IT industry. *IPS.* https://www.ips-journal.eu/work-and-digitalisation/a-bright-future-for-the-western-balkan-it-industry-6539/

Šulin Košar, A. (2017). *Naš Mali avto: avtomobilska industrija [Our small car: Automotive industry].* Statistični urad Republike Slovenije [Republic of Slovenia Statistical Office]. https://www.stat.si/StatWeb/File/DocSysFile/9417/avtomobilska_industrija.pdf

Tasevski, K. (2019). *Change management in Hella Saturnus Slovenia: An analysis of the change of product engineering process (PEP)* [Master's thesis, School of Economics and Business, University of Ljubljana]. University of Ljubljana Repository. http://www.cek.ef.uni-lj.si/magister/tasevski3697-B.pdf

Tsoneva, A. (2021, July 9). Croatia's Rimac to get 200 mln euro ($237 mln) EU grant to develop robotaxi. *SeeNews.* https://seenews.com/news/croatias-rimac-to-get-200-mln-euro-237-mln-eu-grant-to-develop-robotaxi-747235

Vladisavljev, S. (2023, August 8). How did China become the largest investor in Serbia? *CHOICE.* https://chinaobservers.eu/how-did-china-become-the-largest-investor-in-serbia/

Waas, A., Beck, M., Herzberg, R., Hauser, J., Schlehuber, F., Wolk, A., et al. (2021). *At the crossroads: The European aftermarket in 2030.* https://web-assets.bcg.com/36/39/e80d073a4067b-fe89c7482d6db69/the-european-aftermarket-in-2030.pdf

WeBalkans. (2022, February 25). *European Commission launches €3.2 billion investment to advance sustainable connectivity in the Western Balkans.* https://webalkans.eu/en/news/european-commission-launches-e3-2-billion-investment-to-advance-sustainable-connectivity-in-the-western-balkans/

World Bank. (2017). *Looking for skills in the Former Yugoslav Republic Macedonia.* World Bank. https://openknowledge.worldbank.org/handle/10986/27745?show=full

World Bank. (2020). *Republic of North Macedonia : Action plan for recovery of growth and jobs.* World Bank. https://openknowledge.worldbank.org/handle/10986/34755

World Bank Group. (2019a). *Kosovo: A future of green transport and clean air* (final report). https://openknowledge.worldbank.org/bitstream/handle/10986/32283/Kosovo-A-Future-of-Green-Transport-and-Clean-Air.pdf?sequence=1

World Bank Group. (2019b, March 21). *Country partnership framework for the Republic of North Macedonia for the period January 2019–June 2023* (Report No. 135030-MK). https://thedocs.worldbank.org/en/doc/705311555590000225-0080022019/original/NorthMacedoniaCPF2019.pdf

Iris Koleša is a research assistant at the Faculty of Social Sciences, University of Ljubljana, Slovenia, focusing on business internationalisation from managerial, policy and human resources perspectives. She is a recipient of several awards and recognitions for her academic work, including the 2023 Academy of International Business Teaching Innovation Award for the MNC Whispering collection of online teaching materials for international business awarded to the VCL4IB project team. Iris is currently the national representative of Slovenia on the European International Business Academy (EIBA) Board. She has been involved in numerous international and national research projects and continuously presents her findings at both academic and professional events. She is also an experienced communication expert with relevant practical skills in science communication and business support.

Andreja Jaklič, professor and researcher at the Centre of International Relations and Vice-Dean for Research at the Faculty of Social Sciences, University of Ljubljana in Slovenia, has research experience in international trade, foreign direct investment, global value chains and internationalisation strategies. She was a visiting scholar at several universities across Europe and China and received a number of awards for her academic work, including the Academy of International Business Teaching Innovation Award 2023 for the MNC Whispering project. Her publications include over 50 scientific articles in scientific journals (JBR, JWB, IBR, Service Industries, etc.), several monographs and book chapters. She is a co-founder and executive board member of the Academy of International Business Central and Eastern European Chapter, active in editorial boards and served as a board member of the European International Business Academy (EIBA) during 2010–2016.

Service Design, Marketing and Automotive Aftermarket: Perspectives from a Polish Study

Monika Hajdas

Abstract

In this chapter we focus on applying user insights into automotive aftermarket service development, value proposition design and branding strategies. The adoption of user-centered design offers a great potential for the automotive aftermarket research and practice. Industry reports highlight that gaining deeper customer insights is one of the game changing opportunities for automotive aftermarket actors. We investigate this phenomenon by adopting qualitative investigation based on in-depth interviews with female car drivers on a Polish market. Our findings show that there are potential market segments with unmet needs which are yet to be recognized and responded with adequate service and marketing offerings. Our study can encourage owners and managers of automotive aftermarket firms to consider applying user-centered logic in their marketing activities.

1 Introduction

Several major trends are reshaping the automotive aftermarket. Demographic shifts, changing lifestyles and new mobility patterns lead to the transformation of mobility. Shared mobility solutions or mobility as service models are on the rise as young generations are less interested in car ownership. Technological advancements and connected devices lead to the advent of a car as a platform. It can create both opportunities and threats for automotive market players. The chance for tech newcomers

M. Hajdas (✉)
Uniwersytet Ekonomiczny we Wrocławiu, Wrocław, Poland
e-mail: monika.hajdas@ue.wroc.pl

M. Elo, F. Katsardis (eds.), *Automotive Aftermarket*, Management for Professionals, https://doi.org/10.1007/978-3-031-62419-3_12

taking over the market is high, but so are the possibilities of creating meaningful partnerships in this field. Also, the environmental pressures create the demand for sustainable mobility solutions.

The combination of these trends with post-pandemic market recovery poses many strategic and marketing challenges on automotive aftermarket players in the service business, such as repair workshops, operating on B2C markets. What creates the value for customers and other stakeholders, how the value is created and delivered, where and by whom—these questions require strategic reflection and revision. Industry experts highlight the need for strong focus on marketing ability to analyse customer habits, usage and attitudes. In fact gaining deeper customer insights is considered as one of the game changing opportunities for automotive aftermarket actors (McKinsey and Company, 2017). Marketing is also about the ability to apply these insights in the design or redesign of product and service offerings.

In this chapter we focus on the service business and workshop level. We present user-centered design (UCD) (Dell'Era et al., 2020; Joly et al., 2019; LeRouge et al., 2013; Micheli et al., 2018) as a way of thinking about involving users in co-creation of market offerings and service design. We believe the adoption of UCD has a great potential for the automotive aftermarket research and practice, especially for marketing consumer services. User insights, based on combination of thick and big data can lead to incremental or radical innovations in automotive aftermarket products and services.

The main research questions we aim to address in this chapter are the following:

1. What are the concepts and tools of UCD that might be particularly useful for gaining deeper consumer insights and applying them in service and value proposition design on automotive aftermarket?
2. How value can be created and delivered through branding on automotive aftermarket?
3. How UCD concepts and tools can be applied in practice on automotive aftermarket?
4. What are the possible outcomes of applying UCD on automotive aftermarket?

The chapter contributes by introducing learnings from a Polish study addressing consumer marketing and UCD on service at the workshop level. We begin with an overview of user-centered design, service design and with several UCD tools and techniques of researching and synthesizing that lead to identification of user insights. Then we discuss how to apply user insights in value proposition design and innovation. Also, as branding is one of the ways of creating and delivering value for customers and other stakeholders, and as there is an industry call for considering more varied, alternative branding strategies on automotive aftermarket (McKinsey and Company, 2018) we offer a brief overview of key branding concepts and reflect them in the context of automotive aftermarket service business.

2 Theoretical Background

2.1 User-Centered Design for Service Design and Marketing

Gaining deeper customer insights was listed among one of the game changing opportunities for suppliers on automotive aftermarket (McKinsey and Company, 2017). There is also a call for a strong focus on marketing ability to analyse customer habits and apply these insights in the adaptation and creation of new service offerings (Capgemini Consulting, 2010). Therefore, we begin with a brief overview of some *user-centered design* (UCD) methods and tools that can inform the design and the development of automotive aftermarket services and branding.

User-centered design is a problem-solving process in which the needs, desires and limitations of users are investigated, analysed and turned into prototypes, which are further tested and revised according to users' feedback (LeRouge et al., 2013). The UCD concept reflects the marketing orientation in general and the customer orientation in particular (Kohli & Jaworski, 1990) and combines it with a designer way of thinking (Norman, 2013). User-centered design has been widely applied in service design, where it allows for creating new services or improving existing ones (Joly et al., 2019) through better understanding of user experiences (Mahr et al., 2013) and creating corresponding value propositions (Ostrom et al., 2015).

With a rise of UCD, which represents a general principle of bringing the users or customers into the design process (Norman, 2013; Veryzer & Borja de Mozota, 2005; Vredenburg et al., 2002) many conceptual tools have been offered by UCD researchers and practitioners that can facilitate generating user insights (Martin & Hanington, 2012). One of such tools is *persona*, defined as a "fictitious, specific, concrete representations of target users" (Pruitt & Adlin, 2006, p. 11). Persona is a synthesis of knowledge that a firm has about their customers or users, and it is based on a research. Persona helps to go beyond traditional *user profiles*, which categorize, describe and prioritize segments and subsegments of users and their usage of the product, service or a system (LeRouge et al., 2013). A user profile is usually expressed in a brief, verbal and informative form, i.e. "male, average age 45–50, intermediate-level computer user, etc." (LeRouge et al., 2013), whereas a persona is expressed in a narrative form, combining verbal and visual data, which makes it seem like a real person, and provides a story concerning the needs of the user set in a particular product context (Miaskiewicz & Kozar, 2011).

Figure 1 presents the generic framework for a persona development.

The right side of the canvas should be filled with verbally expressed research findings, whereas the left side of the canvas can be supplied with visuals and images representing the persona, their look and feel.

Usually persona tool is followed by additional tools, such as an empathy map and/or user journey. An *empathy map* is a tool that helps to synthesize findings and observations about persona's behaviours, thoughts, concerns and emotions in the context of a particular product category or a usage situation (Kelley, 2018). The framework for an empathy map is presented in Fig. 2.

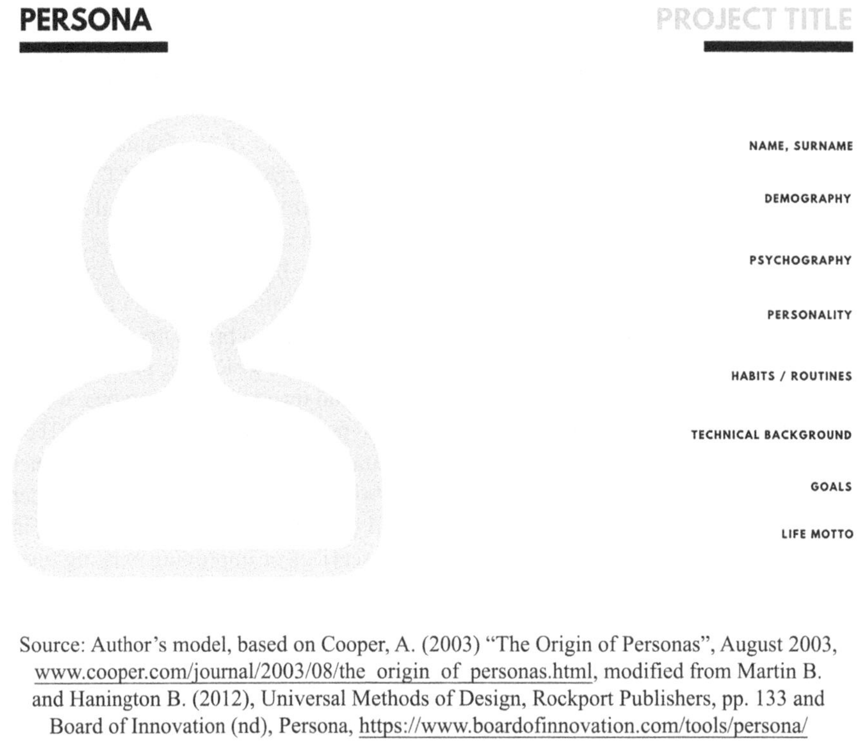

Fig. 1 Persona canvas. *Source*: Author's visual, based on Cooper, A. (2003) "The Origin of Personas", August 2003, www.cooper.com/journal/2003/08/the_origin_of_personas.html, modified from Martin B. and Hanington B. (2012), Universal Methods of Design, Rockport Publishers, pp. 133 and Board of Innovation (nd), Persona, https://www.boardofinnovation.com/tools/persona/

User journey is a visualization of experiences users have when interacting with a product or a service (Martin & Hanington, 2012). It allows to capture the main stages of a buying process but also the smaller moments of interaction, the touchpoints and whether they result in positive or negative experiences (Fig. 3).

Together these three tools form an *extended persona* and help to capture key insights about a selected type of user.

UCD concept may initially seem close to what marketing authors and practitioners know as *market segmentation* (Tynan & Drayton, 1987), which has been applied widely to both B2C and B2B markets. However, user profiles and extended personas go far beyond simple demographics, as they allow to capture the customer's expectations, past experiences and anticipated behaviour (LeRouge et al., 2013).

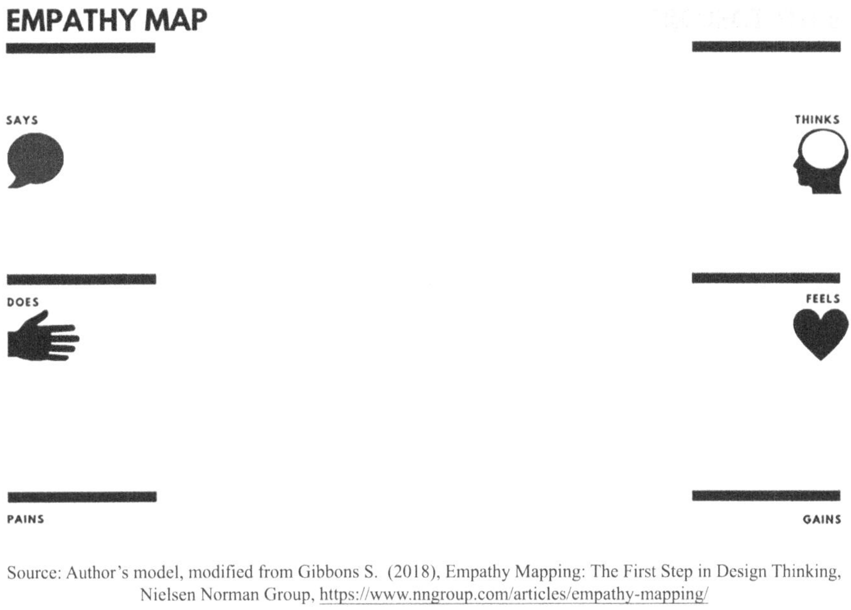

Fig. 2 Empathy map canvas. Source: Author's visual, modified from Gibbons S. (2018), Empathy Mapping: The First Step in Design Thinking, Nielsen Norman Group, https://www.nngroup.com/articles/empathy-mapping/

2.2 Value Proposition Design

Value proposition is one of the most widely used business terms (Anderson et al., 2006) and it relates to a way of strategic thinking that facilitates communication of a firm's ability to offer a superior value package to selected market segments (Payne et al., 2017).

In this section we complement a traditional view of value creation and delivery in dyadic customer-firm relations with some alternative concepts related to value, i.e., how value can be co-created by various actors (Vargo & Lusch, 2004) or by managing network relations (Ramaswamy & Ozcan, 2016). The initial view on value creation relied on an assumption that firms are the value creators and deliverers and customers are the recipients of value (Kotler, 2002; Porter, 1980). More recent views highlight that value can be co-created (or co-destroyed, Bertilsson & Rennstam, 2018) in processes among various stakeholders (Vargo & Lusch, 2004), including institutions such as popular media, social media or cultural conventions (Holt, 2004). The co-creation process is particularly relevant in service design, where customers take part in the service delivery process and co-create value (Luo et al., 2019; Payne et al., 2008).

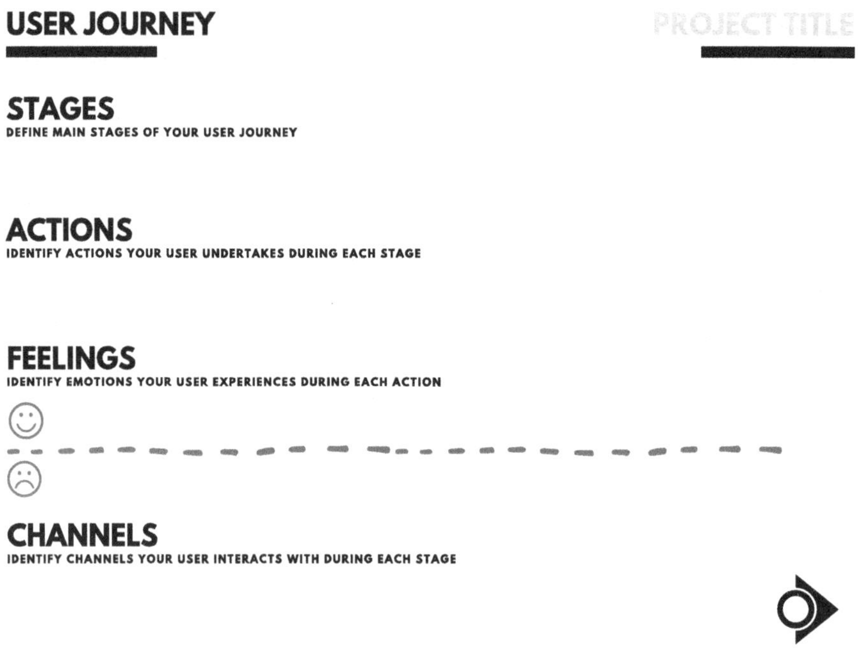

Fig. 3 User journey canvas. Source: Author's visual, modified from Gibbons S. (2017), UX Mapping Methods Compared: A Cheat Sheet, Nielsen Norman Group, https://www.nngroup.com/articles/ux-mapping-cheat-sheet/

One of the widely recognized strategic tools that may foster internal discussions in a firm related to how to apply knowledge about user insights is a Value Proposition Canvas (VPC) (Osterwalder et al., 2014), which can be used by a firm to reflect how it aims to provide value to customers (Fig. 4).

The right side of the VPC refers to a selected customer segment and can be filled with research data, i.e., qualitative findings (thick data) combined with a quantitative data about segment size and usage patterns (big data). It starts with what Osterwalder et al. (2014, p. 12) refer to as *customer jobs*, which are things customers try to accomplish by using particular products or services. A customer job can be a task to complete, a problem to solve or a need to be satisfied.

The left side of the VPC allows for listing all that a firm offers or considers to offer, and the offerings can be tangible (products), intangible (services, brand experiences and stories), digital (forums, platforms) or financial (insurances, financing of a purchase, etc.) (Osterwalder et al., 2014). The products, services and their features should respond to identified customer jobs, create gains and relieve pains. If both sides of the canvas are in line, we achieve a fit (Osterwalder et al., 2014). So the VPC allows to search for ideas that are based on user knowledge and to identify and revise the potential misfits.

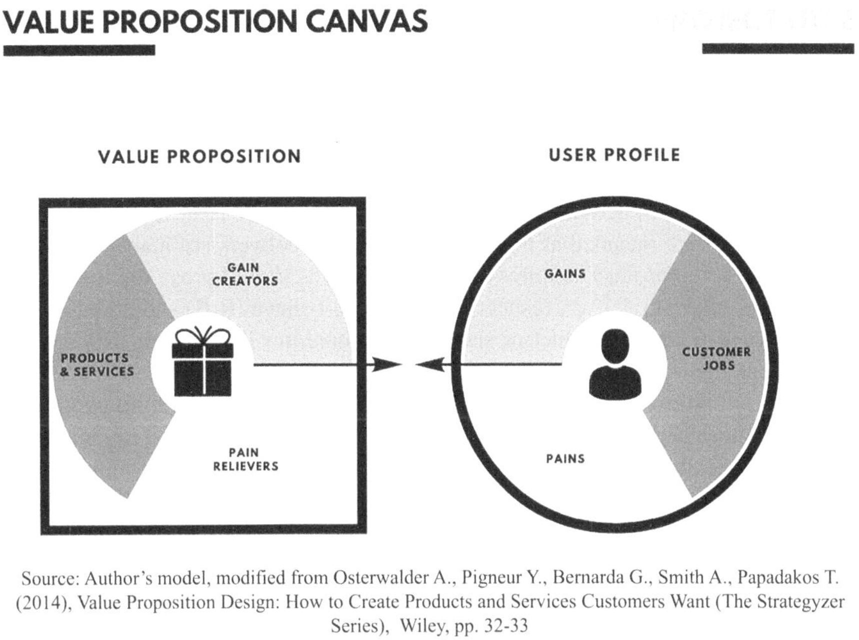

Fig. 4 Value Proposition Canvas. Source: Author's visual, modified from Osterwalder A., Pigneur Y., Bernarda G., Smith A., Papadakos T. (2014), Value Proposition Design: How to Create Products and Services Customers Want (The Strategyzer Series), Wiley, pp. 32–33

2.3 Selected Branding Concepts

Branding is one of the ways of creating and delivering value for customers and other stakeholders. In this section we provide a brief overview of main managerial branding approaches with corresponding illustrative examples from various industries and we discuss them in the context of automotive aftermarket. The value of a brand as a dimension of strategic choices of suppliers on automotive aftermarket varies depending on the region. According to an industry report prepared by McKinsey and Company (2017), the value of a brand is more important in Western Continental Europe compared to the US or the UK, where private labels have a higher market share. Still, according to Capgemini Consulting (2010), branding fosters keeping existing wholesalers or dealers loyal, which seems particularly important due to a flat sales on some markets.

2.3.1 Brand Values and Roles

Brands can deliver different types of values to their customers: functional or rational (Gupta et al., 2018), emotional (Leek & Christodoulides, 2012), hedonic (Kuikka & Laukkanen, 2012), symbolic (Tan & Ming, 2003), or operational (Gupta et al., 2020). Among these types, the automotive aftermarket mainstream brands seem to be focused on functional value for B2C customers supplied by operational value for

B2B customers (i.e., Bosch, Continental). Some brands aim at delivering hedonic and symbolic values as well, these are mainly in-house car tuning brands (Manheimer, 2018) such as Q collection or Q commision by Aston Martin, which involves a personal collaboration with their design team.

Values delivered by brands reflect the role that a brand plays in customers' lives. This role can be functional (problem solving), lifestyle (helping customers create their self-identities) or activist (challenging existing systemic issues) (Koch, 2020). The *functional role* means that the benefits the brand delivers are mainly rational, tangible and set in product-features context. As brands should prove the functional benefits they claim to deliver, respective reasons-to-believe (RtB) should be developed or found within a product or service functionalities or features. RtB can be product ingredients or features, materials used, etc. On B2B markets such RtB are often more complexed and difficult to highlight. A tool that can help to overcome those difficulties and allow customers to see and understand the stated rational benefits are value calculators. These are software applications designed to numerically demonstrate the financial benefit of the offering, i.e., cost savings or increased revenues (Pöyry et al., 2021). Brands that are able to offer such operational value (Gupta et al., 2020) can consider developing value calculators that would show potential customers these advantages. One of the examples of value calculator comes from Goodyear (n.d.). As fuel is one of the main operational costs in transportation industry, no wonder firms are interested in ways of reducing it. Goodyear's interactive calculator allows to estimate how much fuel a fleet can save when using Goodyear tires. The required input information is the fleet size, haul type (regional or long), axel configuration and some fuel usage data.

Although automotive aftermarket as an industry seems to be dominated by functional branding approach, we would like to discuss also the opportunities of other brand roles, as seeing the broader scope of branding options can help firms in making better decisions regarding their differentiation, especially if majority of players follows the same (functional) branding path. It is also in line with McKinsey's (2018) suggestion that players on automotive aftermarket should consider alternative branding strategies.

Let's take a look at an example of another industry, which seems similarly focused on functional branding as automotive aftermarket—pharmaceuticals. Traditionally it has been a typical problem-solving category, where majority of brands followed a similar scheme of branding (functional), naming (Latin-based names), packaging (bullet points with ingredients or benefits) and advertising (experts endorsement and exaggeration in problem presentation). In fact, the replication of this approach by next players created a locked-in marketing convention, which newcomers have obediently followed. However, one brand decided to challenge this marketing convention. Help remedies—a line of basic medicines for most common health issues such as headaches, allergies or cuts—stripped their branding almost to bare bones compared with industry standards and added a lifestyle sparkle interesting particularly for the Millennials generation, which was not a default target group for majority of other pharmaceutical players. The brand name is very simple, *Help*, so is the usage indication: *Help I have a headache* or *Help I've cut myself*. The packaging design is minimalist and pure. All the brand touchpoints reflect the brand mission, which is to make solving simple health issues simple (Dieline, 2009). This example can serve as inspirational benchmark for automotive aftermarket players as it shows that moving beyond conventional functional role can result in brand differentiation, ability to attract particular market segments, overlooked by other brands and gaining additional publicity.

Some firms take the lifestyle aspects of branding further and aim at turning their brands into cultural icons by linking the brand with some relevant and compelling cultural ideas (Holt, 2004), which expand beyond the lifestyle choices. Such approach is called a *cultural branding* (Holt, 2004); it has been successfully applied in consumer markets and it can be as well applied in automotive aftermarket. It requires to link a brand with a selected idea from a culture surrounding the product category. Scholars suggest various cultural categories that might be considered by brand managers, i.e., lifestyle, class and status (McCracken, 2005), gender roles, people's ambitions at work, their dreams for their children, their fears of technology, mid-life crisis, etc. (Holt, 2004).

Although emotional value delivered by a brand may seem more applicable to B2C market segments, scholars emphasize emotional aspects of B2B branding (Leek & Christodoulides, 2012). Emotional values can be delivered to both types of customers (B2C and B2B); however their manifestations seem different on both markets. On B2C markets the emotional value of a brand is reflected rather as a lifestyle or cultural branding concept, whereas on B2B markets it reveals more as a relationship. The topic of "relationship" has been well established in marketing sources and different aspects of relationship in the B2B sector have been described by scholars and practitioners (Melewar, 2009). Emotional branding on B2B markets can also take a form of brand storytelling.

Another role that a brand can play in customers' lives is an *activist* one. An activist brand is the one that goes beyond functional and lifestyle positioning to bring systematic social issues to the public discourse aiming at a structural change (Koch, 2020). This concept is inevitable related to a corporate social responsibility and it requires a brand to take an active role in reshaping social practices (i.e., mobility), transforming industries and related institutions. Recently Interbrand (2022), a global brand consulting company, has published *Brands as acts of leadership* manifesto, in which they have indicated new roles for leading brands, such as dealing with universal, systemic problems, creating a widespread, systemic collaboration devoted to solving these problems and managing the externalities (Interbrand, 2022). Such brand role can be perceived as a purpose-driven branding (Hajdas & Kłeczek, 2021), where brand aims at facilitating a desired social change. As no single market entity can solve the issue of sustainable mobility alone, the leading brand in such arena would have to break down some industry silos and foster an active cooperation of all actors involved in a nexus (Bezerra et al., 2020) of mobility practices.

An outside-the-industry example of a brand that went beyond a functional and lifestyle branding into an activist role is KLM, a Dutch airlines brand that tries to sparkle discussion on responsible flying. As the brand puts it, they are committing to take a leading role in creating a more sustainable future for aviation (KLM, n.d.).

The choice of a brand role is a strategic decision related to brand positioning and competitive strategy. It is worth noticing that brands, as they evolve on the market searching for new differentiation opportunities, may switch from one role to another. Koch (2020) describes an example of Oatly brand, which emerged on the diary market with an oat milk proposition for people with lactose intolerance. The initial brand role was to deliver functional health benefits. Despite the sales increasing steadily, the growth was rather unsatisfactory so the managers decided to switch to a lifestyle concept, supporting urban lifestyle, emerging barista culture and #coffelovers. Recently the brand has made another interesting turn in their role, moving to an activist brand and entering the arena of food politics, focusing on solving global challenges linked to climate change and public health. In a new role, Oatly brand challenges the whole food industry and highlights the problems of animal-based food production. As an activist brand, Oatly undertake several initiatives that go far beyond the scope of traditional marketing activities, i.e., they have sent a petition to the German parliament to make CO2 food labelling a law, initiated a trade organization to promote plant-based food or are part of a EU-funded research project on sustainable food systems where they contribute by testing the possibility of switching from animal-based to plant-based farming business models. Also on automotive aftermarket we can find examples of evolution from one role (i.e., functional) to another (i.e., symbolic). Previous studies showed for example that technological advancements and related architectural layout of cockpit systems in Mercedes-Benz's S-class and BMW's 7-series have evolved due to branding and designer efforts from a functional characteristic to a symbolic identity (Chien & Lin, 2022).

As we already mentioned, neither KLM alone will transform flying nor Oatly alone will transform food, nor will any automotive aftermarket brand alone reshape mobility. But these are definitely good goals to start from (United Nations, 2023) for a brand that wants to act as a leader and is ready to inspire wide, cross-industry partnerships in achieving such goals.

2.3.2 Co-branding and Ingredient Branding

As industry experts suggest that automotive aftermarket players should consider applying multibrand strategies ranging from white-label products to premium supplier brands (McKinsey and Company, 2018), in this section we present selected branding strategies for such consideration.

Co-branding is a term describing co-operations between two or more market entities that connect representations of several brands, such as names, logotypes or product designs (Bengtsson & Servais, 2005). In general it refers to the pairing of two or more brands that cooperate within a marketing field to achieve certain goals (Nasution et al., 2020). Such joint initiatives can take various forms, i.e., cause-related marketing, ingredient branding, or joint sales promotion (Bengtsson & Servais, 2005).

Ingredient branding is a strategy in which key features of one brand are incorporated with a host brand (Desai & Keller, 2002). The well-known and often referred to examples of applying this strategy include Intel, Gore-Tex, or NutraSweet. Scholars have emphasized the various benefits of this approach, such as product differentiation and increased market share (Swaminathan et al., 2012), enhanced market competitiveness and additional economic value (Dalman & Puranam, 2017; Yan & Cao, 2017).

While in the past ingredient branding was quite popular in the automotive industry with examples such as Ford's consumer-facing electronic systems branded by Microsoft (Moon & Sprott, 2016), Recaro car seats or Blaupunkt radio systems (Braind, 2019), nowadays it seems that there are less successful examples of ingredient or co-branding approach in this industry, and if there are, they rather come from out-of-category suppliers (Braind, 2019). One of the examples can be a cooperation between Maserati and Ermenegildo Zegna, which started as suit-style woven fabrics inside the cabin of 100-edition Quattroporte in 2013 and turned into a long-standing trim option, used for the seats, door linings and ceiling headliners on its regular production models of Ghibli, Quattroporte and Levante (Maserati, n.d.).

What is interesting, there seems to be a discrepancy between a low popularity of this strategy and relatively high customers' expectations related to it. Findings from a German automotive market show that 41% of consumers say it is important for them to know that certain car components are coming from a well-known supplier (Braind, 2019). These numbers are even higher in case of premium car brands, such as BMW (57%) or AUDI (56%). More than half of the respondents stated that well-known car components would increase the quality and the value of the car, and 34% would be able to pay more for a car knowing that key components come from well-known suppliers. Findings also show that the ingredient branding could play an important role for components related to sustainability, autonomous driving or e-mobility (Braind, 2019). Definitely suppliers of such components could consider using argumentation based on customer expectations in their B2B communication.

2.3.3 Brand Storytelling

Storytelling has long been recognized as a powerful and effective communication tool in branding (Kasilingam & Ajitha, 2022). Brand storytelling refers to using stories by brands to convey their meaning and context in customers' lives (Key et al., 2021). Among the benefits of brand storytelling we may mention evoking emotional reactions and connections with consumers (Biesenbach, 2018; Dias & Cavalheiro, 2021).

Brand storytelling has long been applied in automotive industry. Some most prominent examples include Land Rover, Honda, Lexus, BMW or Skoda (Lamour, n.d.). But the storytelling has also been successfully applied by other firms, operating on automotive-related markets. Recently, one of the Polish online marketplaces for used cars, Otomoto, has created a campaign "True stories", in which real car sales announcements became the foundations of short documentary movies. The 2-minute spots show unique stories of cars and their owners. The first spot was released before Christmas in 2021. It was based on a sale announcement of

38-year-old Wartburg 353, which has remained in the hands of one family since it was bought by its first owner. The car, which for several decades has always been parked in the same garage, served three generations of owners and has continuously gained sentimental value. Now, after the first owner died, his granddaughter—who has inherited the vehicle—is looking for a new, caring owner for it. Everything we can see in the spot, from the owner of Wartburg, through her family home to the snowy mountain landscape of Lower Silesia is authentic (Otomoto, 2021). According to Otomoto's reports, the spot resulted in the highest number of views and the highest engagement rate among all Otomoto's social media creations in 2021 (Pawłowska & Wróblewski, 2022). The second spot was released by Otomoto soon before Father's Day in 2022. It was based on a sales announcement of 20-year old Toyota 4-Runner, and it told a story of a father and son relationship and their common passion for motorization (Piasecki, 2022).

It is worth mentioning that the emotional car–owner bond results sometimes in a practice of naming the vehicles. Research shows that 42% of American car owners (CBS, 2018) and 32% of UK motorists (Shaw, 2020) name their vehicles. In one of their studies Skoda found 450 unique names given by motorists (Shaw, 2020). Topics like an emotional bond between owner and a car or a shared passion, inherited from father to son or daughter can become storytelling motives for a short-term campaign, as well as foundations of a longer-term cultural branding approach, which we have discussed earlier.

Storytelling can be placed not only in advertisements or TV commercials, which definitely requires more impressive marketing budget, but also in digital channels, social media and can even be squeezed in to 140-characters tweet (Lamour, n.d.), so firms with less impressive resources can also apply it and benefit from.

3 An Empirical Study on Female Car Drivers in Poland

3.1 Research Design and Procedure

In this study we apply the user-centered design way of thinking discussed earlier. Our study is exploratory and based on qualitative inquiry, which allows to address research questions of "how" (Crick, 2021) and is particularly useful when existing theory is insufficient to explain the phenomenon in focus (Graebner et al., 2012). As one of our research questions was "how UCD concepts and tools can be applied in practice on automotive aftermarket" and as previous studies have not explained the application of UCD in the context of this particular market, we opted for a qualitative research approach. We collected the data using semi-structured and in-depth telephone interviews. The study took place in Poland in August and September 2022 and has been carried out by the author of the chapter. It focused on female car drivers who are the occasional customers of car workshops and maintenance stations. The main geographical context of the study was thereby Poland; however, as scholars indicate some methodological challenges related to expanding the findings of qualitative studies beyond the context investigated (Reuber et al., 2022), we have also searched for secondary data from other geographical market that would allow for initial cross-case comparisons. We were able to obtain data from US market in a form of one in-depth interview with a female truck driver conducted and posted online by Stanford University (Stanford d.school, n.d.). We decided to use it for comparison purpose to explore whether the initial findings from the Polish market justify further investigation among other markets.

3.2 Sample and Data Collection

We have applied purposeful sampling (Palinkas et al., 2015) in order to identify and select information-rich cases related to our phenomenon of interest. We have decided to investigate female car drivers and their experiences with car workshops and maintenance stations. Women as car drivers and automotive aftermarket customers remain an underexplored area of research. Historically, the automotive industry, including automotive aftermarket seems to be focused on men as actors and participants. But the closer view on some market statistics shows that excluding women from a target group results in missing an interesting market opportunity in the B2C business. Data reveals that on some markets, women drivers outnumber men (Arbor, 2012). More women are driving than ever before, up from 28% in 1975, to 67% in 2016, presenting a great opportunity for car manufacturers and related service providers (Magnetic Media, 2018). Many automotive aftermarket industry experts call for including women both in marketing strategies (Malik, 2022; Heidbrink, 2022; Tire Business, 2016) and workforce (Deloitte, 2020a). In fact, the rising number of female drivers has been identified by industry experts as one of the key demographic shifts reshaping the automotive aftermarket (Consultancy-me, 2022). It is worth noticing that female drivers are not a homogeneous group and it is safe to assume that they can be segmented according to some psychographic or behavioural criteria. This is where a *user profile* and *persona* tools can help.

For the purpose of this study we have created two user profiles of car maintenance and repair workshops and two corresponding personas of female drivers. The first user profile (UP1, Anna) is based on seven in-depth interviews with female car drivers in Poland. The interviews were semi-structured to make participants feel more at ease and reveal issues that were relevant to them (Adams & Cox, 2008). The questions revolved around *user stories*, which are expressed needs that can inspire innovations (O'hEocha & Conboy, 2010). Interview scenario also included *critical incident technique* (Butterfield et al., 2005), which allows to gain insights about the user's best and the worst experiences when having contact with a product or a service and some *photo studies* (Wagner, 1979; Warren, 2005), which help to gain additional insights into how customers experience retail environments and how they are influenced by the in-store atmosphere (Petermans et al., 2014).

Table 1 User profiles—car maintenance services

User profile	Description
User profile 1 (Poland)	Female, average age 30–40, mainly city car driver, drives a second family car, occasional user of car maintenance services, not interested in mechanics, wants "what needs to be done" for her car, avoids dealing with car services if possible
User profile 2 (USA)	Female, average age 25–30, professional car driver, mechanical engineer, owner of a small transportation firm, truck-driver, self-dependent, heavy user of car maintenance services, wants "the best" for her car, interested in car repairs, would like to know more

The second user profile (UP2, Erica), used for comparative purpose, is based on an interview with a female truck owner conducted in the USA by Stanford d.school (Stanford d.school, n.d.). Table 1 displays both user profiles.

3.3 Findings

In this section we present how persona makes a user profile "come to life" (LeRouge et al., 2013), by reviewing some user stories collected during interviews that were further synthesized into persona 1 (Anna, Poland):

"I know I should take my car to a car maintenance station as I have this scratch on a fender and it should be taken care of before it gets rusty or something but somehow I keep postponing this visit. I am somehow not eager of going there."

"I do take care of such things, quite often in fact. I wish my husband took care of it but he is at work and the working hours of such places cover with the regular office hours. So I have to deal with this."

"Usually it is my husband who takes care of such things but sometimes he can't manage it personally, so he just makes an appointment and asks me to go. I really don't like being in this situation. I just don't feel I belong there. These places are always located in some weird locations, suburbia, abandoned warehouses... Sometimes even my GPS can't find the appropriate address. And then, after struggling with how to get there I finally walk in and see this naked woman poster on the wall. The first thing that comes to my mind is: where am I?? or: am I safe in here?"

"If there are more demanding repairs where they can ask me more complex questions, I try to avoid it. I don't have such knowledge, I have never been interested in it and it makes me embarrassed when they ask questions that are obvious for them and I don't know what's going on. Besides, you have to make some decisions there, for example, you need to change the oil and I don't know if they are pulling me on because I'm a woman."

"I hate going to such places, the experience for me is almost traumatic. Once I had to drive onto the bay, I feared that my car falls into it and these guys there were staring at me. I felt so exposed like they were evaluating my driving skills and only waiting to make jokes about it once I leave."

"I don't know, maybe it's just me but I kind of feel that I enter a man's cave and that I do not belong there."

"It seems to me that these gentlemen in this industry already have some stereotypes about women, blondes, etc. And when a woman comes in they run their imprinted scripts. I once had a comparison between how such a guy talked to my husband and how he talked to me. He already assumed that I was an idiot."

Fig. 5 Photo studies—an experience of a female car driver with a car maintenance workshop environment

> *"There are two types of such places, the first are authorized services with men wearing suits at a frontline. The second are some barns in the middle of nowhere where once – I swear - a barefoot man come out to repair my car! In the first type of places I feel they are going to rip me off. And in the second... I don't even know where to start..."*

One of the interviewees shared a photo from her last visit to a car maintenance workshop. It is presented in Fig. 5.

Let's now also add "life" to the user profile 2, that is Erica (USA). Here are some of her statements from the interview (Stanford d.school, n.d.):

> *"My old truck... at one point I took it in just for a regular oil change and they told me that I needed new tires. Actually I think it was that I had a nail in my tire. And they said oh well, the treads on your tires are really low... Of course I am not going to argue with that, it sounded valid. I didn't want to look dumb...."*
>
> *"And then you know, they are trying to sell you really nice tires. Which is fine because I want the best for my truck. And you know, I have to trust, I have no other choice in terms of what they are trying to sell me."*
>
> *"Seriously I though I hope I would have to pay 40 dol and not 600 dol or how much I would end up paying...".*
>
> *"I always try to be careful and do everything I need to do for my truck cause I don't want to be stranded. Cause I have been stranded before and so I thought well, I need to get this fixed."*
>
> *"Another thing about me is that I am a mechanical engineer and it felt funny that you know (...) I should be able to speak, you know, the car speak and understand like what they are doing and the repairs needed and everything".*
>
> *"And so I am not very good at questioning what they say is wrong with it because I don't want to look dumb or sound dumb. So I am always afraid that anything that I say will let them know that no, I don't know everything about the vehicles and mechanical systems."*

> *"I don't know, maybe it's just me but I've had this feeling that the guy is like: ok, when you're gonna call daddy to get you out of this one? Just because here I am, walking in, a young kid in a really nice truck. And so I wanted like this feeling of independence that I can deal with the situation, I can pay for it, I can figure out what I need to do and just do it."*
>
> *"I find it not fun to get my truck fixed although I am really interested in it. Like I wish they would let me go into the bay and actually explain me step by step what they are doing so I could learn more. But at the same time I feel I had to put on this show like: oh yeah, routine, I know what they are talking about."*

As we can notice, despite probing in different cultural and geographical contexts, both personas share some similar negative assumptions, feelings or experiences with automotive aftermarket services but at the same time their cognitive needs referring to this product category are significantly different, which may require different approaches and solutions.

Insights coming from personas can be transformed into particular design statements. *A design challenge* is another technique from a broad repertoire of UCD frameworks and practices. It refers to finding the right problem to be solved, the one that is based on a deep understanding of the true needs of people and the one that captures the relevant design requirements (Norman, 2013). Some examples of design challenges based on Anna's persona could be: *how to make car maintenance service more women friendly?* Or *more gender inclusive?* Or *how to make a car maintenance station a place women would actually like to go to? More like a coffee shop? Or a spa?* The goal in this stage is to frame users' problems as specific and meaningful design challenges (Norman, 2013). As our examples show, persona can face several problems when dealing with a product or a service, so the design challenges may vary. Therefore it is a common practice in UCD to brainstorm about the problems that a firm would like to consider before brainstorming the actual solutions.

Once the appropriate design challenge is selected, the next step is generating an array of possible solutions (Norman, 2013). Referring to our Anna's persona developed for the purpose of analysis and to the examples of design challenges, solutions don't need to (and probably even should not) evolve around stereotypical "girly aesthetics", yet they could advance and implement more gender neutral approaches. One of the brainstorming techniques involves transferring approaches from other industries (Vullings & Heleven, 2015) by using *analogy*, which has been proven to serve as a powerful creative device (Kao, 2020). This technique allows to observe and learn how firms from other industries solve a similar problem and adapt their way of thinking. Some brands for a while have used a unisex design and language (i.e., Apple or CK One). Some others have recently put gender neutral solutions to their agendas, i.e., Lego removes gender bias from their toys (The Guardian, 2021). These examples can serve as inspirational benchmarks for removing gender bias that seem to preoccupy how some automotive aftermarket products, services or communications are designed or implemented on the repair workshop level.

4 Discussion

We began with an overview of selected marketing concepts related to designing services, value propositions and brands based on user insights. To answer the research question of how these concepts can be applied in practice on automotive aftermarket, we have conducted qualitative study among female car drivers. Based on our findings we can gain deeper understanding how the knowledge about users can be applied in designing (or redesigning) products and services on the automotive aftermarket and reflect the personas developed. As we mentioned in a theoretical section, the value can be co-created (or co-destroyed, Bertilsson & Rennstam, 2018) in processes among various stakeholders (Vargo & Lusch, 2004). For automotive aftermarket firms it means that customers may receive some value that was co-created beyond the firm's boundaries and out of company's control, i.e., on car-related independent online forums, or that value can be destroyed by cultural conventions even prior using a product or a service, like in our Anna and Erica's example, where both personas feel they are treated as secondary car drivers by the workshops' staff, which may reflect the actual staff behaviours and attitudes towards female car drivers or some cultural conventions that had created such impressions in our personas, long before they entered any car workshop for the first time.

How the value is created and delivered, co-created (or co-destroyed), should be the topic of internal and strategic discussions in a firm, i.e., by applying Value Proposition Canvas that has been presented in the theoretical section of the chapter. Some examples of "customer jobs" drawn from our interviews could be*: when I go for a car repair I would like to have all things taken care of without me being involved so I don't feel incompetent* (Anna) or *when I go for a car repair I would like to be included in the process so I could learn more* (Erica). Based on our findings the list of potential solutions would vary for both customer segments, from full service and no contact for Anna—some OEMs already offer a fully digital user journey, from online appointment booking, through status notifications to vehicle pickup and payment (McKinsey and Company, 2018), to seat with a bay-view (an analogy of an open kitchen in a restaurant), some educational features (an analogy of cooking courses) or even participation (an analogy of cooking with a chef) for Erica. The solutions that would meet the expectations of both segments could be transparent pricing communication and more inclusive, less technical language. The examples we use are based on the findings from our study, but our aim is not to offer ready solutions but rather to highlight how value proposition canvas can be used for internal discussions about the value creation and delivery processes. Osterwalder et al. (n.d.) offer several opportunities for innovating value proposition, which we will discuss in the context of the automotive aftermarket.

- Address a *more complete set of jobs*, including related and ancillary jobs; for a B2C customer it could be the possibility of storing tires in a workshop, receiving reminders about car inspections, or an access to relevant online sources, as industry reports highlight that a significant share of car-related information search and purchases has moved online and e-commerce share of parts sales is estimated to increase even by 30% by 2035 (McKinsey and Company, 2017). Also, as a car is becoming a platform for software and applications (McKinsey and Company, 2017) and as 70% of aftermarket experts expect new digital players to gain a significant share of revenues and profits of the automotive aftermarket by 2030 (McKinsey and Company, 2018), automotive aftermarket players could consider creating partnerships with tech companies to be able to co-create the experience of a connected vehicle.

- Switch to *more important jobs*—one of the interviewee in our study told a story, which could exemplify what a more important job potentially could be for a B2C customer:
  "We had some problems with a clutch, it happened on the road. My husband at first wanted to call for assistance as we had an insurance. We were coming back from holidays, travelling with our small kid, it happened 20 km before reaching our home. It was already getting dark, our kid was hungry and tired. We didn't want to wait an hour or who knows how long before the assistance would come. So my husband called his trusted mechanic. The mechanic heard the story and gave him an instant advice: he literally told my husband to kind of 'pump' the clutch until the air is released and the pedal feels tough again. I was sceptical at first but it did work out! After few minutes of 'pumping' we could continue the drive and go home. My husband took the car for the proper repair next day, without us on board"; so the customer job is "when I travel with my family, I want to have a piece of mind that we will get there and back safe and smooth"; so the potential design challenge is how automotive aftermarket firm can address that? Also, as young generations are reported as less interested in car ownership, which will lead to fewer individually owned cars and more mobility solutions (McKinsey and Company, 2017), the mobility will probably become more important customer job to consider for automotive aftermarket players and new business models of shared mobility and mobility as a service will emerge (Deloitte, 2020b). It is worth noticing that some universal human needs, like safety or mobility, can't be met by just a single market player. They should rather be treated as arenas (Interbrand, 2022) for partnerships and networks that go beyond the scope of a single industry, where the players from various industries (manufacturers, service-providers, insurance, state, etc.) create solutions that can provide a seamless customer experience (Jain et al., 2017) throughout the arena.
- Go beyond the functional jobs and help in fulfilling *social or emotional jobs*. Taking into account that vehicles are not only transportation means but also self-identity expression devices (Barbarossa et al., 2015) that create a true interest among both amateurs and professionals, reflected in a number of automotive magazines, TV programs, clubs and both offline and online communities, there are also various social or emotional jobs that customers are trying to "get done". An outside-the-industry benchmark could be a bike culture that has flourished in many countries and resulted in places like Lola bikes in Hague, Netherlands. This concept combines a coffeeshop with bike repairs and ride sharing offers. Another benchmark could be Small vet, an innovative veterinary practice which goes beyond an ordinary pets' medical treatment and also offers well-designed

interiors, 24/7 contact, advisory on pets' health and lifestyle. All this is based on a membership business model, and not on pay-per-visit, which dominates in this industry. These can be inspirational benchmarks for automotive aftermarket players, which are in line with McKinsey's (2018) suggestions that workshops should consider a brick-and-mortar makeover to fit a new customer service mindset.

- Help your customer to *do the job incrementally better*—the responses could include a series of micro improvements to a current value proposition, i.e., decreasing lead times or monitoring the progress of the service (benchmark: status trackers in the last mile delivery industry).

Our findings can also serve as an inspiration for considering selected branding strategies on automotive aftermarket. As we have shown on the examples of our two personas (Anna and Erica), the emotional attitudes of some customer segments towards automotive aftermarket may be negative, so a brand storytelling, if based on understanding the pain points of customers and redesigning the marketing offers accordingly, can be a communication tactic for changing consumer attitudes. In fact, some brands create powerful stories based on bad reviews and harsh opinions they have received from customers and show how they have transformed since. An example can be a Domino's pizza turnaround story, in which employees, after receiving very negative feedback ("tastes like cardboard") decide to completely transform the recipes (Domino's, 2009).

Our findings indicate that another potential for automotive aftermarket actors might be applying cultural branding approach. Let's try to pull out some cultural branding ideas knowing what we already know about our persona, Anna. If she seems somehow unprepared to be a literate customer of car maintenance services, is it her personal trait? Or fault? Hasn't she done her homework properly? Or rather, is it a side effect of cultural upbringing that made her a secondary car driver? If we agree that driving an efficient and reliable car can be a lifesaving issue, shouldn't society prepare young people (including women) better to deal with such issues? This is one of the assumptions of value co-creation concept that we have already discussed, that the value is created and delivered not only in firm–customer dyadic relations, but it is also co-created or co-destroyed (Bertilsson & Rennstam, 2018) in processes among various actors (Vargo & Lusch, 2004), including institutions such as media or cultural conventions. Again, let's look for some outside-the-industry benchmarks for inspiration. Ariel, a washing powder brand, encourages parents in India to involve their young sons in laundry from the early age so they are more prepared to share home duties once they establish their own families (Ariel, 2019). How about fathers involving daughters in conversations about a car repair, oil change and taking them to such places so they are more familiar with the topic once they grow up? Dove, a cosmetic brand, celebrates dads and their ability to care for their kids (Dove, n.d.). How about automotive aftermarket brands encouraging dads to teach their kids some basic knowledge about cars so they are not so easy to rip off once they have to visit a car maintenance workshop someday? Philips brand launches iron for men, so they could be more helpful to their partners once a new born arrives to a family (Philips, n.d.).

As ironing can be demanding for some people, Philips decided for *downskilling*, which means lowering the level of skills needed to use a product or a service (Hajdas & Kłeczek, 2021). Couldn't some automotive aftermarket brands intentionally lower the level of skills needed to discuss a car repair, so Anna feels less intimidated? How about lowering the skills needed to make some basic car repairs? If nowadays men are supposed to be able to change both a tyre and a diaper, as Lynx, a deodorant brand posits in one of their campaigns (Lynx, 2016), shouldn't women be able to effortlessly change a tyre themselves, if they have to or want to do it on their own? How to make changing a tire less physically demanding? How to redesign changing a tire so it feels more like… let's say assembling an IKEA product? This is another example of a design challenge, the one that would empower women (and probably some men too).

We discuss Anna here as we have gathered some insights about her and we speculate about understanding these insights in broader, cultural categories to explain how to transform knowledge about a particular user profile into a cultural branding concept. But definitely, the Anna example does not limit the possibilities of applying cultural branding on automotive aftermarket. And certainly, researching other user profiles, particularly from B2B segment, would reveal other insights that would require brainstorming about other branding concepts.

Following the United Nations Sustainable Development Goals 2030 and bridging gender gaps in this service business, we argue that "the car talk" requires transforming the technical language into a more human, neutral and clear communication. Hence, we could reframe the design challenge into: *how to redesign the way we communicate with our clients so that customers as laypersons can understand what we mean?* Such design challenge can lead to other inspirational benchmarks. In this way each design challenge formulation and reformulation opens a new exploration field for brainstorming about potential solutions.

Referring to our second persona, Erica, one of the potential design challenges could be*: how to empower women as drivers and car owners?* As consumer empowerment has become an important result of marketing actions (Hajdas & Kłeczek, 2021), it can serve as a foundation of purpose-driven branding, which can support solutions building.

In our study, both personas expressed some issues related to the lack of transparent price communication as well as the vague scope of necessary repairs communicated by the staff. It is confirmed by secondary sources, as studies conducted on UK market showed that 29% of female drivers believe they overpaid for car services, simply because they were women (Almeida, 2017). This common concern can also be framed as a design challenge: *how to redesign the way we estimate various repair scenarios to be more precise*, or simply: *how to increase our credibility in terms of pricing so our customers are sure we don't rip them off?* Industry experts highlight that increased price transparency is one of the key trends in the automotive aftermarket (McKinsey and Company, 2017), so this expectation clearly goes beyond the female car drivers segment.

Interesting and yet overlooked,[1] the segment of female car drivers definitely does not exhaust the topic of existing market segments and niches that could become new opportunities for firms operating on the automotive aftermarket. By using the vivid example of this segment we aimed to highlight that user profiles and personas, if based on real research data, can become a way to customize, incorporate and share the research about users in the firm (LeRouge et al., 2013). They can inform the design and the development of brands, products and services by fostering various design challenge formulations. Based on a selected design challenge (or several of them), derived from real user studies firms operating on automotive aftermarket can incrementally or radically innovate their market offerings. It is worth noticing that although embedded in UCD, user profiles and personas can be applied not only to end-customers on automotive aftermarket but also to other stakeholders: employees, B2B partners, suppliers, distributors, etc. A deeper understanding of issues that are relevant to these groups and deciding how a firm can respond can facilitate innovation processes that go far beyond the marketing ones.

5 Conclusions

In this chapter we have presented selected marketing concepts and frameworks that can foster internal discussion about creating and delivering value on automotive aftermarket in times of rapid shifts that are reshaping the market and service business. Our aim was not to offer ready solutions, but rather to reflect challenges, to show a marketing way of thinking and how selected tools, techniques and frameworks can be applied to designing or redesigning market offerings for automotive aftermarket service at the workshop level.

As digitalization is considered one of the most disruptive forces in the automotive industry's 140-year history (Llopis-Albert et al., 2020), digital marketing strategies will gain in importance, with new ways of engaging customers and offering value focused on customer experience being among them (Bain, 2019). It will also foster a digital servitization based on applying digital tools for "the transformational processes whereby a company shifts from a product-centric to a service-centric business model and logic" (Kowalkowski et al., 2017, p. 8). The digital connection will enable dealers, service providers and workshops to interact smoothly with customers after the purchase, either by arranging the next service check or digitalizing consultation (Bain, 2019). Leading car manufacturers and service providers will benefit from using big data analysis to obtain significantly more knowledge about their customers and turning this knowledge into more personalized experiences (Bain, 2019; Smith et al., 2019). On the other hand smaller, independent service providers can try to build closer relations with their customers by acquiring thick

[1] Caroline Criado Perez in her book entitled "Invisible Women: Data Bias in a World Designed for Men" shares findings on how various industries, product and service categories from home appliances to doctors' offices treat men as the default users. So clearly this is not just an automotive aftermarket issue.

data, based on user insights coming from in-depth conversations and observations, similar to ones presented in our study. They can also develop new sales and service formats based on their local market knowledge (Bain, 2019).

We believe the concepts discussed in this chapter together with our findings suggest several courses of action for owners and managers of service businesses in the repair and maintenance context. Our research from Poland illustrates that there are yet target group related potential to be addressed with marketing tools. The gender aspects that Anna and Erica illuminate bring up challenges that can be approached as opportunities for growth and development. Managers could consider revising to what extent their market offerings are free from gender bias and convenient to use despite the gender. Furthermore, the findings can inspire managers to rethink marketing planning and have internal conversations about who their firms' user profiles and personas are, how well they know them, what their pains and gains are, do they deliver the value expected, etc. Our findings can also serve as a pilot study for firms that wish to consider applying user-centered logic in their marketing activities.

User-centered design in research can also support policymakers in understanding market level service elements that are subject to regulation and institutional interests. These relate to consumer issues, e.g. safety, do-it-yourself behaviour and product instructions. Policymaking also defines the certification of skills needed in the repair, maintenance and service process setting the professional requirements and frameworks between consumers and skilled professionals.

There are some limitations related to the findings presented. The interview findings come from an exploratory, qualitative study that is of limited size. The study was conducted in Poland as a pilot and does not represent Poland as a whole. Therefore, the findings cannot be generalized to other contexts, but they offer a vignette for reflecting such concerns and practices more critically in any organization. Future studies employing user-centered design and marketing in other services and processes of automotive aftermarket are suggested to explore regional differences, best practices, diversity and cultural sensitivities.

References

Adams, A., & Cox, A. L. (2008). Questionnaires, in-depth interviews and focus groups. In P. Cairns & A. L. Cox (Eds.), *Research Methods for Human Computer Interaction* (pp. 17–34). Cambridge University Press.

Almeida, L. (2017). 1 in 3 women felt they were overcharged in garages, Motorverso, https://www.motorverso.com/1-3-women-felt-overcharged-garages/

Anderson, J., Narus, J., & Van Rossum, W. (2006). Customer value propositions in business markets. *Harvard Business Review, 84*(3), 91–99.

Arbor, A. (2012). Women drivers outnumber men, but still drive less, University of Michigan News, https://news.umich.edu/women-drivers-outnumber-men-but-still-drive-less/

Ariel. (2019). The #ShareTheLoad journey. https://www.ariel.in/en-in/share-the-load/the-share-the-load-journey

Bain. (2019). A survival guide for Europe's Car Dealers, https://www.bain.com/insights/a-survival-guide-for-europes-car-dealers

Barbarossa, C., Beckmann, S. C., Pelsmacker, P. D., Moons, I., & Gwozdz, W. (2015). A self-identity based model of electric car adoption intention: A cross-cultural comparative study. *Journal of Environmental Psychology, 42*, 149–160.
Bengtsson, A., & Servais, P. (2005). Co-branding on industrial markets. *Industrial Marketing Management, 34*, 706–713.
Bertilsson, J., & Rennstam, J. (2018). The destructive side of branding: A heuristic model for analyzing the value of branding practice. *Organization, 25*(2), 260–281. https://doi.org/10.1177/1350508417712431
Bezerra, B., dos Santos, A. L., & Delmonico, D. V. B. (2020). Unfolding barriers for urban mobility plan in small and medium municipalities—A case study in Brazil. *Transportation Research Part A., 132*, 808–822.
Biesenbach, R. (2018). *Unleash the power of storytelling: Win hearts, change minds, get results*. Eastlawn Media.
Braind. (2019). Automotive Supplier shy away from Ingredient Branding—but 41% of consumers say it is important to them! 25 Feb, 2019, https://www.braind.co/blog/automotive-study-ingredient-branding
Butterfield, L. D., Borgen, W. A., Amundson, N. E., & Maglio, A. T. (2005). Fifty years of the critical incident technique: 1954-2004 and beyond. *Qualitative Research, 5*, 475–497.
Capgemini Consulting. (2010). *The Aftermarket in the Automotive Industry. How to Optimize Aftermarket Performance in Established and Emerging Markets*. Institute of Technology Management, University of St. Gallen.
CBS. (2018). Survey: 42 percent of americans name their vehicles, CBS News, https://www.cbsnews.com/pittsburgh/news/42-percent-americans-name-their-vehicles/
Chien, WC. And Lin, CH. (2022). Technology in automotive brands: Function or fashion?. In: Kurosu, M. (eds) Human-computer interaction. Theoretical approaches and design methods. HCII 2022. Lecture Notes in Computer Science, 13302. Springer. https://doi.org/10.1007/978-3-031-05311-5_28
Consultancy-me. (2022). Demographic shifts and trends reshaping the automotive aftermarket, 21st April, https://www.consultancy-me.com/news/4933/demographic-shifts-and-trends-reshaping-the-automotive-aftermarket
Crick, J. M. (2021). Qualitative research in marketing: What can academics do better? *Journal of Strategic Marketing, 29*(5), 390–429.
Dalman, M. D., & Puranam, K. S. (2017). Consumer evaluation of ingredient branding strategy. *Management Research Review, 40*, 768–782.
Dell'Era, C., Magistretti, S., Cautela, C., Verganti, R., & Zurlo, F. (2020). Four kinds of design thinking: From ideating to making, engaging, and criticizing. Creativity and Innovation Management.
Deloitte. (2020a). Women in automotive industry study, https://www2.deloitte.com/uk/en/pages/manufacturing/articles/2020-women-in-automotive-industry-survey.html
Deloitte. (2020b). Future of automotive sales and aftersales, https://www2.deloitte.com/de/de/pages/consumer-industrial-products/articles/future-of-automotive-sales-and-aftersales.html
Desai, K. K., & Keller, K. L. (2002). The effects of ingredient branding strategies on host brand extendibility. *Journal of Marketing, 66*, 73–93.
Dias, P., & Cavalheiro, R. (2021). The role of storytelling in the creation of brand love: The PANDORA case. *Journal of Brand Management, 29*, 58–71.
Dieline. (2009). Before & after: Help remedies, https://thedieline.com/blog/2009/7/20/before-after-help-remedies.html?
Domino's.(2009).Domino's®Pizzaturnaround,https://www.youtube.com/watch?v=AH5R56jILag
Dove. (n.d.). Dove Men+Care celebrates everyday fathers, whose masculinity and strength comes from their ability to care, https://www.dove.com/au/men-care/Fathers2.html#!
Goodyear. (n.d.). How much fuel can your fleet save with fuel max™? https://comercial.goodyearca.com/en/Calculadora-de-Ahorro-de-Combustible
Graebner, M. E., Martin, J. A., & Roundy, P. T. (2012). Qualitative data: Cooking without a recipe. *Strategic Organization, 10*(3), 276–284. https://doi.org/10.1177/1476127012452821

Gupta, S., Foroudi, P., & Yen, D. A. (2018). Investigating relationship types for creating brand value for resellers. *Industrial Marketing Management, 72*, 37–47.
Gupta, S., Gallear, D., Rudd, J., & Foroudi, P. (2020). The impact of brand value on brand competitiveness. *Journal of Business Research, 112*, 210–222.
Hajdas, M., & Kłeczek, R. (2021). The real purpose of purpose-driven branding: Consumer empowerment and social transformations. *Journal of Brand Management, 28*(4), 359–373.
Heidbrink, S. (2022). Women and the automotive aftermarket, https://www.vehicleservicepros.com/shop-operations/collision-repair/customer-relations/article/21166075/women-and-the-automotive-aftermarket
Holt, D. (2004). *How brands become icons*. Harvard University Press.
Interbrand. (2022). Brands as acts of leadership, https://interbrand.com/thinking/brands-as-acts-of-leadership/
Jain, R., Aagja, J. P., & Bagdare, S. (2017). Customer experience—a review and research agenda. *Journal of Service Theory and Practice, 27*, 642–662.
Joly, M. P., Teixeira, J. G., Patrício, L., & Sangiorgi, D. (2019). Leveraging service design as a multidisciplinary approach to service innovation. *Journal of Service Management, 30*, 681–715.
Kao, C. (2020). How figurativity of analogy affects creativity: The application of four-term analogies to teaching for creativity. Thinking Skills and Creativity.
Kasilingam, D., & Ajitha, S. (2022). Storytelling in advertisements: understanding the effect of humor and drama on the attitude toward brands. *Journal of Brand Management*.
Kelley, T. (2018). Build Your creative confidence: Empathy maps, IDEO, https://www.ideo.com/blog/build-your-creative-confidence-empathy-maps
Key, T. M., Keel, A. L., Czaplewski, A. J., & Olson, E. M. (2021). Brand activism change agents: strategic storytelling for impact and authenticity. *Journal of Strategic Marketing*.
KLM. (n.d.). Fly responsibly, https://flyresponsibly.klm.com/gb_en#home
Koch, C. (2020). Brands as activists: The Oatly case. *Journal of Brand Management*, 1–14.
Kohli, A. K., & Jaworski, B. J. (1990). Market orientation: The construct, research propositions, and managerial implications. *Journal of Marketing, 54*(2), 1–18.
Kotler, P. (2002). *Marketing management*. Prentice Hall.
Kowalkowski, C., Gebauer, H., Kamp, B., & Parry, G. C. (2017). Servitization and deservitization: Overview, concepts, and Definitions. *Industrial Marketing Management, 60*(2017), 4–10.
Kuikka, A., & Laukkanen, T. (2012). Brand loyalty and the role of hedonic value. *Journal of Product & Brand Management, 21*, 529–537.
Lamour, T. (n.d.). *How storytelling is shaping the automotive industry*. Purple Publish. https://purplepublish.com/en/storytelling-in-automotive-industry/
Leek, S., & Christodoulides, G. (2012). A framework of brand value in B2B markets: the contributing role of functional and emotional components. *Industrial Marketing Management, 41*, 106–114.
LeRouge, C. M., Ma, J., Sneha, S., & Tolle, K. M. (2013). User profiles and personas in the design and development of consumer health technologies. *International Journal of Medical Informatics, 82*(11), e251–e268.
Llopis-Albert, C., Rubio, F., & Valero, F. (2020). Impact of digital transformation on the automotive industry. *Technological Forecasting and Social Change, 162*, 120343–120343.
Luo, J., Wong, I. A., King, B., Liu, M. T., & Huang, G. I. (2019). Co-creation and co-destruction of service quality through customer-to-customer interactions. *International Journal of Contemporary Hospitality Management*.
Lynx. (2016). Men in progress. Masculinity, https://www.youtube.com/watch?v=aTjMYRH1uEY
Magnetic Media. (2018). Women drivers: A missed opportunity for motors brands, Magnetic Media, 2nd March, https://magnetic.media/research/studies/talking-to-women-drivers
Mahr, D., Kalogeras, N., & Odekerken-Schröder, G. (2013). A service science approach for improving healthy food experiences. *Journal of Service Management, 24*(4), 435–471.
Malik, A. (2022). How the aftermarket can market to women, Auto Service World, https://www.autoserviceworld.com/how-the-aftermarket-can-market-to-women/

Manheimer, J. (2018). The 15 best in-house car tuning brands, Hiconsumption, https://hiconsumption.com/best-in-house-car-tuning-brands/
Martin, B., & Hanington, B. (2012). *Universal methods of design. 100 ways to research complex problems, develop innovative ideas, and design effective solutions*. Rockport Publishers.
Maserati. (n.d.). Crafters of style. Maserati and Ermenegildo Zegna. https://www.maserati.com/gb/en/brand/partners/ermenegildo-zegna
McCracken, G. (2005). *Culture and consumption II: Markets, meaning, and brand management*. Indiana University Press.
McKinsey and Company. (2017). The changing aftermarket game—and how automotive suppliers can benefit from arising opportunities, Advanced Industries, June 2017.
McKinsey and Company. (2018). Ready for inspection—automotive aftermarket in 2030, June 2018.
Melewar, T. C. (2009). The notion of relationship in branding. *Journal of Brand Management, 17*, 95–96.
Miaskiewicz, T., & Kozar, K. A. (2011). Personas and user-centered design: How can personas benefit product design processes? *Design Studies, 32*, 417–430.
Micheli, P., Wilner, S. J., Bhatti, S. H., Mura, M., & Beverland, M. B. (2018). Doing design thinking: Conceptual review, synthesis, and research agenda. *Journal of Product Innovation Management*.
Moon, H., & Sprott, D. (2016). Ingredient branding for a luxury brand: The role of brand and product fit. *Journal of Business Research, 69*, 5768–5774.
Nasution, R. A., Arnita, D., & Purnama, S. Q. (2020). Believe it or not: The effect of involvement on the credibility of image transfer through co-branding. *Journal of International Consumer Marketing, 33*, 578–594.
Norman, D. A. (2013). *The design of everyday things: Revised and expanded edition*. Basic Books.
O'hEocha, C., & Conboy, K. (2010). The Role of the User Story Agile Practice in Innovation. Computer Science, LESS.
Osterwalder, A., Pigneur, Y., Bernarda, G., & Smith, A. (2014). *Value proposition design: How to create products and services customers want*. Wiley.
Osterwalder, A., Pigneur, Y., Bernarda, G., & Smith, A. (n.d.). *Six ways to innovate from the customer profile*, Strategyzer, www.strategyzer.com/vpd
Ostrom, A. L., Parasuraman, A., Bowen, D. E., Patrício, L., Voss, C. A., & Lemon, K. (2015). Service research priorities in a rapidly changing context. *Journal of Service Research, 18*(2), 127–159.
Otomoto. (2021). OTOMOTO przedstawia „Prawdziwe historie" Zobacz nasz świąteczny spot!, https://www.youtube.com/watch?v=y3Z8mO3_nas
Palinkas, L. A., Horwitz, S. M., Green, C. A., Wisdom, J. P., Duan, N., & Hoagwood, K. E. (2015). Purposeful sampling for qualitative data collection and analysis in mixed method implementation research. *Administration and Policy in Mental Health and Mental Health Services Research, 42*, 533–544.
Pawłowska, P., & Wróblewski, J. (2022). Otomoto i Madogz prezentują "Prawdziwe historie", Media Marketing Polska Online, https://mmponline.pl/artykuly/257511,otomoto-i-madogz-prezentuja-prawdziwe-historie
Payne, A. F., Storbacka, K., & Frow, P. (2008). Managing the co-creation of value. *Journal of the Academy of Marketing Science, 36*(1), 83–96.
Payne, A., Frow, P., & Eggert, A. (2017). The customer value proposition: evolution, development, and application in marketing. *Journal of the Academy of Marketing Science, 45*, 467–489.
Petermans, A., Kent, A., & Cleempoel, K. V. (2014). Photo-elicitation: Using photographs to read retail interiors through consumers' eyes. *Journal of Business Research, 67*, 2243–2249.
Philips. (n.d.). Steam iron. https://www.philips.com.sg/c-p/GC4490_02/azur-steam-iron
Piasecki, P. (2022). Otomoto z kolejną odsłoną cyklu "Prawdziwe historie", Media Marketing Polska Online, https://mmponline.pl/artykuly/259075,otomoto-z-kolejna-odslona-cyklu-prawdziwe-historie
Porter, M. E. (1980). *Competitive strategy: Techniques for analyzing industries and competitors*. The Free Press.

Pöyry, E., Parvinen, P., & Martens, J. (2021). Effectiveness of value calculators in B2B sales work—Challenges at the sales-call level. *Journal of Business Research, 126*, 350–360.
Pruitt, J., & Adlin, T. (2006). *The persona lifecycle: Keeping people in mind throughout product design*. Morgan Kaufmann.
Ramaswamy, V., & Ozcan, K. (2016). Brand value co-creation in a digitalized world: An integrative framework and research implications. *International Journal of Research in Marketing, 33*, 93–106.
Reuber, A. R., Alkhaled, S., Barnard, H., Couper, C., & Sasaki, I. (2022). Something borrowed, something new: Challenges in using qualitative methods to study under-researched international business phenomena. *Journal of International Business Studies, 53*(9), 2147–2166.
Shaw, N. (2020). Most popular names people give their cars in the UK, Cambridge News, https://www.cambridge-news.co.uk/news/uk-world-news/most-popular-names-people-give-18296081
Smith, W. S., Coleman, S., Bacardit, J., & Coxon, S. (2019). Insight from data analytics with an automotive aftermarket SME. *Quality and Reliability Engineering International, 35*, 1396–1407.
Stanford d.school. (n.d.). Truck owner POV, Truck owner POV on Vimeo.
Swaminathan, V., Reddy, S. K., & Dommer, S. L. (2012). Spillover effects of ingredient branded strategies on brand choice: A field study. *Marketing Letters, 23*, 237–251.
Tan, T., & Ming, M. T. (2003). Leveraging on symbolic values and meanings in branding. *Journal of Brand Management, 10*, 208–218.
The Guardian. (2021). Lego to remove gender bias from its toys after findings of child survey, by Hellen Russel, Mon 11th October, https://www.theguardian.com/lifeandstyle/2021/oct/11/lego-to-remove-gender-bias-after-survey-shows-impact-on-children-stereotypes
Tire Business. (2016). Study: Women vehicle owners determining aftermarket strategy, https://www.tirebusiness.com/article/20160610/NEWS/160619997/study-women-vehicle-owners-determining-aftermarket-strategy
Tynan, A. C., & Drayton, J. (1987). Market segmentation. *Journal of Marketing Management, 2*(3), 301–335. https://doi.org/10.1080/0267257X.1987.9964020
United Nations. (2023). Global Sustainable Development Report (GSDR) 2023, https://sdgs.un.org/gsdr/gsdr2023
Vargo, S. L., & Lusch, R. F. (2004). Evolving to a NEW DOMINANT LOGIC FOR MARKETING. *Journal of Marketing, 68*, 1–17.
Veryzer, R. W., & Borja de Mozota, B. (2005). The impact of user-oriented design on new product development: An examination of fundamental relationships. *Journal of Product Innovation Management, 22*(2), 128e143.
Vredenburg, K., Mao, J., Smith, P., & Carey, T. (2002). A survey of user-centered design in practice. In *Proceedings of the conference on human factors in computing systems processing (CHI)* (pp. 471–478). ACM Press.
Vullings, R., & Heleven, M. (2015). *Not invented here: The key strategies for cross-industry innovation*. BIS Publishers.
Wagner, J. (1979). *Images of information*. Sage.
Warren, S. (2005). Photography and voice in critical qualitative management research. *Accounting, Auditing & Accountability Journal, 18*(6), 861–882.
Yan, R., & Cao, Z. (2017). Is brand alliance always beneficial to firms. *Journal of Retailing and Consumer Services, 34*, 193–200.

Monika Hajdas is an associate professor at the Wrocław University of Economics and Business. She also holds a postgraduate degree in Trendwatching & Future Studies as well as Design Management. She is a certified academic tutor and Design Thinking facilitator. Her research interests include innovation management and brand management. She has published her work in *Psychology & Marketing, Journal of Brand Management, Journal of Marketing Theory and Practice, Journal of Retailing and Consumer Services, Journal of Family Business Strategy,* and

International Journal of Management Education, among others. Monika combines scientific work with business practice. For the last 20 years she has advised in and facilitated dozens of projects for global and local firms from various industries: finance, pharmaceuticals, retail, real estate, FMCG, education, HoReCa, luxury goods, interior design, graphic design, building materials, oil industry, local government.

The Relevance and Perception of African Business Potentials: German Perspectives and Strategies on Internationalization to Africa

Tilo Halaszovich and Erik Deitersen

Abstract

Africa has been labelled the next big market for many years. From a German perspective, these expectations are not yet met by the activities of German firms in many industries. The German engagement in Africa is, at best, limited so far. Nevertheless, the growing African automotive aftermarket is becoming increasingly important for German suppliers, due to the increasing demand for spare parts and the need to find markets where internal combustion engines are still in high demand for the next decade.

Against this background, this chapter sheds light on the hurdles and strategies of firms doing business in Africa. From academic and practical perspectives, the approaches of German firms are analysed and evaluated. The findings and cases consistently highlight the relevance of local sales partners and their specific knowledge in successfully addressing the African automotive aftermarket.

1 Introduction

For many years, Africa has been discussed as market and source of supplies for German companies. Likewise, the German government, most noteworthy the former chancellor Dr. Angela Merkel, has been pushing Africa on the political and economic agenda. Nevertheless, only slightly less than 1000 German companies have invested

T. Halaszovich (✉)
ICN Business School, Nancy, France
e-mail: tilo.halaszovich@icn-artem.com

E. Deitersen
africon GmbH Management Consultancy, Bremen, Germany
e-mail: info@africon.de

M. Elo, F. Katsardis (eds.), *Automotive Aftermarket*, Management for Professionals, https://doi.org/10.1007/978-3-031-62419-3_13

in Africa. Focusing on Sub-Sahara Africa (SSA), this figure is even less: 220 German companies are engaged in foreign direct investment (FDI) in the region. With more than 1000 German companies involved in mostly export activities, trade is more common from Germany towards SSA. But even regarding trading activities, Africa as a whole represents less than 2% of all German exports according to the German Statistisches Bundesamt. Trade and FDI activities, thereby, are mostly centred around mechanical engineering, construction, services, and the automotive sector (Halaszovich & Mattfeld, 2020). More than 70% of all German subsidiaries in SSA are tasked to handle marketing and sales, followed by (technical) services (33%). Unlike FDI in other regions, especially in developing Asia, production activities are only present in slightly more than 10% of German subsidiaries. Hence, doing business in Africa is considered a sales and export business, while the region is not considered as a sourcing location. A finding that resonates with the need for context-specific approaches towards Africa, both in business and in research (Zoogah, 2008).

From these figures, it becomes clear that Africa's reputation in Germany as an economically viable region is contested. The reasons for this perception are manifold and range from stereotyping as continent of political unrest and economic weaknesses to more nuanced perceptions of difficulties with institutional factors, such as corruption (Mattfeld & Halaszovich, 2022). Interestingly, many firms active in Africa report that—even with their experience—doing business in Africa is difficult and not manageable for every firm (Halaszovich, 2019). To overcome issues faced in the region, almost all firms report that access to local networks, i.e. having reliable local partners, is a key prerequisite (Mattfeld et al., 2022). In line with this, prior cooperation and contacts are even named as the main reason for country selection in SSA by German managers (Halaszovich, 2019).

Based on recent studies of German firms in SSA (Halaszovich, 2019; Halaszovich & Mattfeld, 2020), the following patterns can be identified:

1. Country selection: German firms centre their activities in a rather small set of leading countries in the region. South Africa, thereby, is by far the most attractive country (about 35% of all activities). Beyond South Africa, Nigeria and Kenya are prime destinations (each with about 7% of all activities).
2. Country portfolios: Only about 20% of all German firms are active in just one country. More than half of the firms operate in two to five countries. The remaining firms even operate in more than five countries.
3. Motives for engagement: The main motives for engagement in SSA by German firms are centred around market access and increasing sales. Sourcing is named as the least important motive. In line with these motives, local subsidiaries are mostly tasked with marketing and sales (about 70% of all subsidiaries) and service (about 30% of all subsidiaries). Only about 10% have production facilities.

Taken together, these patterns reveal that from a German perspective, an engagement in Africa requires a lead market strategy, supported by a portfolio of selected second-tier countries. Hence, individual countries in the region are not perceived as viable stand-alone markets. Moreover, low-wage production or sourcing of natural resources—dominant motives for western firms in many developing regions of the world—do not drive German international business activities towards Africa.

Looking at the top-5 industries, German firms in SSA are active in mechanical engineering (39%), services including logistics and consulting (13%), construction (10%), chemicals (8%), and automotive and automotive aftermarket (7%) (Halaszovich & Mattfeld, 2020).

The purpose of this chapter is to discuss German business actors in the automotive aftermarket and their approaches to Africa. We analyse the relevance of local partners and networks in serving fragmented markets in SSA, both from an academic perspective and real-world cases. Thereby, we aim to answer the following research questions:

1. What is role of local partners and networks in the operations of German suppliers in the automotive aftermarket in Africa?
2. Which knowledge is essential to operate successfully in these markets? And which side, i.e. German company or local partner, should provide which kind of know-how?

Currently, doing business in Africa—and especially an inclusion of the local side in these transactions—is an emerging and still under-researched topic. With our contribution we try to highlight its relevance and stimulate more focus on the underlying mechanisms.

The remainder of the chapter is structured as follows: First, we outline the structure of the African automotive aftermarket. Next, strategies of German companies in this market are discussed. Thereafter, we refer to the International Business theory to outline the specific risks and obstacles of foreign firms doing business in developing countries. An illustrative case highlights the approach and issues of a German automotive aftermarket supplier in Africa. Before we conclude with a discussion of our findings, we discuss the results of a recent report on the German "Mittelstand" in Africa.

2 The African Automotive Aftermarket

Unlike the stable activities in most industries, Africa has become increasingly more relevant for German automotive aftermarket suppliers over the last two decades. This is mainly driven by rapid population growth, a growing middle class, and an increasing number of vehicles on the road—with significant room for further growth. Additionally, suppliers tied to the internal combustion engine (ICE) are exploring markets in which sales of their products will likely still grow over the upcoming 1–2 decades. Overall, Africa is often perceived as one of the last areas globally where significant growth is possible for German aftermarket suppliers.

The automotive market in Africa, which is by definition the main driver for the aftermarket, is characterized by 50–60 million vehicles (excl. motorcycles) on the road. The three largest countries in this regard are Nigeria (11.8 million vehicles), South Africa (11.4 million vehicles), and Egypt (6.1 million vehicles). Besides the number of vehicles, is the vehicle age of particular interest for the automotive aftermarket, with a higher age increasing the need for maintenance and part replacements. In North Africa, vehicles are about 12–14 years old. In southern Africa, the vehicle age can be even higher (17 years in Nigeria, 15–16 years in Ghana or Ivory Coast). While the number of cars and their age make SSA an interesting market from a European perspective, the automotive market in this region is heavily dominated by Asian vehicle brands, especially by Toyota. More than 50% of all light vehicles in SSA are Toyotas. In Tanzania and Ethiopia the share is even 70%. Compared to SSA, European car brands are stronger in the Maghreb region: 30% Dacia in Morocco, 25% Renault in Algeria. Focusing on the automotive aftermarket in SSA, the market has grown by 7% per year between 2015 and 2021, with the leading suppliers coming from China, the UAE (as a trading hub), or the USA (mostly with second-hand parts).

From a German perspective, the automotive aftermarket is dominated by a small number of large multinational enterprises (MNEs), such as Bosch, Schaeffler, or ZF, who are dominant in serving the aftermarket as suppliers. Small and medium-sized enterprises (SMEs) do exist in the German system but have only limited market power and are often contractually bound to German carmakers. Serving highly fragmented overseas markets, such as SSA, is therefore less common among German SMEs and mostly carried out by MNEs.

3 Strategic Approaches of German Suppliers

Strategic approaches of German suppliers to address the African automotive aftermarket are basically centred around two schemes: Establishing local production or export business models.

Some German companies produce components in South Africa and some North African countries. These factories supply to local vehicle factories, as well as to global (after)markets. Especially in the latter case, global suppliers make use of affordable labour costs in Northern Africa, to produce products that are relatively manually produced.

The much larger number of German aftermarket suppliers however focuses purely on exports into the African region. They heavily rely on networks of external distributors. Own subsidiaries, especially outside South Africa, are very rare—Bosch's various entities in Africa being a notable exception.

These networks of distributors consist of two principal types of distributors: the first type are local country importers that buy parts and sell them in their respective country/countries of operation. The second type are regional distributors, which sell parts across a range of countries. Quite often, these regional distributors are export houses based in Dubai or Europe. They typically buy large quantities of parts from various global suppliers, consolidate them in a central location and then sell smaller shipments to African (and other) customers. Hybrids and variations of both types of companies exist.

German suppliers commonly rely on a mix of both regional distributors and country importers. Larger markets in North and South Africa are more commonly served through dedicated country importers. In many cases, several such importers are appointed per country. While relative to regional distributors, these country importers tend to represent smaller accounts with higher risks of payment defaults, the more direct contact to larger markets is often perceived to justify the additional efforts and risks—among other factors. Regional distributors on the other hand are more commonly relied upon for smaller markets in the Sub-Sahara region. Here, German suppliers often face challenges related to the indirect contact to the final target markets (e.g. information relating to customer preferences, market insights, successes and failures are not reliably passed on from the regional distributors to the suppliers). Ultimately, country importers typically allow for a more dedicated development of a market for a supplier, which however must be justifiable by higher obtainable results.

Taken together, serving the African automotive aftermarket is of strategic importance to German suppliers, yet they are facing a trade-off between accessibility to the markets via dedicated distributors and the risk and obstacles associated with establishing and maintaining these local connections.

4 Foreignness and Local Connections: A Theory Perspective with Explanatory Power

The general activities of German firms in Africa as well as the activities within the automotive aftermarket highlighted the importance of local connections from the practitioners' point of view. This raises the question why local connections even outweigh economic considerations for German firms' activities in Africa.

Doing business in Africa is characterized by numerous opportunities and also challenges for companies from developed countries. From a scientific perspective, three dominant themes can be identified (Halaszovich, 2020; Halaszovich & Lundan, 2016):

(1) Competitiveness of Foreign Firms

Companies from developed countries are generally perceived to be more competitive compared to their local competitors. This competitive edge is based on the knowledge, skills, and resources MNEs usually control. As dominant actors in their markets, these firms are believed to outperform local competitors in areas related to their products and services, e.g. by offering innovative and high-quality products, by leveraging superior managerial and marketing skills as well as by having access to international networks and resources.

While many of these assumptions can indeed explain the global success of large multinational companies in the automotive industry, such as Bosch or Schaeffler, it remains questionable if the same logic applies to the automotive aftermarket in Africa.

(2) Challenging Institutions and Institutional Voids

It is a well-established concept in the international business literature that firms doing business abroad are affected by the differences between the situation in their home country and their host countries. These so-called distances affect the ability of a firm to operate in a host country in multiple ways. Of particular interest with regard to developing countries, the concept of institutional distance has been discussed in the literature (c.f. Kostova et al., 2020). Institutional distance describes the difference between the formal institutions and the informal institutions between the home and host countries. Increasing institutional distance increases the uncertainty and risk of doing business in the host country. Empirical research over the last two decades revealed that this kind of distance indeed negatively affects internationally active firms. These consequences range from a negative impact on the performance of the firm, especially of its operations in the host country, to the entry mode choice where increasing distance leads to a reduction of ownership shares, and the location choice as such. Regarding the latter, firms actively avoid countries with high institutional distance and prefer locations with a lower distance.

In addition to the differences between the institutions in the home and host countries, developing countries are often characterized by flawed institutions. These flaws are routed in missing or untransparent regulations, which in turn leave room for informal regulations. These informal rules can often involve bribery or corruption and create discriminatory decisions, such as favouring second-best businesses in a transaction. In the international business literature, these phenomena are called institutional voids. More generally, a country with high institutional voids lacks transparent and fair market conditions. Informal markets are strong but cannot be accessed by external parties.

All of these conditions obviously are also affecting the German engagement in Africa, either indirectly or directly. The perception of the institutional distance between Germany and many of the African markets might be one of the core reasons why only very few German companies engage in this region at all, as their performance expectations do not justify these activities proactively. Those firms that are active in Africa clearly prefer export over FDI, which is in line with a low commitment regarding the entry mode. And finally, the very selective location choice observed among German firms in Africa reflects the effect of institutional distance and voids.

(3) Liabilities of Foreignness

Companies venturing abroad are generally affected by additional burdens, often summarized as liabilities of foreignness. "The concept of liabilities of foreignness (LOFs) describes the additional costs that multinational enterprises have to face relative to their indigenous competitors when operating in foreign markets." (Denk et al., 2012, p. 322). Researchers identified three sources, the so-called hazards, driving LOFs.

1. Unfamiliarity hazard: As outlined above, firms doing business abroad face a different and often unfamiliar environment in the host country. A key challenge is to obtain information from and about the host market as well as the firm's ability to interpret this information correctly. If either the information is biased or incomplete or the firm lacks the ability to process the information, incorrect assessments can occur (Caves, 1971; Eden & Miller, 2004; Elango, 2009; Petersen & Pedersen, 2002).
2. Relational hazard: Foreign firms are new and rather external participants in the host country. As such, they lack established network and have only limited knowledge about and access to local networks. As outsiders they may also face a lack of trust by local business partners (Eden & Miller, 2004; Ring et al., 1992). This phenomenon is also described as liability of outsidership (LOS) by Johanson and Vahlne (2009). The lack of relational networks results in higher internal and external transaction costs for foreign firms (Caves, 1971).
3. Discrimination hazard: As outsiders, foreign firms are often perceived as threads to the host country's environment or might be negatively attributed. Therefore, foreign firms may suffer from discrimination by local stakeholders, e.g. the government, consumers, or competitors (Balabanis & Diamantopoulos, 2004; Henisz & Williamson, 1999).

While LOFs are a key factor for international business failure, little is known how MNEs can potentially overcome these issues (Halaszovich, 2020). Reviewing the international business literature, experience can be identified as key to overcome LOFs. Yet, gaining experience takes time to learn and requires the organizational will to accept failures in the process. The authors of this chapter, therefore, tried to quantify the relevance of local partners, i.e. insiders from the host countries' institutions and voids, for the performance of foreign firms in developing countries (Halaszovich, 2019; Halaszovich, 2020; Halaszovich & Lundan, 2016; Halaszovich & Mattfeld, 2020). In these studies and business reports, a consistent picture emerged: Those foreign firms that rely on local partners outperform foreign firms that rely on their own capabilities or international networks. Establishing and managing local connections plays a key role in doing business in developing countries.

In the next chapter, we dive into more details by outlining the findings from a report on the German "Mittelstand" in Africa (Halaszovich, 2019).

Case—The Trade-Off Between Country Importers and Regional Distributors

The following example can visualize this trade-off between country importers and regional distributors: Around the year 2020 a large European supplier changed its distribution strategy in East Africa from several smaller country distributors to one larger regional distributor. As a result, the supplier suffered a significant and relatively immediate decline in absolute sales in the region.

On the other hand, the previously painful payment delays were reduced. It subsequently took the supplier several years to return to the sales levels obtained prior to the adjustment—while local markets had meanwhile seen additional growth, that the supplier was not able to fully participate in.

Independently of regional distributors and country importers, the mode of cooperation between suppliers and distributors is relatively similar. The distributor and the supplier often jointly agree on a sales strategy, which will include target market segments, exact products to sell, price positioning and other factors. The distributor buys stock from the supplier, before marketing and selling such to end-users and other buyers (e.g. smaller retailers, workshops). The supplier supports these efforts through marketing funds or activities, incentive programs, training, and other initiatives. Supplier staff, mostly based in Europe or Dubai, visits the distributors (in regular intervals) to monitor and support activities, as well as to develop the overall relationship.

Hurdles and challenges in this process generally come from *internal* and *external* directions.

Internal challenges include factors under the direct control of the supplier and/or the distributor. A common example is the limited attention that German suppliers place on African markets, which commonly results in limited sales to the region. In some instances, stereotypes and a lack of information about the African continent prevent German suppliers from launching any direct activities in the region. In other cases, sales managers that are primarily responsible for other regions such as Southern Europe, are awarded the additional task to manage African markets. Due to the geographical distance, cultural barriers, as well as often smaller market sizes, these sales managers often place a stronger priority on their original/other regional responsibilities. Business dealings in Africa often take a longer time to succeed, sometimes causing resignation on the side of German suppliers. Another related internal problem is the lack of support in the form of, for example, marketing, visits, training, or product pricing provided by the supplier to local distributors.

External challenges include price-sensitive buyers, especially in Sub-Saharan countries. With many old vehicles in these countries and relatively low-income levels, many vehicle owners are not willing or able to afford good quality spare parts—such as most German suppliers are offering. Lower-cost parts from Asian producers are often the preferred choice, with counterfeits of "Western" brands often creating additional hurdles. Other external challenges are a lack of skills and formal education of mechanics in many countries, as well as fragmented market structures with few larger formal entities. Poor fuel quality poses technical challenges to some suppliers of ICE-related parts. Many German suppliers suffer from "boom and bust" market cycles, which are relatively common in many emerging markets and create planning challenges for German suppliers.

5 The German "Mittelstand" in Africa—Dos and Don'ts in Local Network Connections

The German "Mittelstand" is a globally well-known and rather unique group of companies in Germany that share features of SMEs and MNEs alike, one might refer to them as small, sometimes even mini MNEs. Some of these companies are market leaders in their niche market and many are routed in an engineering tradition, leveraging the reputation of "Made in Germany" in their international markets. The "Mittelstand", thereby, can be regarded as the backbone of the German economy, while many of these firms are often not well known. This featured earned them the nickname "Hidden Champions"—leading but unknown companies. Due to their economic relevance, innovation capabilities, and potential for job creation, the German GIZ (Gesellschaft für Internationale Zusammenarbeit mbH) tasked one of authors of this chapter in 2019 to study their activities in SSA (Halaszovich, 2019). The main aim of the study was to identify patterns in their approach to the market. From this study, several insights were gained—highlighting the relevance of local partners and mutual knowledge exchange between German and African firms.

5.1 Sources of Knowledge

In line with the self-perception of engineering driven companies, the majority of German firms in the study reported that technical knowledge should come from the German side and that this know-how is essential for successful activities in SSA. The picture became more nuanced when we asked the German firms about knowledge related to marketing and sales. Here, the firms favoured local knowledge over their German home country experiences, even though marketing knowledge from Germany was still considered an important asset (see Fig. 1).

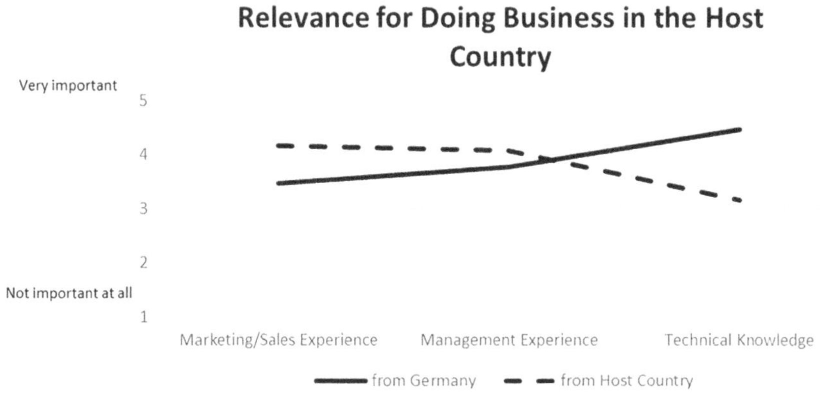

Fig. 1 Relevance of knowledge (*source*: Halaszovich, 2019)

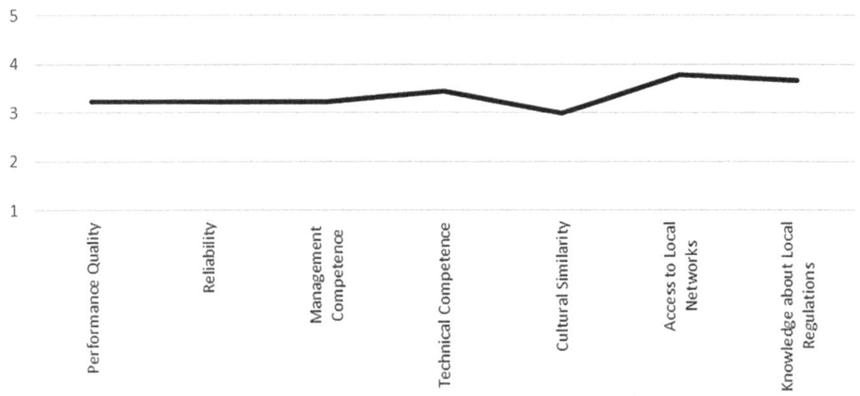

Fig. 2 Evaluation of local sales partners (*source*: Halaszovich, 2019)

In addition to this rather rough differentiation of knowledge categories, the firms have been asked to evaluate their assessment of capabilities of local partner firms from SSA. A value of 1 indicates very weak capabilities, 5 indicates very strong capabilities.

As can be seen in Fig. 2, local sales partners are perceived as rather competent companies in all categories. But most noteworthy, these local companies excel in their access to local networks and knowledge on local regulations. These capabilities perfectly match with the discussion on LOFs and the hurdles of foreign companies doing business in developing countries. It is therefore not surprising that the study also found statistically supported evidence that those German firms that collaborate closely with local sales partners outperform those who don't do it or engage in only loose collaborations.

A pattern that was also observed among the German firms was that larger companies work more closely with international sales partners, comparable to companies located at the Dubai hub as outlined in the case study above. Here, the study found a positive effect on firm performance only in very early growth stages of a market entry. In the long run, when companies try to establish themselves in the market, international sales partners turned out to negatively affect performance.

Taken together, it is the particular knowledge about local rules and regulations and their enhanced ability to access local business networks that make local sales partners valuable collaborators for German firms. An advantage that international sales partners cannot sustain beyond an early growth period.

5.2 Transfer of Knowledge

If technical knowledge from Germany and market knowledge from SSA are the two cornerstones of successful activities in the region, one would expect that each and every partnership includes tailor-made processes for know-how exchange, facilitating mutual learning processes of all parties involved. Surprisingly this is not the

Fig. 3 Knowledge exchange between partners (*source*: Halaszovich, 2019)

case. From all firms in the study, only 68% actively provide knowledge to their local business partners. Almost one third of the German firms do not share any their know-how with their local partners. Slightly more German firms, 73%, try to learn from their local partners. Taken together, more than 25% of the German firms active in SSA do not engage in any knowledge exchange—despite the relevance of these exchange processes for successful operations in SSA.

Among those firms with exchange processes, the study measured the extent of transfer in different categories towards and from the local partners, i.e. technical knowledge, quality management, marketing and sales, management, and cultural knowledge. Figure 3 shows the share of firms engaged in the different categories.

It is not surprising that literally all German firms, that are active in knowledge exchange, provide technical knowledge trainings for their local partners and more than 50% also provide trainings on quality management. These activities underline the fact that the main competitive advantage of the German "Mittelstand" is based on their engineering capabilities. More surprising is the fact that German firms also favour technical knowledge and quality management in their learning processes from local partners. This pattern contradicts the assessment of the importance of local technical know-how as depicted in Fig. 1. One might assume that this discrepancy stems from the experience of German firms, i.e. realizing that local technological know-how can help to adjust the products to local needs. Yet, this conclusion is rather unlikely as it has been the very same companies explaining that local technological know-how is not important that here reported their knowledge transfer activities. Therefore, this pattern uncovers a certain myopic approach of the German "Mittelstand": Companies driven by engineering feel most comfortable when talking about technology—even when the benefits are fairly limited. This is further stressed by the fact that only 50% of the German firms actively ask for local marketing and sales knowledge (the most important kind of knowledge according to Fig. 1) and even less than 50% try to enhance their cultural knowledge.

As expected from theory, the study found strong statistical support for the relevance of local marketing and sales know-how for the performance of German firms in SSA. In early stages of market cultivation, gaining cultural insights from local partners also showed a strong positive effect. Yet, this effect seems to diminish over time. Interestingly, the effect of providing technical knowledge to local partners—the one activity all German firms active in knowledge exchange were engaged in—was found to be effective only in early stages of market cultivation but not in the long run.

Summarizing the study on the German "Mittelstand" in Africa, it can be concluded that local sales partners are instrumental to the success of foreign firms in SSA. Their specific profile allows these local firms to access networks and leverage their knowledge on local markets and regulations. To benefit fully from local partners, foreign firms also need to invest in mutual knowledge exchange programs. A high risk, especially for engineering-based companies—like many companies in the automotive aftermarket, is to focus too strongly on technological knowledge. While the people involved in the exchange processes might feel more comfortable when discussing technical specifications, the exchange programs should focus strongly on aspects of local marketing and sales as well as cultural differences.

6 Discussion and Conclusion

When entering the African automotive aftermarket, foreign companies have many strategic options to choose from. Simplified, they have to decide on which country or set of countries to enter, and how they want to enter the market. International Business theory suggests that entering rather unstable markets with weak institutions should be done by means of exporting rather than foreign direct investments. A pattern that was clearly approved by our studies on the German "Mittelstand" and the AA case study.

Exporting, thereby, can take different forms, such as direct exporting with or without a local sales partner or indirect exporting via international sales partners. From our empirical insights, direct exporting with local sales partners and indirect exporting with international sales partners are the dominant forms for German firms, including suppliers in the AA market. Our findings illustrate that, despite higher burdens, direct exporting with local sales partners is the most promising approach. This is especially true if a foreign firm wants to benefit from local market growth as in the case of the German AA supplier who failed to exploit these local dynamics with an international sales partner from Dubai.

And finally, the relevance of mutual knowledge exchange in the right domains, i.e. favouring market knowledge over technical knowledge, became clearly visible. Those foreign firms that acknowledge local knowledge in this regard are able to overcome institutional voids and leverage their superior quality even in difficult markets such as the African AA. These firms dedicate resources to the region and move closer to local customers, to ensure a holistically competitive offering. They decide where to place the right priorities, which distribution partners are needed to develop markets and what kind of support these need to succeed. These companies consequently record growing sales figures in this region with significant future growth potential.

In the rather near future, the emerging middle class in Africa will create new opportunities and challenges for firms to establish themselves in the service market. While this process might align the African market with global standards, it is likely that it will also create further needs for an "African way" to serve the needs of these demanding customers. Foreign and local firms will be challenged to adjust not only their products but also their go-to-market approach accordingly.

The African diaspora and their potential as bridge builders between the markets (Mattfeld & Halaszovich, 2022) is increasingly becoming a relevant topic in research and practice (Bodomo, 2010; Rana & Elo, 2017; Riddle et al., 2008). While some members of the diaspora are already in their second or third generation in Europe, an ever-increasing number of immigrants from the African continent try to establish themselves in Europe. And we are just at the beginning to understand and leverage the potential of these people for the benefits of companies and societies alike (Kothari et al., 2022).

References

Balabanis, G., & Diamantopoulos, A. (2004). Domestic Country bias, Country-of-origin effects, and consumer ethnocentrism: A multidimensional unfolding approach. *Journal of the Academy of Marketing Science, 32*(1), 80–95. https://doi.org/10.1177/0092070303257644

Bodomo, A. (2010). The African trading community in Guangzhou: An emerging bridge for Africa–China relations. *The China Quarterly, 203*, 693–707.

Caves, R. E. (1971). International corporations: The industrial economics of Foreign Investment. *Economica, 38*(149), 1–27. https://doi.org/10.2307/2551748

Denk, N., Kaufmann, L., & Roesch, J.-F. (2012). Liabilities of foreignness revisited: A review of contemporary studies and recommendations for future research. *Journal of International Management, 18*(4), 322–334. http://linkinghub.elsevier.com/retrieve/pii/S1075425312000543

Eden, L., & Miller, S. R. (2004). Distance matters: Liability of foreignness, institutional distance and ownership strategy. *Advances in International Management, 16*, 187–221. https://doi.org/10.1016/S0747-7929(04)16010-1

Elango, B. (2009). Minimizing effects of "liability of foreignness": Response strategies of foreign firms in the United States. *Journal of World Business, 44*(1), 51–62. https://doi.org/10.1016/j.jwb.2008.03.012

Halaszovich, T. F. (2019). Deutscher Mittelstand in Afrika - Motivation, Ausgestaltung und Einbettung deutscher Investitionen in afrikanischen Zielländern.

Halaszovich, T. F. (2020). When foreignness becomes a liability: the effects of flawed institutional environments on foreign versus domestic firm performance in emerging markets. *European Journal of International Management, 14*(1), 118–143. https://doi.org/10.1504/EJIM.2020.103803

Halaszovich, T. F., & Lundan, S. M. (2016). The moderating role of local embeddedness on the performance of foreign and domestic firms in emerging markets. *International Business Review, 25*(5), 1136–1148. https://doi.org/10.1016/j.ibusrev.2016.02.003

Halaszovich, T. F., & Mattfeld, S. (2020). Impact of Covid-19 on export and investment activities of German companies in Sub-Saharan Africa.

Mattfeld, S. & Halaszovich, T. F. (2022). Die Rolle der afrikanischen Diaspora bei der Internationalisierung deutscher Unternehmen in Afrika. Study for the Deutsche Afrika Stiftung.

Henisz, W., & Williamson, O. E. (1999). Comparative economic organization: Within and between countries. In Business and Politics (Vol. 1, Issue 3, pp. 261–277).

Johanson, J., & Vahlne, J.-E. (2009). The Uppsala internationalization process model revisited: From liability of foreignness to liability of outsidership. *Journal of International Business Studies, 40*(9), 1411–1431. https://doi.org/10.1057/jibs.2009.24

Kostova, T., Beugelsdijk, S., Scott, W. R., Kunst, V. E., Chua, C. H., & van Essen, M. (2020). The construct of institutional distance through the lens of different institutional perspectives: Review, analysis, and recommendations. *Journal of International Business Studies, 51*(4), 467–497. https://doi.org/10.1057/s41267-019-00294-w

Kothari, T., Elo, M., & Wiese, N. (2022). Born as a citizen and Reborn as an Alien: Migrant superdiversity in global business. *AIB Insights, 22*(3), 1–6.

Mattfeld, S., Piaskowska, D., & Halaszovich, T. F. (2022). Digital outsidership: The practical and affective effects of digitalization in international business relationships between Kenyan and German firms. In D. Dikova & E. Ipsmiller (Eds.), *Research handbook on innovation in international business* (pp. 177–205). Edward Elgar Publishing.

Petersen, B., & Pedersen, T. (2002). Coping with liability of foreignness: Different learning engagements of entrant firms. *Journal of International Management, 8*(3), 339–350. https://doi.org/10.1016/S1075-4253(02)00068-6

Rana, M. B., & Elo, M. (2017). Transnational diaspora and civil society actors driving MNE internationalisation: the case of Grameenphone in Bangladesh. *Journal of International Management, 23*(1), 87–106.

Riddle, L., Brinkerhoff, J. M., & Nielsen, T. M. (2008). Partnering to beckon them home: Public-sector innovation for diaspora foreign investment promotion. *Public Administration and Development: The International Journal of Management Research and Practice, 28*(1), 54–66.

Ring, P. S., Van De Ven, A. H., Smith, P., & Van De Ven, A. H. (1992). Structuring cooperative relationships between organizations. *Strategic Management Journal, 13*(7), 483–498. https://doi.org/10.2307/2486599

Zoogah, D. B. (2008). African business research: A review of studies published in the Journal of African Business and a framework for enhancing future studies. *Journal of African Business, 9*(1), 219–255.

Tilo Halaszovich is Associate Professor of International Marketing at the ICN Business School, Nancy (France). Tilo is a senior editor of the European Journal of International Management, a member of the Editorial Review Board of the Management International Review journal and serves on the board of the European International Business Academy. His research focuses on competitiveness of firms in developing countries, institutional distance, and the role of technology and entrepreneurship in international business. He has published in leading journals, including Journal of Business Research, International Business Review and European Journal of International Management.

Erik Deitersen is a Director with africon GmbH. Since 2014, he has been working with companies from around the world to assess markets and devise growth strategies in Africa. As the head of africon's automotive practice, he has worked on and accompanied more than 30 automotive projects in the region.

Part IV

Perspectives on Innovation, Business Ecosystems and Global Challenges

Innovation Capabilities in the Automotive Aftermarket: Case Study from Automechanika Frankfurt 2022

Anders Leth Villumsen and Claire Croucher Petersen

Abstract

The automotive sector is currently undergoing a remarkable transformation, shifting away from traditional individual mobility powered by combustion engines to more complex and interconnected systems. In this evolving sector, the automotive aftermarket has an essential role in providing services to maintain the global vehicle fleet; these encompass repairs, optimisation, and enhanced user experiences. By extending the longevity of vehicles, the aftermarket plays a crucial role in shaping the sector. However, current scientific discussions of innovation are predominantly centred around products developed and produced by vehicle manufacturers, disregarding this role. This chapter aims to fill this gap by exploring the innovation capabilities within the independent aftermarket using a case study of five innovations showcased at the Automechanika Frankfurt 2022 trade fair. Our findings reveal that the aftermarket exhibits diverse degrees of innovation capabilities across a wide range of product offerings. Furthermore, this chapter demonstrates how the current innovations focus on making the whole mobility sector more environmentally sustainable for future generations.

1 Introduction

Mobility and access to goods, services, and social networks rely strongly on transportation. The sector enables the economy to function, people to go to work, attend school, and advance in their life (ITF, 2021), involving a wide range of actors, from those who design, develop, and manufacture vehicles to those who maintain, repair,

A. L. Villumsen (✉) · C. C. Petersen
Department of Technology and Innovation, University of Southern Denmark, Odense M, Denmark
e-mail: Avill18@student.sdu.dk; Clvas18@student.sdu.dk

M. Elo, F. Katsardis (eds.), *Automotive Aftermarket*, Management for Professionals, https://doi.org/10.1007/978-3-031-62419-3_14

and offer after-sale services and parts. Automotive mobility is undergoing multiple transformations and innovations. Fordism and Toyotaism (Monden, 2011; Ohno, 1988; Rubenstein, 2001) are just two examples of disruptive process innovations that are commonly mentioned as examples from this sector and used in teaching and other sectors. Consequently, when one thinks of innovation in the context of the automotive sector, the first things that come to mind are these disruptive process innovations or incremental innovations performed on commodities by vehicle manufacturers.

This one-sidedness also exists in academic research literature, where most contemporary research is predominantly focused on inventions created by vehicle manufacturers rather than the automotive aftermarket (AAM). The AAM may be described by the transactions taking place in an aftermarket, which contain two distinct characteristics: They are used in conjunction with a primary product and purchased after the primary product (Shapiro & Teece, 1994, p. 139). Thus, the AAM is the sector that services and maintains the vehicles after they have been manufactured and sold, encompassing all actors in the value chain from parts producers to digital service providers.

The service side of the sector may be divided into two distinctive types: dealer networks and independent non-dealer networks, as depicted in Fig. 1. It is the original equipment manufacturer (OEM), i.e., the organisation and value chain that manufactured the original vehicle, that forms the dealer networks. Whereas other independent actors in the independent aftermarket (IAA), e.g., spare parts manufacturers, international trading groups, and independent workshops, form diverse non-dealer networks (Kempf et al., 2018; Dombrowski & Engel, 2014). Bosch car service-network is an example of the latter (Wang & Kimble, 2016). Please note that Fig. 1 provides a simplified overview of the sector, and in reality, there are additional actors involved and the structure more complex.

The primary purpose of this chapter is to address the nature of innovation capabilities in the automotive aftermarket sector and how these capabilities can be

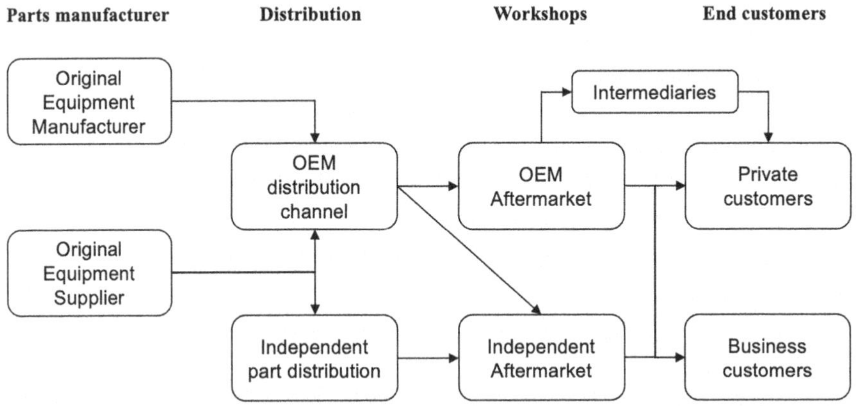

Fig. 1 Automotive aftermarket overview (adapted from Dombrowski & Engel, 2014)

utilised to ensure a sustainable future for future generations. Innovation capabilities are defined by Aas & Breunig (Aas & Breuning, 2017, p. 8) as "a firm's ability to identify new ideas and transform them into new or improved products, services, or processes that benefit the firm". We focus on selected independent aftermarket actors to examine their capabilities and use a case study approach to explore their innovations presented at Automechanika Frankfurt 2022. Our contribution to the current literature lies in filling the research gap on innovation capabilities within this sector. In addition, the cases shed light on the role of AAM in generating innovations and communicating them. We seek to answer the three following research questions:

- How are innovation capabilities expressed and communicated in the AAM?
- What is emphasised when presenting key innovations?
- How do innovation capabilities affect the development of products in the mobility sector?

This chapter begins by outlining various theories of innovation that will serve as the theoretical framework for investigating the innovation capabilities in the automotive aftermarket. It then delves into some broad perspectives on automotive innovation and explains the significance of the findings. Following this, a case study of five selected items illustrating current innovation development in the sector is presented, and lastly, implications and future recommendations are discussed.

2 Innovation

There are multiple ways to define innovation, but it is commonly considered to consist of two key components: something new, original, or improved, and something that generates and captures value (Dance, 2008). These terms are, however, broad and may cover a wide range of products, processes, and organisational structures. The OECD offers a more defined definition and differentiates innovation between the two following constructs:

- Product innovation: a new or improved good or service that differs significantly from the firm's previous goods or services and that has been introduced on the market (OECD, 2018, p. 70).
- Process innovation: a new or improved business process for one or more business functions that differs significantly from the firm's previous business processes and that has been brought into use in the firm (OECD, 2018, p. 72).

Although the OECD definition remains broad, it does provide a framework for understanding the concept and its application in various contexts. With a better understanding of how innovation may be regarded, it is possible to assess it more precisely.

2.1 Degrees of Innovation

According to existing literature, innovation is a difficult construct to measure because it is multifaceted and context-dependent (Fischer, 2001; Garcia & Calantone, 2002; Tidd, 2001). As a result, the degree of novelty can be determined by the adopting unit, marketplace, sector, or industry (Lawson & Samson, 2001; Rogers, 2003). There are numerous methods for determining the degree, such as using hierarchical decision models. That evaluates the degree to which innovation is achieved by using a set of criteria such as novelty, complexity, transferability, scalability, and potential for success (Daim, 2016). Another method is to classify the innovativeness or degree of innovation based on the relation between market novelty and the technology used to create or develop the innovation (Abernathy & Clark, 1985; Henderson & Clark, 1990). This method creates a four-quadrant innovation matrix, as depicted in Fig. 2.

The four quadrants in Fig. 2 describe four unique categories:

- *Incremental or sustaining innovation* is an offering with improved performance using existing technology. Minor improvements made on existing products, such as an improved longevity of spare parts, e.g., lightbulbs or windshield wipers, are examples within the sector.
- *Disruptive innovation* occurs when a new technology is introduced into an existing market that threatens to replace or disrupt the traditional approach; in the AAM, this could be mobility as a service (MaaS), which is the concept of providing mobility on demand (MOD) services such as ride-hailing, car-sharing, and bike-sharing (Flügge, 2017).
- *Radical innovation* occurs when novel technology is introduced to customers and a new market is created. This can also be described as a paradigm shift if it occurs at the sector level (Yin et al., 2017). This could be the development of autonomous driving technology in the sector.

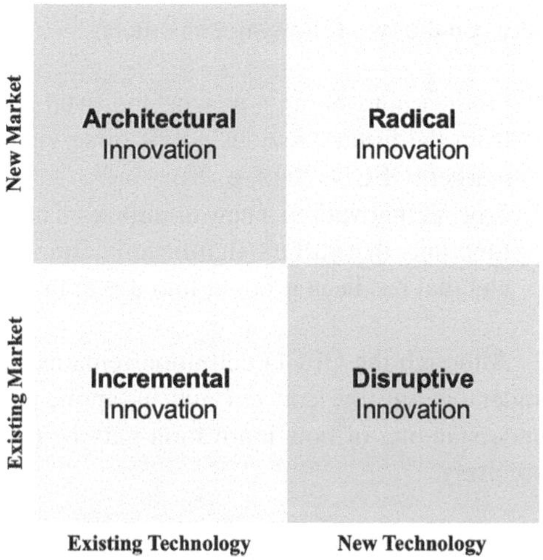

Fig. 2 Four degrees of innovation matrix (adapted from Henderson & Clark, 1990)

- *Architectural innovation* occurs when existing technology is introduced to a new market. An example in the AAM would be the introduction of digital sales channels to end customers, allowing them to buy spare parts from the manufacturer, thus excluding the repair shops.

It is possible to delve deeper into the creation and capabilities of innovation using this method of assessing the innovation outcome, or rather, the innovativeness.

2.2 Innovation Process and Capabilities

The generic innovation process can be defined as the process of transforming ideas into reality and capturing value from them (Tidd & Bessant, 2021), and this ambidexterity of exploitative and exploratory operations is required for a company to survive and prosper (Tushman & O'Reilly, 1996). Apart from the generic process, there are numerous other entrepreneurial routes, such as Bricolage (Baker & Nelson, 2005; Albert, 2019; Radjou et al., 2012), or Reverse Innovation (Govindarajan & Ramamurti, 2011), that the innovator can choose to follow.

The chosen innovation process or route is closely related to an individual firm's ability to continuously transform knowledge and ideas into new products, processes, and systems for the benefit of the firm and its stakeholders, or to maintain or increase market share as defined by Lawson and Samson (2001, p. 384). These functions are closely related to the firm's resources, which may be defined as assets owned by the firm, whereas capabilities describe a firm's ability to deploy and transform its resources into something beneficial (Ulaga & Reinartz, 2011).

The capabilities relate to a firm's contentious configuration, or, more accurately, reconfiguration of its internal and external competencies to succeed (Teece et al., 1997). Consequently, it ultimately refers to a firm's ability to replicate successful innovations, which is required to ensure positive future performance of the firm (Tushman & O'Reilly, 1996).

Figure 3 illustrates the dimensions of innovation novelty, market characteristics, and human capital. This can be applied to evaluate the level of innovation and innovation capabilities for companies (Aas & Breuning, 2017). The main states within the figure can be outlined as follows:

- *An incremental and static* state would necessitate the ability to use existing resources and human capital to incrementally improve products and services already delivered to existing markets or recurring customers. An example of this could be to optimise, manufacture, and sell spare parts to workshops.
- *A radical and dynamic* state would entail the ability to adapt or acquire the human and social capital required to develop new products for novel market demands. An example of this could be to make joint ventures between companies.
- *An incremental and dynamic* state would involve the ability to adapt or acquire human and social capital, to improve existing products for novel market demands.

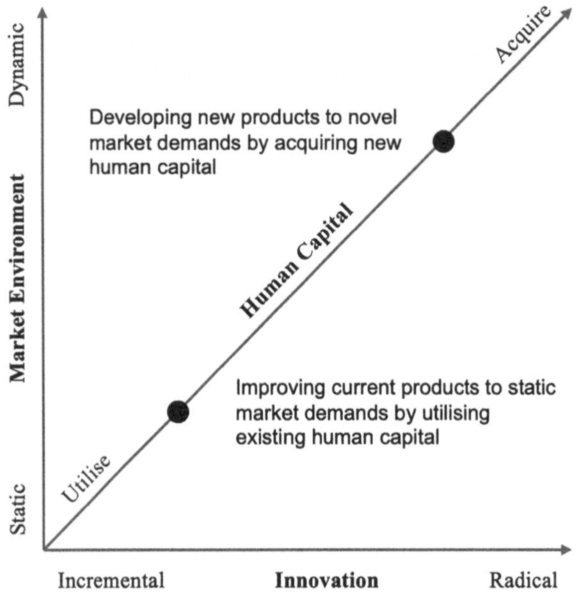

Fig. 3 Innovation capabilities (adapted from Aas & Breuning, 2017)

This could be when optimising internal organisational processes used for service offerings.

- *A radical and static* state would necessitate the ability to develop new products, services, or processes using existing resources. An example of this could be an in-house research and development department developing a new product.

3 General Elements Influencing Innovations Within the Automotive Aftermarket

Technological developments have created new mobility opportunities and improved the AAM's ability to maintain and service the current global car parc. This plays an essential component in the road towards a more sustainable impact on our planet, as outlined in the 17 Sustainable Development Goals (SDGs) developed by the United Nations (2015).

3.1 The Age of the Global Car Parc

One of the main trends in the AAM is the increasing age of the car parc. In Western Europe, vehicles are now, on average, 11.8 years old, while in Central and Eastern Europe, they are considerably older (ACEA, 2022). Older vehicles are more likely to require maintenance and replacement parts, which significantly impacts the aftermarket and the relationship between OEM and IAA supply channels. When a car is still under warranty, owners tend to take it to an authorised repair facility. However, as vehicles age, customers become inclined to shift to the IAA (BCG, 2021). Innovation is present not only during the initial product development of vehicles but also throughout their usage phase, involving services and products consumed during their lifetime.

In Western Europe, the proportion of vehicles older than eight years has increased from 50% in 2011 to 65% in 2019, and the industry anticipates that this trend will continue, with the share reaching as high as 75% by 2030 (BCG, 2021). Other contributing factors to the age are impacts from the COVID-19 pandemic and the war in Ukraine, which have decreased sales of new vehicles (Gu, 2022). However, the revenue in the largest European automotive market, Germany, has experienced an expansion between 1990 and 2020 (DAT, 2021). Similarly, the car maintenance and repair industry has also experienced an increase in recent years (excl. 2019 and 2020; ZDK, 2021), and it is estimated that 63% of revenue streams stem from the aftermarket (CC, 2010).

3.2 Investments

In 2020, the European share of R&D investments in the automotive sector was globally the largest, accounting for 44.8% of the combined global expenditures in the sector (EC, European Commission [EC], Executive Agency for Small and Medium-sized Enterprises, and Bremer, 2021a, p. 14; ZEW, 2022). Within the EU, it accounted for 33.6% of the total R&D expenditure, amounting to € 62 bn, compared to China's € 13 bn (9.3%) and the US's € 19 bn (5.6%; EC, 2021a, p. 7). Moreover, it represents 6.1% of EU employment and 7% of the GDP (EC, 2021b, p. 10)

3.3 Internet of Things

Internet of Things (IoT) is transforming the automotive industry, specifically connectivity, which is seen as an enabler for other services and technologies that drive innovation in the sector, creating large amounts of data and enabling new possibilities and challenges (EC, 2020). This includes new vehicles that can be autonomous self-driving vehicles that learn and adapt to dynamic environments and evolve with the environment (EC, 2020). But also, an enhanced in-vehicle experience with improved navigation solutions, telematics, and driver assistance systems that add comfort and efficiency to car ownership (EP, 2021).

3.4 Smart Mobility

Novel business models based on access-based consumption have emerged in recent years. In these models, consumers gain temporary access to vehicles for their journey from point A to point B. Instead of selling cars individually, manufacturers using these models offer mobility packages, transforming themselves into service providers of smart mobility (Bardhi & Eckhardt, 2012; Flügge, 2017). A concept within smart mobility is mobility as a service (MaaS), where customers can use multiple mobility providers from a single platform, this may include car sharing schemes, public transportation modes, and bikes (EC, 2021). Mobility on demand (MoD), which is similar to MaaS, is on the rise. Car sharing, which allows customers to use cars when needed without owning them, shifts the ownership of vehicles from individual users to the service providers, and is creating new possibilities in the AAM (EC, 2021; Flügge, 2017).

3.5 Sustainability and Circular Economy

The Green Deal moving the EU towards zero emissions by 2035 (EC, 2019) has created a greater focus on the circular economy and sustainability within the European Union. This transition process can be considered a key enabler of innovations within the sector. In a circular economy, businesses aim to reduce waste via the improved design of materials, products, systems, and business models, and the idea of end-of-life is replaced by restoration (Moreno et al., 2016). Sustainability may, in a business context, be divided into three distinct accounts: environmental, economic, and social (Elkington, 1998).

Innovation has a large role in driving a sustainable future. This is, for instance, reflected in efforts spent optimising the fuel economy of cars as well as the increased innovation of non-fuel-reliant vehicles. Legislation and consumers demand more sustainable solutions and alternatives, which are realised by the AAM industry through innovation and optimisation driven by technological progress. Sustainable development and innovation within the aftermarket are, furthermore, enabled through repairability, reparation, and remanufacturing of the vehicles.

3.6 Repairability, Reparation, and Remanufacturing

Repairability may be described as the technical ability for a product to be repaired, encompassing the enabling framework for the repair, covering areas such as documentation given by the manufacturer, ease of disassembly of the product, availability of replacement parts, and the relationship between the price of a spare component and the original product (ADEME, 2018; Ministére de la Transition Écologique, 2020).

The process of repairing may be divided into two parts: calibration, in which the user detects object changes and adapts practices around them; and realignment, in which the user attempts to adjust the material capacities of objects to return them to a usable state (Godfrey et al., 2021). As car parcs age, the OEMs within the aftermarket try to attain their consumers by increasing repair and use phase services (BCG, 2021).

When repairing is not possible, remanufacturing can be a circular alternative. It refers to the process of restoring a spent product to its original state through a process of product rebuilding, which requires less material and energy and reduces CO_2 emissions compared to manufacturing new parts (Subramoniam et al., 2009; Sundin, 2004). These remanufactured parts may then be used as a substitute for virgin parts.

4 Research Approach

We conducted a comparative case study (Ebneyamini & Moghadam, 2018; Morgan et al., 2017) to examine the innovation capabilities of five AAM companies. The AAM companies were selected based on their participation at Automechanika Frankfurt 2022, which is the largest trade show for the AAM with 2.804 companies from 70 countries (Automechanika Press, 2022). The selection criteria were the following: AAM company, manufacturing company with physical or digital products, exhibition booth, and presentation of an innovation. Our research team consisted of two people, both with engineering degrees. During the trade fair, we met with numerous AAM exhibitors and collected photographs, videos, and observations. We conducted vis-à-vis interviews and related other materials with the assistance of the organisation Talents for Automotive Aftermarket (TALENTS4AA).

The companies used in the case study are Dinex A/S, Nissens A/S, Robert Bosch GmbH, Schaeffler Automotive Aftermarket GmbH & Co. KG, and ZF Friedrichshafen AG. A brief overview of each company is presented below.

Bosch Bosch is a German company founded in 1886. The company operates in multiple sectors: mobility solutions, consumer goods, industrial technology, energy, and building technology. They provide repair shops with technology and solutions for auto diagnosis and repairs, as well as a variety of vehicle spare parts, in the aftermarket. Their sector of mobility solutions generated € 45 bn in revenue in 2021, accounting for 58% of their total revenue (Bosch, 2022).

ZF ZF is a German company founded in 1915. They employ a total of 157,549 people across their 206 locations. The firm operates in the automotive aftermarket, where they deliver driveline, chassis technology, active and passive safety technology, and fleet management tools. Their revenue in the aftermarket was € 3 bn in 2021, accounting for 8% of their total revenue (ZF, 2021).

Nissens Nissens is a Danish company founded in 1921 that develops and manufactures engine cooling and climate system components for the aftermarket, with a focus on thermal and manufacturing. They had 550 employees and 22 sales distribution centres worldwide in 2021, with a total revenue of € 215 million (Chr. Augustinus Fabrikker, 2022; Nissens, 2022).

Schaeffler Schaeffler is a German company founded in 1946. They currently employ 82,981 people across 185 locations around the world. The majority of the company's operations are in the automotive technologies, automotive aftermarket, and industrial sectors. Their aftermarket product portfolio includes repair solutions for engines, transmissions, and chassis. The automotive aftermarket generated €1.8 bn in revenue in 2021, accounting for 13% of their total revenue (Schaeffler, 2021).

Dinex Dinex is a Danish company founded in 1982. It started out as a spare part exhaust system for heavy-duty vehicles, with a focus on quality, agility, and collaboration. With over 2000 employees and customers on three continents, the company has established itself as a global actor in aftermarket and as an OEM for exhaust and emission solutions. In 2021, the company's revenue was € 270 m (Dinex, 2022).

The case data was analysed using content analysis and reflecting theory. Each case was analysed and compared employing the following categorisation: product, photo, description of the product, category of product, innovation degree, and innovation type.

The results are shown in Table 1, which highlights five of the innovations we were shown at Automechanika.

5 Findings and Discussion

In our case study, we chose five different products, each taking an offset in current developments within the sector. These five cases and their products illustrate rich and different types of innovations, highlighting the multiple roles and impacts that AAM companies have as innovators and contributors to future mobility.

In most cases, the setting for showcasing innovative products consisted of presenting the artefact and displaying essential information. Most of the information was typically provided by a competent technician or salesperson able to address product-related inquiries. This choice ties in with the primary objective of most organisations at the fair, which, for us, seemed to be strengthening their current customer relations and exploring new potential customer relations. The presence of a technician or salesperson enabled a direct, one-on-one interaction that was more participatory and offered a customised presentation of their innovations. This choice is expected in a B2B environment where most organisations frequently have a lengthy purchasing timeframe, necessitating a personal relationship and idiosyncratic investments in individual relationships. This choice also enabled them to differentiate their information flow about innovations between their customers, potential customers, and competitors.

Choosing which details to emphasise for each specific innovation is dependent on the receiver and their proximity to the organisation, i.e., a customer's salesperson would be interested in different details than a journalist. Therefore, the findings for

Table 1 Case study overview

Company	Product	Photo	Description	Categories	Innovation degree	Innovation type
Dinex A/S	Dinex reconditioned RX filters		Reconditioned DPF exchange filters are offered as an exchange service, where customers send their used filters and receive a reconditioned filters ready to install	• Tangible • Remanufacture • Sustainability • Service	Incremental New to company and market	Service Process
Nissens Automotive A/S	EGR spare part		The EGR cools the engine using a coolant that flows through it. Using an optimised design and durable materials it extending the lifespan of the sparepart	• Tangible • Durability • Sustainability • Product	Incremental New to market	Product
Robert Bosch GmbH	ESI[tronic] 2.0 Diagnostic Software		Diagnostic software for the maintenance, diagnosis and repair of vehicles using specific vehicle data	• Intangible • Software • Platform • IoT	Disruptive New to company and market	Service Process
Schaeffler AG	INA Engine KIT		The INA repair kit offers a complete solutions containing all engine parts to be replaced for the individual vehicles	• Tangible • Repair • Sustainability • Product	Incremental New to company	Product
ZF Friedrichshafen AG	Autonomous Shuttle		The Autonomous Shuttle system is offered by ZF as an end-to-end service partner. Aiding the customer in all the steps from construction to maintenance	• Tangible • Intangible • IoT • MaaS MOD	Radical New to company and market	Service Product

Note: all product photos are obtained from the companies respective webpages, (Dinex, 2020; Nissens, 2021; Bosch, 2022; Schaeffler, 2022; ZF, 2022) each innovation explained and reflected below are what we perceived as engineering students when being presented to the artefacts at the trade fair.

The reconditioned RX filter exchange system is a system that allows Dinex's customers to exchange spent RX filters for remanufactured ones. Their system retrieves used filters, reconditions ('cleans'), and tests them before reinserting them into the system (Dinex, 2020). This allows a single filter, which would otherwise be discarded, to have multiple life cycles. Using the system not only saves the customer money, but it also extends the filter's environmental footprint over a longer lifespan. However, the system's structure cannot be considered novel, as similar systems, such as glass bottle deposit systems, have existed for decades. However, in this context, their model can be viewed as a step towards servitisation for the company, shifting their focus from product-centric to service-centric (Baines et al., 2009). As a result, their value proposition has shifted from selling a product to selling services or usage in a product service system. Their current focus, however, is still primarily product-oriented, despite this particular system being result-oriented (Tukker, 2004).

The EGR spare part developed and manufactured by Nissens can be considered a classical and well-executed example of incremental innovation. As it is an optimised design, containing more durable materials than its predecessor, thus being an improved version of products already present on the market. The level of incremental novelty can be discussed further with the incorporation of coolant in the sparepart.

Bosch's diagnostic software is an example of how technology, particularly the concepts of connectivity and IoT, can be used to optimise processes for individual independent repair shops. The system's next phase will include digital assistance that will aid the individual mechanic via augmented reality (AR). The system gives the shops access to a database containing information about each specific vehicle and model, allowing them to optimise their workflow. This is especially useful for independent repair shops, which must service multiple brands to compete with OEMs.

The Schaeffler engine repair kit is also an incremental innovation that facilitates the repair of complex technologies, indicating a shift away from the practice of replacing large modules when only minor components require replacement. Furthermore, their various kits include all of the components needed to repair specific engines from individual brands, simplifying the process and creating new opportunities for additional service offerings for their customers.

The autonomous shuttle system developed by ZF can be considered a radical innovation in how mobility is bought and perceived. Their system is to a great extent service-centric, with ZF as the main service provider, delivering a complete solution to their customers (municipalities or large organisations) covering the whole lifecycle of their vehicles in a closed ecosystem. They have developed an autonomous electric vehicle fleet that can be used as an integrated part of an MaaS system, and the MoD, creating new possibilities within the sector (ZF, 2022).

6 Discussion

The traditional innovation philosophy of the automotive sector has focused on contentious quality improvement, reducing production costs, and implementing novel technologies and processes. Yet, the aftermarket today is also, to a great extent, impacted by sustainability. This attentiveness is twofold: as many of the services provided already revolve around replacing, restoring, and maintaining vehicles, ultimately extending their longevity, it inherently aids the transition towards more sustainable consumption patterns. The other is as a result of legislation directing European firms towards the goal of zero emissions within the European Union (EC, 2019). Sustainability was an overarching theme we experienced throughout the trade fair and was emphasised by all firms used in this case study. It is also worth noticing that every company in this case study also emphasises this aspect on their websites.

The anticipated increase in supply and demand for electric and hybrid vehicles, combined with a decrease in sales of conventional combustion engines, creates new opportunities for the aftermarket. This necessitates that the sector transform some of its innovation capabilities to accommodate the change and perhaps consider moving from being product centric to service-centric; a process most of the companies presented in this case study have already initiated.

The findings from the case study illustrate that the supplying companies within the AAM have the required capabilities to innovate and create novel offerings to partners and end users. Furthermore, their capabilities exist on various levels and in various configurations, creating innovations of a similar degree to their vehicle manufacturing counterparts.

7 Conclusion

This chapter aims to explore the innovation capabilities of the independent automotive aftermarket. Our findings show that innovations within the sector are developed by multiple actors of various sizes and product categories, ranging from manufacturers supplying parts to original vehicle manufacturers and the aftermarket to those focusing solely on the aftermarket.

The innovations in this case study were classified as radical, disruptive, or incremental in varying degrees. The dissimilarity in degrees of innovation demonstrates the sector's current ability to develop much-needed solutions to cope with the environmental impact of the current car parc. This is especially important given the current developments in the mobility sector.

IoT influenced companies such as Bosch and ZF, whose focus at the Automechanika fair was more technology driven. One of the largest driving forces of innovation at the Automechanika, if not the largest, was sustainability. Sustainable innovation came into play, whether it was driven by a focus on materials, reuse, replaceability, or repairability, among other factors. All companies in the case study emphasised the sustainable aspect of mobility when communicating their innovations. Whether it was highly relevant to their innovation or not, sustainability dominated their marketing communication at the Automechanika 2022 fair. This reflects the driving force behind most innovations in 2022 and beyond. User demands and policymakers require more efficient cars that are produced, maintained, and repaired with a smaller environmental impact. The AAM has historically played a role in creating a more sustainable vehicle market by prolonging the lifetime of vehicles and reducing emissions, and it sets a high standard for OEMs to match. The competition among producers of parts further drives development and innovation to meet future demand and shortages, ultimately resulting in technological and sustainable development.

The study is limited to European companies and showcases one exhibition; hence, it does not offer any longitudinal or deeper analysis. This limitation does not allow insights for other types of innovations, like frugal innovation, and it excludes developing countries and bottom-of-the pyramid markets. For further studies, we recommend investigating other geographical areas than Europe, especially focusing on the differences in the aftermarket sector in Asia, Africa, and America. We believe that examining different types of innovations in developing and emerging markets as well as developed markets may hold the potential for rethinking R&D practices, innovation and creativity, and drive business development.

Acknowledgements We would like to thank Maria Elo and Fotios Katsardis for their invaluable assistance, guidance, and support during the research and writing process for this chapter. We would also like to thank all the individuals and organisations, especially Talents4AA, who provided us with the resources and data. Your assistance has been greatly appreciated.

References

Aas, T. H., & Breuning, K. J. (2017). Conceptualizing innovation capabilities: A contingency perspective. *Journal of Entrepreneurship, Management and Innovation, 13*(1), 7–24. https://doi.org/10.7341/20171311

Abernathy, W. J., & Clark, K. B. (1985). Innovation: Mapping the winds of creative destruction. *Research Policy, 14*(1), 3–22. https://doi.org/10.1016/0048-7333(85)90021-6

ACEA. (2022). Vehicles in use Europe 2022. Acea.auto https://www.acea.auto/files/ACEA-report-vehicles-in-use-europe-2022.pdf

ADEME (Agence de l'Environnement et de la Maltrise de l'Energie). (2018). *Panorama de l'offre de réparation en France: actualisation 2018.* ADEME. https://librairie.ademe.fr/dechets-economie-circulaire/1014-panorama-de-l-offre-de-reparation-en-france-actualisation-2018.html

Albert, M. (2019). Sustainable frugal innovation - the connection between frugal innovation and sustainability. *Journal of Cleaner Production, 237*, 117747. https://doi.org/10.1016/j.jclepro.2019.117747

Automechanika Press. (2022, September 17). A strong show: The international automotive aftermarket is back in Frankfurt. Automechanika. https://automechanika.messefrankfurt.com/frankfurt/en/press/press-releases/automechanika-frankfurt/final-report.html

Baines, T. S., Lightfoot, H. W., Benedettini, O., & Kay, J. M. (2009). The servitization of manufacturing: A review of literature and reflection on future challenges. *Journal of Manufacturing Technology Management, 20*(5), 547–567. https://doi.org/10.1108/17410380910960984

Baker, T., & Nelson, R. E. (2005). Creating something from nothing: Resource construction through entrepreneurial bricolage. *Administrative Science Quarterly, 50*(3), 329–366. https://doi.org/10.2189/asqu.2005.50.3.329

Bardhi, F., & Eckhardt, G. M. (2012). Access-based consumption: The case of car sharing. *The Journal of Consumer Research, 39*(4), 881–898. https://doi.org/10.1086/666376

BCG [Boston Consulting Group]. (2021, March). At the crossroads: The European aftermarket in 2030. BCG. https://web-assets.bcg.com/36/39/e80d073a4067bfe89c7482d6db69/the-european-aftermarket-in-2030.pdf

Bosch. (2022). Bosch ESI[tronic] Software for Automotive Workshops. Boschaftermarket.com https://www.boschaftermarket.com/gb/en/diagnostics/ecu-diagnosis/esitronic-software/

CC [Capgemini Consulting]. (2010). *The aftermarket in the automotive industry: How to optimize aftermarket performance in established and emerging markets.* Capgemini. https://www.capgemini.com/wp-content/uploads/2017/07/tl_The_Aftermarket_in_the_Automotive_Industry.pdf

Chr. Augustinus Fabrikker. (2022). Nissens A/S. Augustinusfabrikker.dk. https://augustinusfabrikker.dk/en/our-ownerships/k-nissen-international-s/

Daim, T. U. (2016). *Hierarchical decision modelling: Essays in honor of Dundar F. Kocaoglu.* Springer International Publishing.

Dance, J. (2008, May 22). *Summary of "Innovation": 30+ Definitions.* Fresh Consulting. https://www.freshconsulting.com/insights/blog/what-is-innovation/

DAT. (2021). Revenue of the used and new car markets in Germany from 1990 to 2020 (in billion euros) [Graph]. *Statista.* Retrieved November 3, 2022, from https://www.statista.com/statistics/486834/used-and-new-car-market-revenue-germany/

Dinex. (2020). Reconditioned Euro 6 DPF Filters. Dinex.net. https://www.dinex.net/aftermarket/recon-rx-solution

Dinex. (2022). Dinex History. Dinex.net. https://www.dinex.net/about-us/history

Dombrowski, U., & Engel, C. (2014). Impact of electric mobility on the after sales service in the automotive industry. *Procedia CIRP, 16*, 152–157. https://doi.org/10.1016/j.procir.2014.01.022

Ebneyamini, S., & Moghadam, M. R. S. (2018). Toward developing a framework for conducting case study research. *International Journal of Qualitative Methods, 17*(1), 160940691881795. https://doi.org/10.1177/1609406918817954

Elkington, J. (1998). Partnerships from cannibals with forks: The triple bottom line of 21st-century business. *Environmental Quality Management, 8*(1), 37–51. https://doi.org/10.1002/tqem.3310080106

European Commission [EC]. (2019). *A European Green Deal.* European Commission. https://ec.europa.eu/info/strategy/priorities-2019-2024/european-green-deal_en

European Commission [EC], Executive Agency for Small and Medium-sized Enterprises, & Bremer, V. (2021b). *Advanced technologies for industry : AT watch : technology focus on sustainability in the automotive industry in Europe.* Publications Office. https://data.europa.eu/doi/10.2826/94114

European Commission [EC], Executive Agency for Small and Medium-sized Enterprises, Heimberger, H., Horvat, D., & Kroll, H. (2020). *Advanced technologies for industry : sectoral watch : technological trends in the automotive industry.* Publications Office. https://data.europa.eu/doi/10.2826/13579

European Commission [EC], Joint Research Centre, Grassano, N., Hernández, H., & Guevara, H. (2021a). *The 2021 EU industrial R&D investment scoreboard : executive summary.* Publications Office. https://data.europa.eu/doi/10.2760/248161

European Parliament [EP], Directorate-General for Internal Policies of the Union, Brown, D., Flickenschild, M., & Mazzi, C. (2021). *The future of the EU automotive sector.* European Parliament. https://data.europa.eu/doi/10.2861/680215

Fischer, M. M. (2001). Innovation, knowledge creation and systems of innovation. *The Annals of Regional Science, 35*(2), 199–216. https://doi.org/10.1007/s001680000034

Flügge, B. (2017). *Smart mobility - connecting everyone: Trends, concepts and best practices.* Springer Fachmedien Wiesbaden. https://doi.org/10.1007/978-3-658-15622-0

Garcia, R., & Calantone, R. (2002). A critical look at technological innovation typology and innovativeness terminology: A literature review. *The Journal of Product Innovation Management, 19*(2), 110–132. https://doi.org/10.1111/1540-5885.1920110

Godfrey, D. M., Price, L. L., & Lusch, R. F. (2021). Repair, consumption, and sustainability: Fixing fragile objects and maintaining consumer practices. *The Journal of Consumer Research.* https://doi.org/10.1093/jcr/ucab067

Govindarajan, V., & Ramamurti, R. (2011). Reverse innovation, emerging markets, and global strategy. *Global Strategy Journal, 1*(3–4), 191–205. https://doi.org/10.1002/gsj.23

Gu, L. (2022). Inflation, weak growth and uncertainties cloud global *auto sales outlook.* Scotiabank.com. https://www.scotiabank.com/ca/en/about/economics/economics-publications/post.other-publications.autos.global-auto-report.september-22%2D%2D2022.html

Henderson, R. M., & Clark, K. B. (1990). Architectural innovation: The reconfiguration of existing product technologies and the failure of established firms. *Administrative Science Quarterly, 35*(1), 9–30. https://doi.org/10.2307/2393549

ITF. (2021). *ITF Transport Outlook 2021.* OECD Publishing. https://doi.org/10.1787/16826a30-en

Kempf, S., Heid, B., Padhi, A., Tschiesner, A., Cornet, A., Mohr, D., Camplone, G., Nauclér, T., Baumgartner, T., & Hanser, R. (2018). Ready for inspection – the automotive aftermarket in 2030. Mckinsey.com. https://www.mckinsey.com/industries/automotive-andassembly/our-insights/ready-for-inspection-the-automotive-aftermarket-in-2030

Lawson, B., & Samson, D. (2001). Developing innovation capability in organisations: A dynamic capabilities approach. *International Journal of Innovation Management, 05*(03), 377–400. https://doi.org/10.1142/s1363919601000427

Ministére de la Transition Écologique. (2020). The anti waste law in the daily lives of the French people what does that mean in practice? [online] pp. 22–25. https://www.ecologie.gouv.fr/indice-reparabilite#scroll-nav__6

Monden, Y. (2011). *Toyota production system* (4th ed.). Productivity Press.

Moreno, M., De los Rios, C., Rowe, Z., & Charnley, F. (2016). A conceptual framework for circular design. *Sustainability (Basel, Switzerland), 8*(9), 937. https://doi.org/10.3390/su8090937

Morgan, S. J., Pullon, S. R. H., Macdonald, L. M., McKinlay, E. M., & Gray, B. V. (2017). Case study observational research: A framework for conducting case study research where observation data are the focus. *Qualitative Health Research, 27*(7), 1060–1068. https://doi.org/10.1177/1049732316649160

Nissens. (2021). EGR Valves. Nissens.com. https://showroom.nissens.com/egr-valves

Nissens. (2022). *About* Nissens automotive. Nissens.com https://nissens.com/en-gb/about-nissens/about-nissens-automotive.aspx

OECD/Eurostat. (2018). *Oslo Manual 2018: Guidelines for Collecting, Reporting and Using Data on Innovation* (4th ed.). The Measurement of Scientific, Technological and Innovation Activities, OECD Pub. https://doi.org/10.1787/24146137

Ohno, T. (1988). Toyota production system: Beyond large-scale production. *Productivity Press.* https://doi.org/10.4324/9780429273018

Radjou, N., Ahuja, S., & Prabhu, J. C. (2012). *Jugaad innovation: think frugal, be flexible, generate breakthrough growth.* Jossey-Bass.

Rogers, E. M. (2003). *Diffusion of innovations* (5th ed.). Free Press.

Rubenstein, J. M. (2001). *Making and selling cars: Innovation and change in the U.S. automotive industry.* Johns Hopkins University Press.

Schaeffler (2021). Schaeffler Annual report 2021. Schaeffler.com https://www.schaeffler.com/remotemedien/media/_shared_media_rwd/08_investor_relations/reports/2021_ar/2021_schaeffler_annual_report_en_y3w5pl.pdf

Schaeffler. (2022). INA Engine Kit. Aftermarket.schaeffler.com https://aftermarket.schaeffler.com/en/news-ina-engine-kit

Shapiro, C., & Teece, D. J. (1994). Systems competition and aftermarkets: An economic analysis of kodak. *Antitrust Bulletin, 39*(1), 135–162. https://doi.org/10.1177/0003603X9403900104

Subramoniam, R., Huisingh, D., & Chinnam, R. B. (2009). Remanufacturing for the automotive aftermarket-strategic factors: Literature review and future research needs. *Journal of Cleaner Production, 17*(13), 1163–1174. https://doi.org/10.1016/j.jclepro.2009.03.004

Sundin, E. (2004). *Product and process design for successful remanufacturing*. Linköpings Universitet.

Teece, D. J., Pisano, G., & Shuen, A. (1997). Dynamic capabilities and strategic management. *Strategic Management Journal, 18*(7), 509–533. https://doi.org/10.1002/(SICI)1097-0266(199708)18:7<509::AID-SMJ882>3.0.CO;2-Z

Tidd, J. (2001). Innovation management in context: Environment, organization and performance. *International Journal of Management Reviews : IJMR, 3*(3), 169–183. https://doi.org/10.1111/1468-2370.00062

Tidd, J., & Bessant, J. (2021). *Managing innovation: Integrating technological, market and organizational change* (7th ed.). Wiley.

Tukker, A. (2004). Eight types of product-service system: Eight ways to sustainability? Experiences from SusProNet. *Business Strategy and the Environment, 13*(4), 246–260. https://doi.org/10.1002/bse.414

Tushman, M. L., & O'Reilly, C. A. (1996). The ambidextrous organizations: Managing evolutionary and revolutionary change. *California Management Review, 38*(4), 8–30. https://doi.org/10.2307/41165852

Ulaga, W., & Reinartz, W. J. (2011). Hybrid offerings: How manufacturing firms combine goods and services successfully. *Journal of Marketing, 75*(6), 5–23. https://doi.org/10.1509/jm.09.0395

United Nations. (2015). The 17 Goals, United Nations. *United Nations*. https://sdgs.un.org/goals

Wang, H., & Kimble, C. (2016). How external factors influence business model innovation: A study of the Bosch group and the Chinese automotive aftermarket. *Global Business and Organizational Excellence, 35*(6), 53–64. https://doi.org/10.1002/joe.21712

Yin, E., Ansari, S., & Akhtar, N. (2017). Radical innovation, paradigm shift and Incumbent's dilemma the case of the auto industry. *Future Studies Research Journal, 9*(1), 138–148. https://doi.org/10.24023/FutureJournal/2175-5825/2017.v9i1.301

ZDK. (2021). Revenue of the car maintenance and repair industry in Germany from 2004 to 2020 (in billion euros) [Graph]. *Statista*. Retrieved November 03, 2022, from https://www.statista.com/statistics/589634/revenue-car-services-germany/

ZEW. (2022). Expenditure on innovation in the car industry in Germany from 2011 to 2022 (in billion euros) [Graph]. *Statista*. Retrieved November 03, 2022, from https://www.statista.com/statistics/589497/car-industry-germany-innovation-spending/

ZF. (2021). Annual report 2021. Zf.com https://www.zf.com/master/media/en/corporate/m_zf_com/company/bonds_relations_/financial_reports/annual_report/2021_2/ZF_Annual-Report_2021.pdf

ZF. (2022). Autonomous Shuttle. Aftermarket.zf.com. https://aftermarket.zf.com/en/aftermarket-portal/autonomous-shuttle/

Anders Leth Villumsen holds a master's and bachelor's degree in Product Development and Innovation and has extensive experience working in diverse interdisciplinary teams across multiple product categories. His research interests include sustainability, circular economy, innovation management, and product development and optimisation.

Claire Croucher Petersen work experience is within supply chain management, product design, and optimisation, and her research interests encompass sustainability and innovation management. She possesses a bachelor's and a master's degree in Product Development and Innovation, which provides her with an engineering background within the fields of product design and management, innovation, and product commercialisation.

How Can the Nordic Automotive Aftermarket Provide Opportunities Counteracting the Challenges of Disadvantaged Entrepreneurs?

David Breitenbach, Susanne Sandberg, and Maria Elo

Abstract

This chapter explores how the automotive aftermarket could provide potential opportunities for the challenges refugee and immigrant entrepreneurs face in the Nordic countries. As the automotive industry has a vast societal impact, including geographical, economical, and employment effects, the aim of this paper is to explore the Automotive Aftermarket (AA) as a potential gateway for disadvantaged newcomers, e.g., refugees and immigrants in the Nordics, to become entrepreneurs and reach prosperity. Through qualitative research it is shown that the automotive aftermarket industry has excellent market conditions for disadvantaged entrepreneurs through low entry barriers and increasing demand side. The industry is also characterised by strong network influences, while its institutional infrastructure might be difficult to navigate. Additionally, this paper also found that language has a central piece in overcoming refugee entrepreneurs' challenges in social integration. In summary, the paper concludes that raising awareness of

D. Breitenbach (✉)
PatentRenewal.com ApS, Copenhagen, Denmark

S. Sandberg
School of Business and Economics, Linnaeus University, Kalmar, Sweden
e-mail: susanne.sandberg@lnu.se

M. Elo
University of Southern Denmark (SDU), Odense, Denmark

Belt and Road Institute of International Business, School of Economics, Shanghai University, Shanghai, China

Abo Akademi University, Turku, Finland

Turku school of Economics, University of Turku, Turku, Finland

Institute Arbeit und Technik, Gelsenkirchen, Germany
e-mail: melo@sam.sdu.dk

M. Elo, F. Katsardis (eds.), *Automotive Aftermarket*, Management for Professionals, https://doi.org/10.1007/978-3-031-62419-3_15

the opportunities the automotive aftermarket provides in the Nordics could be beneficial both for the industry's social impact and newcomers who wish to contribute to the development of the entrepreneurial landscape of AA.

1 Introduction

The twenty-first century's entrepreneurial landscape has been significantly influenced by globalisation, digitalisation, technological advancements, rapid urbanisation, and many other phenomena, spurring enhanced mobility and an inherently global and accessible world (Triandafyllidou, 2018). For many individuals, travelling or living abroad or just simply interacting with people from foreign countries have become a natural part of their lives. As studies show, this could actually be a transformative experience for many, since individuals tend to open up for change, discovering new ways of living and learning how to thrive in a new environment when their well-known environment and previous reference points fundamentally change (Adam et al., 2018; Kiss et al., 2020; Zhang & Chun, 2018).

However, migration is not a new phenomenon by any means, since humans have been migrating for a long time (Massey et al., 1987)—one number mentioned is 50–60,000 years, where there has been a particular intensification during the last 25 years (Triandafyllidou, 2018). Academic research distinguishes several categories of migration, and among others, the following categories might be the most relevant for this particular study:

1. *International migration* refers to the movement of people from their home country into another. The reason for leaving is often work- or study-related or family reunification. International migration is often described with the terms of emigration (leaving one's country of origin) and immigration (moving to a new country) (IOM, 2022).
2. *Internal migration* meaning the movement of people within a specific country or geographic area. Internal migration can be driven by a variety of factors, including economic opportunities, social and cultural factors, and environmental conditions (Massey et al., 1987).
3. *Forced migration* describes the forced movement of individuals, making them leave their homes due to conflict, persecution, or violence. Forced migration can include refugees, asylum seekers, and internally displaced persons (IDPs) (UNHCR, 2022).
4. *Return migration* includes those who have previously migrated to a new location and are returning to their country of origin or another location where they have previously lived in (Massey et al., 1987).

Out of the given four types, forced migration is getting significantly more focus in the last couple of years from researchers, policymakers, and businesses (UNHCR, 2022). According to the United Nations High Commissioner for Refugees, the number of people forced to flee their homes, due to political conflicts, human rights violations, or general violence worldwide, have doubled in the last decade, growing from more than 45 million people to 89.3 million in 2021. Close to 30 million people of those have been refugees either under UNHCR or UNRWA's mandates (UNHCR, 2023). According to the UNHCR's estimations, the number of refugees is expected to grow continuously, while the number of opportunities and options available to them are expected to get increasingly dire (UNHCR, 2022).

As refugees are forced to disrupt their lives and migrate from their country of origin (COO) to a new country of residence (COR), their lives will inevitably change. Not only the lives of refugees will transform, but research shows that host societies also go through a certain transition as a result of the influx of refugees. For example, the local economy of a certain settlement, city, or municipality might experience a significant shift as it experiences the arrival of refugees, and new models of integration policies might be required so that the disadvantaged can reach their potentials as valuable assets of/to their new COR (Elo & Hieta, 2017; Elo, Zubair & Zhang, 2022; Sandberg et al., 2019).

Although migration is considered a relatively well-researched area in business and economics (Bonacich, 1973; Portes & Manning, 2008), discussions have begun to separate refugees from immigrants only recently (Harima et al., 2021). In fact, even the term 'refugee' is distinct from 'migrant', as refugees are fleeing their country of origin (COO) often due to a violent conflict, while migrants are rather seeking better economic opportunities in their new COR (IOM n.d.; Scherschel 2011).

This paper's focus will be those disadvantaged and often underprivileged that enter the field of entrepreneurship as a direct consequence of their forced migration: *refugee entrepreneurs*. The potential has already been recognised by many researchers exploring different aspects of the challenges they might face (Alexandre et al., 2019; Alrawadieh et al., 2018; Bizri, 2017; Harima et al., 2021; Lee, 2018; Sandberg et al., 2019; Shneikat & Alrawadieh, 2019; Turner, 2020). More concretely, research has shown that refugees tend to overcome their challenges by developing distinct survival attributes (Mueller, 2014), start-over intentions (Mulholland & Ryan, 2014) and complex social capital (Pedrini et al., 2016; Zolin et al., 2016). Upon analysing the current landscape of refugee entrepreneurship literature further, this work found that current research is dominated by country-specific studies; however, industry-specific analysis is lacking in general. There are a few exceptions on impactful sectors, such as tourism and hospitality (Alrawadieh et al., 2018; Shneikat & Alrawadieh, 2019); however, automotive aftermarket as a context of research is still underdeveloped. Thereby this paper addresses this gap by discovering the opportunities the automotive aftermarket could provide for the challenges refugees could potentially face during their entrepreneurial journey.

The term *automotive aftermarket* refers to all car-related market activities following the initial sales of new vehicles, such as second-hand sales, car maintenance and repair services, or sales of car parts (Bowring, 1990). As the automotive industry is shifting, the automotive aftermarket closely follows in its tracks; having vast societal impact, including geographical, economical, or employment effects (Mathivathanan et al., 2018). This paper explores the relevance of this market to refugee entrepreneurs. The study will be conducted in the Nordics as its context.

Nordic businesses may also benefit from the intrapreneurship of newcomers who do not have the capitals to set up a business, but do have particular skills and networks to grow their employer's business entrepreneurially (Elo & Vincze, 2019).

Therefore, the research question that this paper aims to answer is the following:

> *"How can the Nordic automotive aftermarket provide opportunities for the challenges of refugee entrepreneurs?"*

The aim of this research paper is to explore the automotive aftermarket as a potential gateway for refugees in the European Nordics to become entrepreneurs and reach prosperity. As the industry needs and wants entrepreneurial talent and offers a low entry barrier opportunity for entrepreneurship for newcomers, this paper aims to discover how refugees' previous knowledge and experience could be valued in use. Overall, this paper addresses opportunities in the automotive aftermarket for disadvantaged entrepreneurs. Additionally, this paper contributes to the theory discussion by extending theory on "entering" in economic integration with the automotive aftermarket as the entry point for disadvantaged people to entrepreneurship and a better life.

In order find answers to the research question and contribute to the academic discussions on refugee entrepreneurship, this paper will first introduce a theoretical overview of the topic, detailing its definition, its influencing factors and its typology. Following this, the paper's methodology will be presented, introducing the research question, research design, data collection, and analysis. In Sect. 4, the collected data will be presented from two different perspectives: firstly, from the refugee entrepreneurs, and secondly, from industry experts. Then the paper will conclude with the discussion of the findings and the academic contributions and implications.

2 Theoretical Framework

In this section, the paper will introduce the differences between (im)migrant and refugee entrepreneurship, then identify the specific challenges refugees might face based on their typology. The section will be finished by a short overview on the Automotive Aftermarket with the aim of introducing its relevance to this study.

2.1 Differences Between (Im)migrant and Refugee Entrepreneurship

The topic of refugee entrepreneurship is relatively new to research. Even though there were early theories on immigrant entrepreneurship, like the Middleman Minority theory (Bonacich, 1973) or Migrant Enclaves (Portes & Manning, 2008), self-employment of refugees in particular has not been the focus of attention until just recently (Wauters & Lambrecht, 2008). One of the earliest researches on refugee entrepreneurs belongs to Gold and Kibria (1993) who conducted their research on Vietnamese refugees and Soviet Jewish refugees in the US and argued that refugee entrepreneurs can be characterised by traumatic events, less extensive network conditions and have restricted access to resources than other entrepreneurs (Gold, 1988; Gold & Kibria, 1993).

In general, researchers have predominantly considered migrants as agents of entrepreneurship in both the COO and COR (Portes & Shafer, 2006; Rath, 2003; Waldinger, 1989). On the other hand, the majority of this research have not made a notable distinction between refugee entrepreneurship and immigrant entrepreneurship. Recent research such as Harima et al. (2021) and Sandberg et al. (2019) reference Wauters and Lambrecht's (2008) contribution as a turning point, publishing some of the first influential studies arguing that refugee entrepreneurship should be researched distinctively from immigrant entrepreneurship.

One significant characteristic of refugees is their *reason for migration*. While immigrant entrepreneurs are leaving their home countries and engage into entrepreneurial activities abroad in order to survive or prosper economically (Vinogradov & Isaksen, 2008), refugees are individuals who have fled their countries due to political conflicts, human rights violations or violence (UNHCR, 2022), thus they did not leave by own will but were forced to migrate. This leads us to the second point of differences; *access to resources* in both COR and COO. Refugees, as they are forced to migrate, may have limited access to legal protections and support in COR, as their legal status differs from others. Oftentimes, they face difficulties to get a job or to gain access to social benefits (Wauters & Lambrecht, 2008). As refugees leave their home country often unprepared and on the run, they rarely know consciously where they will end up, and what to expect in that specific geographical area, country, or municipality. Empirical evidence suggests that most refugees are forced to leave their most precious resources, such as valuable assets or financial capital, behind while fleeing. Combining this with discrimination (Kupferberg, 2003), language barrier (Wauters & Lambrecht, 2008) and lack of knowledge about the COR's culture and business practices (Gericke et al., 2018), leaves no other choice to them than to either study or retrain themselves (Mulvey, 2015), be underemployed (Vinokurov et al., 2017) or turn to self-employment (Mawson & Kasem, 2019). Additionally, the network conditions of refugee entrepreneurs are undeniably less developed than for immigrant entrepreneurs. Unsurprisingly, as a result of the lack of conscious planning, refugees' social network in the host country is much more likely to be less extensive (Gold & Kibria, 1993).

Lastly, the *relation to their home country* has significant differences as well. While immigrant entrepreneurs decided to leave their home countries on their own, most refugees under normal circumstances would have never left their homes. Additionally, immigrant entrepreneurs often seek to return to their home country and employ return migration, in a higher social status and with more significant economic power (Dustmann, 1996), while refugees might never be able to go back to their previous homes even if they want to, as they were persecuted there and their homes might also be ruined due to war (Wauters & Lambrecht, 2008).

A summary of the distinctive characteristics of refugee entrepreneurs can be found in Table 1.

Table 1 Summary of main differences between a refugee and an immigrant entrepreneur

	Refugee entrepreneur	Immigrant entrepreneur
Reason for migration	Forced migration due to persecution	Pursuit of economic success or survival
Access to resources	Limited access to mobilise resources in COR or COO	More likely to mobilise resources in both COR and COO through networks
Network conditions	Less extensive networks	Opportunity to build up relationships prior to their arrival in COR
Relation to home country	Reluctant to return to their homeland	Higher chance of returning home

Source: Own design based on (Adam et al., 2018; Gold, 1988; Gold & Kibria, 1993; Sandberg et al., 2019; Wauters & Lambrecht, 2008)

2.2 The Barriers of Entrepreneurship for Refugees

There are several reasons refugee entrepreneurship has gained the attention of the research community. One of these is the constantly growing interest for the topic of identifying and overcoming the refugee entrepreneurs' initial challenges as newcomers in their COR.

Once a refugee starts their entrepreneurial journey, there are several challenges waiting for them that might hinder their desire for wanting to become entrepreneurs in their COR. One approach to group these challenges was created by Wauters and Lambrecht (2008) who identified three categories: (1) Opportunity structures, (2) Group characteristics, and (3) Institutional and social environment. In their model, opportunity structures include market conditions and access to entrepreneurship, while group characteristics represent human capital and social networks. Finally, institutional and social environment include all legal and socio-economic obstacles.

For *market opportunities*, Wauters and Lambrecht (2008) found that many refugees (and immigrants also) often start their entrepreneurial journey from a negative motivation, basically as an emergency solution as they are unable to find a job. They found evidence on refugees in Belgium who wish to be independent from welfare programmes and aim to survive on their own, e.g. via entrepreneurship. There is a significant relevance for these market opportunities stemming from negative motivation, as other empirical evidence suggest that refugees often face a phenomenon known as *"Refugee trauma"* (Pugliese, 2009). These previous and current negative experiences combined may have a negative effect on refugees' entrepreneurial motivations and might deactivate their human capital for a short period of time (Williams & Shepherd, 2016). Interestingly, it has been demonstrated that refugees can indeed overcome their previous traumatic experiences with resilience (Goodman et al., 2017; Nanda & Khanna, 2010) which they primarily showcase through their entrepreneurial activities (Pieloch et al., 2016; Shepherd et al., 2020), which can be enabled by opportunities in the COR (Wauters & Lambrecht, 2008).

An interesting aspect of opportunity structures is the choice of sector, since refugee entrepreneurs tend to choose a sector with low entry barriers that require limited skills and initial resources, as they face lack of financial capital and assets (Gold &

Kibria, 1993; Wauters & Lambrecht, 2008). Elo, Täube and Volovelsky (2019) showed that this particularly concerns first-generation refugees who tend to stay in lower-level investment and lower-level tech businesses, such as service and retail, while their children seem to take off and do tech, IT, digital and other more complex and more international businesses. Hence, the generation who enters as a refugee or otherwise vulnerable or disadvantaged migrant (Elo, Zubair & Zhang, 2022) really benefits from easy entry businesses that they grow and expand—often in surprising circumstances. Another aspect that makes their access to entrepreneurship limited is the lack of ability to prove their professional skills as they are unable to bring their official certificates and diplomas with themselves upon fleeing their home country or their documents are not equivalent with the required ones in their COR (Wauters & Lambrecht, 2008). This is an example of the meaningfulness and importance of the formal status and identity of the migrant-refugee that enable them to act in legitimate and formal manner as entrepreneurs (Elo et al., 2022).

The second area most refugees face challenges in is their *social and economic integration* into their COR (Harima et al., 2019; Shneikat & Alrawadieh, 2019) i.e. group challenges (Wauters & Lambrecht, 2008). Studies have shown that entrepreneurial activities might be one of the best ways of overcoming this challenge, as these activities not only contribute to the integrative efforts of refugees into the COR's economic and societal systems but also provide opportunities for job creation and socio-economic value creation (Shneikat & Alrawadieh, 2019). However, it has also been found that in order to professionally integrate refugees into their COR and enable them to run their own business, they require knowledge of languages (Sandberg et al., 2019), Wauters & Lambrecht, 2008), as this enables them to study and communicate efficiently with their (potential) customers.

Another significant aspect of successful entrepreneurs is their social embeddedness (Barberis, 2017), that can also contribute positively to the integration of refugees (Wauters & Lambrecht, 2008). However, it has been found that in many cases refugees tend to find themselves alone in their COR, as they neither integrate with the local community, nor are able to find their compatriots (Wauters & Lambrecht, 2008). On the other hand, studies have also proven that many refugees indeed often have the desire to start their own firms (Alexandre et al., 2019; Mawson & Kasem, 2019), thus being more opportunity than necessity-driven than previously considered (Sandberg et al., 2019). What is more interesting, by starting their entrepreneurial journey might also help refugees not only to face their previous traumatic experiences, but they can create economic and socio-economic values in the host country, such as creating innovation, providing products and services to the local market, creating jobs and paying their taxes (Shneikat & Alrawadieh, 2019). Moreover, by engaging into entrepreneurial activities studies proved that they are essentially improving the image of their fellow refugees as well (Turner, 2020) and promote counter-xenophobic instruments in society (Harima et al., 2021), spread their COO's culture (Tavakoli, 2020), or contribute to political activism by voicing their opinion (Lee, 2018).

The last category of barriers refugees might face when starting their entrepreneurial journey is *institutional infrastructure* (Harima et al., 2021; Wauters &

Lambrecht, 2008). Refugees as newcomers often lack knowledge about local customs, legal obligations, and institutions in their COR and the paperwork required might discourage them to initiate their entrepreneurial journey, fearing losing themselves in the spiderweb of bureaucracy (Wauters & Lambrecht, 2008). To be able to start a business the refugee entrepreneur most often is in need of capital, i.e. to gain funding from financial institutions, such as banks (if they provide capital guarantee), or from friends, family, and/or other investors (Wauters & Lambrecht, 2008). Here Sandberg et al. (2019) highlighted that it tended to be difficult for entrepreneurs with refugee background to receive formal funding from banks, instead they had to rely on family and friends in the COO, or other areas where these connections resided, to be able to start and prosper their business in the COR.

The funding situation is one example of the problems, when refugees arrive to their host countries, they have a constraint of access to generally available support systems and infrastructure (Mohammed et al., 2016; Sandberg et al., 2019). This category also includes societal and market-related obstacles (Harima et al., 2021), like discrimination (Kupferberg, 2003), cultural differences in doing business (Gericke et al., 2018) or market-specific needs in product and service design (Harima et al., 2021).

A summary of the identified barrier categories of refugee entrepreneurship is presented in Fig. 1, along with identification of potentially overlapping elements.

With basis in the three identified barrier categories, each pinpointed into specifics (e.g. market opportunities into choice of sector), it is visible/has been observed that there are potential overlaps between the categories. For example, between choice of sector and social embeddedness there is an overlap in terms of the market specific needs. In the same fashion, cultural differences in doing business is in the overlap between social embeddedness and knowledge about local customs, law, and institutions. Lastly, the ability to prove skills and certification verge on the border between institutional knowledge and choice of sector. These are all important factors in terms of handling the barriers described—but the most important note is to handle the language barrier. Without that skill, the barriers might be too overwhelming for the refugee entrepreneur.

2.3 Typology of Refugee Entrepreneurs

One criticism of recent refugee entrepreneurship research was that most studies employ a heterogenic approach to refugees, refusing to take into account that refugees can be different in many aspects (Harima et al., 2021). Harima et al. (2021) built a multiple embeddedness model that enabled them to sort the refugee entrepreneurial types into several subcategories:

1. Type 1—includes those refugees that are not able to use their previous qualifications due to institutional obstacles and turn towards opportunities from their homeland socio-economic resources.

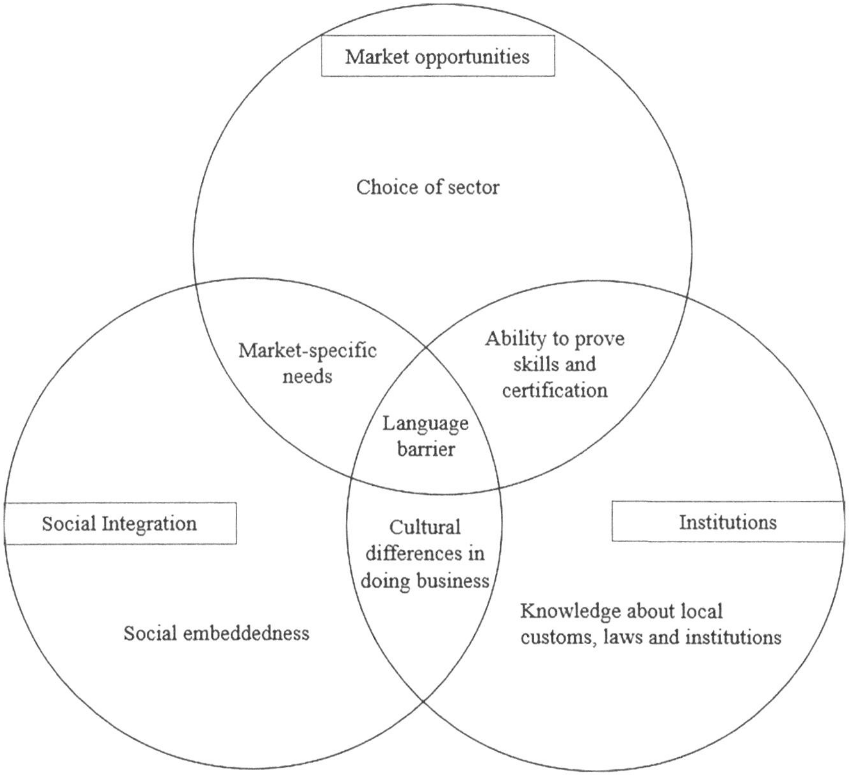

Fig. 1 Identified categories of refugee entrepreneurship barriers with potential overlaps. *Source*: Own design based on (Barberis, 2017; Gericke et al., 2018; Harima et al., 2019; Harima et al., 2021; Wauters & Lambrecht, 2008)

2. Type 2—refers to those who use their experience and knowledge of the differences between their home countries and their host countries and leverage this through their transnational networks.
3. Type 3—takes into account that refugees can also act as social entrepreneurs, as their previous traumatic experiences drive them to help fellow refugees or asylum seekers that live in the same host country.
4. Type 4—shows those who are successful in transferring their previous qualifications and human capital from their homeland to their host countries.
5. Type 5—those entrepreneurs belong in this type who recognise the difficulties in their homeland through the lens of the new institutional environment in their host countries and tend to attempt solving those problems in their homelands remotely or through a social enterprise.
6. Type 6—refers to those who can employ "globally applicable, substantial human capital" (Harima et al., 2021).

3 Research Methodology

The aim of this research was to explore the challenges faced by refugee entrepreneurs in the Nordic automotive aftermarket and connect these obstacles with the opportunities the industry provides. The research design applied for this study was a systematic review of the existing literature on refugee entrepreneurship, including academic articles, books, statistics, secondary interviews with stakeholders, and other relevant sources on the topic. The research design was extended by incorporating semi-structured interviews (Kvale & Brinkmann, 2009) on a purposive sample of refugee entrepreneurs and other industry stakeholders in the Finnish, Swedish, Danish, and Norwegian automotive aftermarkets. A qualitative approach was chosen in order to get more in-depth information about the interviewees which is preferable when studying entrepreneurship (Dana & Dana, 2005), in particular for cases of forced migrants, such as refugee entrepreneurs (Rodgers, 2004).

The research question for this study was *"How can the Nordic automotive aftermarket provide opportunities for the challenges of refugee entrepreneurs?"* The study explores and identifies specific challenges faced by refugee entrepreneurs and attempts to understand the strategies and coping mechanisms used by these individuals to overcome these obstacles. It explores the connection with the available opportunities in the automotive aftermarket industry, which provides one pathway to success and prosperity.

In order to conduct the research, an interview guide was developed for the semi-structured interviews (Merriam, 2009) with the participants. The guide included open-ended questions designed to elicit in-depth responses from the participants about their perspectives on refugee entrepreneurs in the Nordic automotive aftermarket. Then, a purposive sampling approach (Merriam, 2009) was used to identify a diverse group of refugee entrepreneurs as well as other stakeholders in the Nordic automotive aftermarket. Potential interviewees were selected based on their experience as refugee entrepreneurs or stakeholders/experts in this industry, as well as their willingness to participate in the study.

Several refugee entrepreneurs within the automotive industry were contacted through organisations, charities, and NGOs and invited to participate in the study. However, the majority of potential participants were unresponsive or did not wish to participate in this research. As a solution, and in order to be consistent, four cases (sufficient starting point according to Eisenhardt (1989)) of refugee entrepreneurs within the automotive aftermarket were identified and compiled from extant research (Hakkarainen, 2018; Ihanainen-Alanko, 2022; KirkkonummenSanomat, 2020; Prokirkkonummi, 2022; Saksa, 2016). They are presented and analysed as secondary empirical material in combination with primary empirical material from industry expert interviewees, a method inspired, e.g. from Schönwälder and Weber (2023).

The expert interviews took the perspective of industry stakeholders. Here two interviewees took part in the study, resulting in two expert interviews as primary empirical material to be presented and analysed in this paper.

1. The Sales and Marketing Manager of a Swedish Automotive firm—referred to as Interviewee 1 in later sections.

2. One of the Ambassadors of the independent automotive aftermarket in the Nordics—referred to as Interviewee 2 in later sections.

The interviews were conducted online as this was the most convenient for the participants, each interview lasting about 60 minutes. They were audio-recorded and transcribed for analysis, and the transcribed data from the interviews was coded using a thematic analysis approach (Braun & Clarke, 2006), which involves identifying common themes and patterns within the data. The themes that emerged from the data were used to answer the research question and meet the objectives of the study.

4 Cases of Business Entrepreneurs with Refugee Background in the Nordic Automotive Aftermarket

In this section the findings of the paper's research will be presented in two parts. First, the entrepreneurial perspectives will be introduced, then the findings of stakeholders' interviews.

4.1 Entrepreneurial Perspectives

The first story will be presented in Text Box 1 about a refugee entrepreneur, called Alan Karim Mohammad.

Text Box 1 Story of Alan Karim Mohammad

Alan Karim Mohammad was born in Sulaymaniyah, Iraq, in the Kurdish region during the Persian Gulf War. The boy has grown up in a middle-class family, where his father supported him and his mother from abroad and the mother took care of the household. Alan's brothers worked in the construction industry, while his little sister still went to school. As his father worked abroad, Alan has always dreamt about going abroad like his dad, discovering all the opportunities that the *'Golden land of Europe'* could offer.

Due to the war Alan was forced to leave his hometown and take off to find a safe place. On his way through Europe Alan built up his network from fellow refugees. Eventually he ended up in Finland without his passport, without any identification, just with a mobile phone in his pocket. After sharing his life story with an interpreter, he got his hands on a temporary passport, only to state that the person carrying the document has not been fully identified. He began his integration by starting to learn Finnish and tried to get to know as many new people as possible.

Alan applied for several training programmes; however, he fell short in most cases due to the language barrier. After many tries Alan got into a vocational school for auto mechanics; however, even after completing his training he couldn't get a job due to lack of experience:

"They always said that I was a beginner. At least five years of work experience was required."

After several declines for different job positions Alan has requested a loan from his father and brothers for €15,000 and bought a small basement workshop called AutoSky. He then started engaging into minor jobs, like car washing, changing tires or light bulbs but also doing small repairs. One of Alan's former vocational school friends joined him and younger student from the school attended his workshop for internship purposes.

"It was free labour for me. By doing it myself, I learn something new every day".

After his first year in business, Alan didn't earn anything, but used the revenue to get the tools he needed. In his second year, he was already able to pay back his father and brothers with interest without applying for any start-up money or grants. In his view, these opportunities required too much administrative work.

The next challenge Alan had to face was caused by growth. He soon needed to find a suitable office space in a good location in order to keep his business alive and continue its growth. He applied for a loan at Nordea, successfully received it and started renovating a former bus depot. For the renovation Alan had requested his brother Daban to join him, as he had extensive experience in the construction industry. Daban now owns 2/10th of AutoSky Oy and is a partner in his brother's workshop.

Alan believes that his entrepreneurial spirit comes from his childhood. His father in particular has encouraged and helped his sons to move forward in life from an early age. When asked, Alan highlighted a few characteristics he believes an entrepreneur needs to have in order to succeed:

"Determination, goal orientation, hard work and good social skills".

Another interesting aspect of Alan's journey is his integration into the Finnish society. As he describes his journey, he highlighted the role of his partner and their son, Romeo.

"I've been trying to adapt to Finnish culture. It wasn't easy, but when you are in daily contact with Finns, you always see and hear and learn."

Alan's journey continues until today, leading and working at his car repair shop and living in a happy family in Finland.

Source: Own work based on (Hakkarainen, 2018)

In Alan's story, there are numerous challenges of entrepreneurship mentioned; such as limited language barriers or instructional obstacles (Wauters & Lambrecht, 2008). Moreover, the story has demonstrated that language knowledge and social embeddedness (e.g.; having a Finnish family or getting to know fellow refugees) (Harima et al., 2021) contributed significantly to Alan's societal integration into the Finnish society aligning with theory. Interestingly, Alan has managed to secure a bank loan, that is unusual in most refugee entrepreneurs' cases, as most of them are unable to provide capital guarantee from such institutions (Sandberg et al., 2019; Wauters & Lambrecht, 2008). Alan has also expressed in his story that he experiences strong cultural differences in Finland, which he overcomes through his network.

The second story can be found in Text Box 2, introducing Lulzim Verbocvi who has started his own workshop in Kirkkonummi, Finland.

Text Box 2: Story of Lulzim Verbocvi

Lulzim was born in Kosovo and moved to Finland in his twenties in 1992 due to the break of Yugoslavia. Naturally, his mother tongue is Albanian; however, since leaving Kosovo he now speaks fluently in Finnish, Swedish, and English. Living in Rajakummu, Lulzim has a family including his wife and three children.

After arriving to Finland, Lulzim has started his entrepreneurial journey quite early with a Pizza Service back in 2003, which he has run together with his brothers. Nowadays, he also has a car repair shop, called Lulin Autohuolto, a car wash and a construction company in Kirkkonummi and he employs approximately 12 people.

Lulzim's car repair shop has become a success and he managed to be acquired by Mekonomen, a Swedish automotive aftermarket company.

Lulzim doesn't only contribute to his host country with his business activities, but is also considered a proactive member of the local area. He has applied for municipal elections, and is considered a sworn supporter of entrepreneurship. In his campaign, he wished to contribute to an entrepreneur-friendly Kirkkonummi by providing the necessary resources in terms of permits and infrastructure for new entrepreneurs in due time, so they can start their businesses with a reduced bureaucratic burden.

As he describes in his election campaign; family-centredness and children are also close to his heart, and he is a sworn supporter of child services. He relies strongly on his family and considers this a key support mechanism as one doesn't need to rely on so much external support.

Source: Own design based on (KirkkonummenSanomat, 2020; Prokirkkonummi, 2022)

Lulzim's story demonstrates well how one can become a social entrepreneur, proactively contributing to the interests of other fellow refugees and immigrants, as seen in the case of Type—3 refugee entrepreneurs (Harima et al., 2021). Another interesting aspect of Lulzim's story is his network embeddedness, as it not only enabled him to sell his business but also supported him in his integration into his new COR.

Text box 3 introduces the story of Razak Sahib, who arrived in Finland from Iraq as he was forced to leave his war-torn home. Interestingly, Razak became an employee in this story; however, he got hired by a fellow refugee entrepreneur Al-Ani Omar.

Text Box 3 Story of Razak Sahib

Razak Sabib lived with his parents, two sisters, and two brothers in their home in Baghdad, Iraq. He is a car mechanic by profession and didn't speak any foreign languages, only his mother tongue, Arabic.

Razak had strong roots in the automotive aftermarket, as he already started working at his father's car repair shop when he was only 10 years old. As he grew, he gained extensive experience and by the time he was 20 he already had his own car repair workshop in Iraq. Razak's life was going well before the war, as he had enough work to take care of himself and his bride. The sudden change of events came, when in 2003 US forces arrived to Iraq with the aim of evicting Saddam Hussein from power and looking for weapons of mass destruction. After 8 years, the American troops left; however, war hasn't stopped. A civil war had developed between the two Muslim groups, Sunnis and Shias during the years of occupation, although they had lived side by side in harmony before the war. Razak's car repair shop disappeared in chaotic circumstances like thousands of other businesses. He explained that he also lost his fiancée due to their religious identity, as Razak was Shia, while his bride was Sunni Muslim. As the civil war escalated, more and more people were being killed and disappeared without a trace. Razak was also one of many who has been persecuted by the Shiites. As he felt that his life was in danger, he fled to Finland through Turkey and Europe.

As Razak describes, things took a slower pace after arriving to Finland as the bureaucracy required a longer process and language barriers were proving to be challenging. He needed to wait 6 months for the Finnish Immigration Service to fully investigate his identity, how he came to Finland and explore his background, and soon after he had his papers, he was faced with the fact that he cannot join a car repair shop without knowing Finnish. As a fellow refugee entrepreneur explained to Razak;

"The problems start when a foreigner cannot talk to the customer about what is wrong with the car and what should be repaired. The pace of work is so busy that others don't have time to give advice. That's what it boils down to."

He soon realised that it might be challenging to join a Finnish car repair shop; however, as a professional who has not been afraid of risks and work since their childhood he decided to turn to his local diaspora for help. He found out that most of the car repair shops in the area are already in the hands of foreigners and there are many Iraqi car repair shops in Finland.

That is the way he met with Al-Ani Omar, a mechanic by his trade, who opened his own car repair shop after doing unpaid work for a whole year.

"I was doing unpaid work for two six-month periods, but I received support from the state and the municipality so that I was able to get by."

After Al-Ani Omar completed his training, he got a permanent job, worked hard and soon applied for start-up money for his own workshop.

This is the workshop where Razak found his place for now, planning to restart his own business, that was destroyed by the war in Iraq.

As Razak describes, language is still the most significant barrier in his life in Finland;

"People look at me in the city and know that I am an asylum seeker. I don't know if it's racism, fear, suspicion or something else. I can't explain to them who I am. I would like to tell them that I am not a nuisance or a danger to them."

Razak does not want Finns to think of refugees as economic immigrants, since in reality they are just desperately seeking for a *"normal life without war"*.

Source: Own design based on (Saksa, 2016)

In Razak's story, we are presented with Al-Ani Omar, a Type-4 refugee entrepreneur (Harima et al., 2021), who has managed to transfer their previous qualifications from Iraq to Finland through training and language learning.

Razak's story also demonstrates a case of refugee trauma (Pugliese, 2009), as Razak was forced to flee his home and his own business to avoid persecution. In Razak's case, we see strong suggestions on how social embeddedness can support refugees to find jobs, initiate entrepreneurial activities and develop one's resilience (Goodman et al., 2017; Nanda & Khanna, 2010), perhaps assisting them to eventually overcome their refugee trauma.

Lastly, Text Box 4 demonstrates the story of Ali Al-Mayahi, who arrived in Finland as a refugee from Iraq at only 13 years old.

Text Box 4 Story of Ali Al-Mayahi

Ali Al-Mayahi was born in Iraq and was forced to flee his home country while being only 13 years old. Soon after his arrival to his host country Ali graduated from primary school and then went on to study a profession. First, he studied to be a cook for less than 6 months, then he moved to study construction for less than a year.

Ali had already demonstrated early signs of initiative, as he stopped studying and started working as an employee at a pension's office, then at a slaughterhouse. As he describes, one of his main motivations was lack of capital.

"I stopped studying because I was short of money and because I couldn't really sit still in school. I got a job as a personal assistant in Pirkanmaa where I worked in a pension's office. Then I moved to Eastern Finland and worked in a slaughterhouse for less than a year."

While working at the slaughterhouse Ali bought second-hand cars with the goal of refurbishing and then selling them. He was so fascinated by the opportunities in this industry, that he soon started dreaming about having his own workshop.

"I have always wanted our service to be such good quality and affordable so the customer could recommend to their friends."

In 2022, Ali's dream had come true, as he founded his own workshop, called Tigers Cars Oy, offering car painting, maintenance, repair interior decoration and towing. He is currently operating with two full-time employees, both with Moroccan background. While Ali himself doesn't have any training in the industry, he aims to study and understand the processes better. Helping him at his side, his colleagues have already gained extensive experience and a degree in automotive engineering in Finland.

Interestingly, as he describes his workshop's competitive advantage stems from the industry practices in his home country, Iraq.

"In Finland, if a car has a fault, you have to pay for the inspection first. In Iraq you only pay for the repairs done, not for the inspections. We want to do the same here."

The entrepreneur also plans ahead; as he envisions future improvements in technology, he believes that second-hand car markets will thrive and the recycling of car parts will be more efficient, providing new opportunities for the automotive aftermarket sector.

When asked about his advice on how to become a successful entrepreneur he provided the following two advices:

1. *"Don't be afraid to start a business, but plan carefully."*
2. *"Avoid borrowing, save instead."*

Following his own advice, Ali dreams big, envisioning a nationwide network of workshops in every major Finnish town and city.

Source: Own design based on (Ihanainen-Alanko, 2022)

Ali's entrepreneurial journey can be referred to as a Type-2 refugee entrepreneur's case (Harima et al., 2021), as he is utilising his knowledge and previous experience to identify market-specific differences between his home country, Iraq, and his host country, Finland. Ali attempted to leverage this through his entrepreneurial journey, and proved those studies right that claimed many refugees often have the desire to start their own firms—thus becoming an entrepreneur driven by grasping an opportunity rather than necessity, such as discrimination or inability to get a job due to issues with supporting their skills (Alexandre et al., 2019; Mawson & Kasem, 2019; Sandberg et al., 2019).

4.2 Expert Perspectives

In this section the paper will introduce the perspectives of other stakeholders in the industry regarding the barriers refugee entrepreneurs might face in the Nordic automotive aftermarket. The section will first introduce the experts' opinions on market opportunities, followed by their takes on social integration and institutional barriers.

4.2.1 Market Opportunities

When it comes to market opportunities (Wauters & Lambrecht, 2008) both interviewees described that young people are more likely to be open-minded and interested in joining this industry compared to middle-aged entrepreneurs without any expertise. They also added that among the potential societal factors influencing one's decision in their job search, capital and family situation are the biggest influencing factors for the middle-aged entrepreneurs.

> I think it's a matter of age. Younger people are of course more open or and more interested to both joining this industry but also learning new technologies. As we see the youth understand better and better that this workplace is not so dirty as it used to be. It will still take many years to get to a clean workplace, we're talking at least 10 more years before the electrical cars or the hybrid cars are dominating in a workshop. So, this dirty piece will still be there with oil and grease and so forth, but if you look at the older generation, they are not as interested in learning new things. At least if they are workshop owners, everything depends on their salaries, saving money, and the family situation. (Interviewee 1)

Interviewee 2 pointed out the role of early network impressions on entrepreneurs in the automotive aftermarket when asking about what influences one to choose the automotive industry as their area of business.

> *Well, I think this depends on how you start your life. If your father is working with cars, then you are automatically getting into this business. I don't believe that school is the main or first influencer in most cases since they don't work as much with the workshops. Mostly, it's from friends who are generally interested in cars or from the family they grew up with.* (Interviewee 2).

A different approach, but certainly relevant to societal factors comes from Interviewee 1's perspective, who shares that often refugee entrepreneurs can start their journeys through the help of their families.

> *...some people obviously then move into garages or workshops, and it is very often related to their family or they are supported by their family. It could be the support of their father; it could be their siblings, brother-in-law and so on.* (Interviewee 1).

Another interesting aspect raised by the interviewees was the entrepreneurial opportunity based on geographical location (Wauters & Lambrecht, 2008).

It's kind of obvious what kind of job is available depending on where they live. So, if they live on the countryside, then either they become a farmer, or there is a small industry in the village. Perhaps there are workshops or grocery stores or something where they can work. But some people obviously then move into garages or workshops.... But on the city side, it's slightly different, even though it's a lot of word of mouth. So, if you have people in your neighbourhood who already work in some kind of workshops, whether it's manufacturers or brands, workshops, or independent workshops, then then they start to do a two-year high school, which is I think, still the most common education to become a mechanic. (Interviewee 1).

As the demand on the countryside is relatively lower than in a city, many entrepreneurs face the problem of getting the right inventory to these areas on time in the needed quality. According to Interviewee 1, this could lead not only to a decline in the quality of workshop service but to a negative reputation for refugees.

"I will not say it's a huge problem, but it is a problem. And it's especially difficult on the countryside. The biggest problem happens, when small workshops start to work on mechanic tasks without the required expertise and tools. For example, if they start to change an oil filter, and they buy filters that are not appropriate, not approved or came from somewhere which is not legal, then it could lead to a huge problem." (Interviewee 1).

4.2.2 Social Integration

One of the first topics that have been highlighted during the interviews was the role of languages. Both interviewees expressed that this is a topic that is often considered black or white; however, language can have many aspects in the case of refugees and should not be viewed as a simple matter. For example, Interviewee 1 described that, although the lack of language knowledge in the Nordics might seem discriminatory in certain cases, it really does matter when we take social integration or the ease of doing business into consideration.

"For example, if the refugees, potential customers or even workshop owners don't speak English or a common language - or at least not good enough - then it will take too much time and power and effort to conduct business with each other or to communicate in the team. This also could prove difficult when business owners wish to make their teams more productive or when potential customers attempt to describe the exact problems with their car. This is not against the refugees, but language barrier can definitely make business challenging." (Interviewee 1).

Interviewee 2 emphasised the perspective on languages when speaking about the required investment of tutoring in a workshop.

"It's [language differences – the authors] *not black and white. So, in most cases, it depends on what kind of skills do you bring with you as a refugee, because especially if you come to the countryside, to a small workshop with maybe three, four or five mechanics, they normally don't have the time to educate you as a new person, and in certain cases, especially if you don't speak the common language, they will not be patient enough."*

Another aspect the interviewees emphasised the role of language was in relation to storytelling. For example, Interviewee 1 emphasised that the ability for a refugee to introduce their skills and expertise to others is significantly impactful, especially when they are aiming to integrate in the COR.

> *"If a refugee can also sell themselves that can be very powerful. If they can tell others what is their history, talk about their knowledge and expertise, share what they have done in their past and what they can do now that is powerful. If this is paired up with a strong language knowledge like English, Swedish, Finnish or Danish then more or less immediately they will definitely become a valuable source for our industry. The good thing is that I believe there are many good opportunities today in the Nordics to learn these languages compared to other countries"* (Interviewee 1).

However, in order to reach a professional language knowledge efficiently, refugees are in need of support either from their compatriots or from an organization or authority.

> *"I agree, yes, it is difficult for a refugee to sell themselves without any support. Maybe it's slightly easier to start their journey in brand workshops, because there you already have many nationalities. So, the odds that the refugee can speak with someone, or someone can translate is much higher there. However, in many cases support organisations and authorities are needed to help these people to sell themselves."* (Interviewee 2).

Lastly, Interviewee 2 also emphasised the fact that almost half of the workshops have immigrants employed in Sweden, who can translate, train, and help their compatriots.

> *"If I say immigrants, and there you have also a spectrum of ages, because some of them, especially male, they already used to work with cars in their country where they are coming from. So, they are super good experts. And when they then approach, let's say, a garage or a workshop, and say, this is my resume, and they say wow, you are more than welcome. Please start tomorrow. If you travel around in Stockholm, Gothenburg or Malmö and go in workshops, you will realise that almost 50% are immigrants in the workshop, and they can help each other in many aspects"* (Interviewee 2).

4.2.3 Institutional Infrastructure

The last aspect the interviewees talked about was institutional background and market conditions.

Regarding institutions, Interviewee 2 highlighted that getting initial funding in the Nordics might be less challenging than perhaps in other geographical areas, as there is public support available. However, in order to get the actual funding, the refugee needs to have all their paperwork in order. As this might take some time, the timeline of getting these funds is often uncertain.

> *"To start the company as a refugee you can get some help from most Nordic governments, in case you have the right background and fill in your papers the right way. Unfortunately, this can be quite a lengthy process, but once you have your company, and you are in business then it's about convincing a chain or concept owner to secure a loan for getting the required inventory. This can be for example, three months of products that is needed, including tools and products needed for doing service on cars and you can get 90-, 120- or 180-days credit. Basically, it is a commercial agreement with the chain that you're talking to.*

> *For this you have a much higher chance in a city, where demand is generally higher."* (Interviewee 2).

Interviewee 2 as also expressed that in certain parts of the Nordic countryside refugees might not be most welcome, referring to the possibility of discrimination (Kupferberg, 2003).

> *"What we see is that there is a difference in the perspective on refugees depending on if it's on a countryside or city, but for different reasons. So, in general, I think refugee entrepreneurs can definitely be a valuable opportunity, but it depends on how one takes care of themselves and their business. This is a customer-facing job, where you work with people. You can easily grow your network, but you can also easily find dissatisfaction around you"* (Interviewee 2).

5 Discussion

While current academic literature has expanded on refugee entrepreneurship in recent years (see e.g.: Bizri, 2017; Lee, 2018; Sandberg et al., 2019; Wauters & Lambrecht, 2008), and many barriers of refugee entrepreneurship have been identified (Harima et al., 2021; Wauters & Lambrecht, 2008), research has been dominated by country-specific studies (see for example: Alexandre et al., 2019; Harima et al., 2021; Lee, 2018; Mawson & Kasem, 2019; Tavakoli, 2020; Turner, 2020) and was only attempting (Alrawadieh et al., 2018; Shneikat & Alrawadieh, 2019) to create industry-specific opportunity roadmaps for refugee entrepreneurs towards entrepreneurial success.

Upon analysing the findings of this paper, the specific challenges and opportunities of the Nordic automotive aftermarket will be presented in this section in three categories: market opportunities, social integration and institutions and market.

Regarding the *market opportunities* of the automotive aftermarket, empirical evidence has shown that the choice of sector can vary. Some choose it as they have previous qualifications and knowledge about the industry (Saksa, 2016), some have seen the opportunities arising from the market-specific differences (Ihanainen-Alanko, 2022), and others find their entrepreneurial opportunity due to the low entry barriers (Hakkarainen, 2018). As the studied entrepreneurs with a refugee background have chosen to engage in the automotive aftermarket industry, this supports the findings of Elo et al. (2019) that first-generation refugees tend to engage in low-entry barrier industries as it tends to be the most accessible when coming new to the COR.

All of the cases align with already existing theory (Gold & Kibria, 1993; Harima et al., 2021; Wauters & Lambrecht, 2008); however, there are a few common aspects also in this regard in the presented cases. For one, all refugee entrepreneurs had limited skills and initial resources, as they faced a general lack of financial capital and assets (Gold & Kibria, 1993; Wauters & Lambrecht, 2008). Secondly, in none of the cases were the entrepreneurs able to prove their skills and expertise through certificates and qualification. Expert interviews suggest that the automotive aftermarket is considered a rather practical industry, where demonstration of skills might also be acceptable, if the refugee is able to communicate those efficiently (Interviewee 1 and 2).

Interestingly, empirical data has also shown differences between entrepreneurial opportunities in the countryside and around the city as it seems that both funding opportunities as well as business opportunities are mainly focused on central cities in the Nordics (Interviewees 1 and 2).

As for *social integration*, the empirical data suggests that most of the introduced refugees struggled at one point with integrational issues (Hakkarainen, 2018; Ihanainen-Alanko, 2022; Saksa, 2016). All empirical data demonstrated that the most significant element in social integration is language knowledge, as this has enabled refugees to study and increase their social embeddedness (Barberis, 2017) in the host society. Interestingly, the data has also proven that the automotive aftermarket industry is able to support refugees in overcoming this particular challenge, as more than half of the Nordic workshops employ or are owned by immigrants or refugees (Interviewee 1). This enables refugee entrepreneurs to increase their network more efficiently and find members of their diaspora that can help them ingrate into their new COR (Saksa, 2016). As a plus, expert interviews suggest that options for refugees for acquiring the local language knowledge are widely accessible and affordable (Interview 1). This paper has also found that as the automotive industry is a customer-facing industry, it provides the opportunity for refugee entrepreneurs to increase their social embeddedness and learn on the job how to navigate the differences in doing business between their COO and COR (Interviewee 2).

Lastly, *institutional and market-related barriers* were analysed. Based on empirical evidence it has been found that bureaucracy indeed poses a barrier of refugee entrepreneurship (Wauters & Lambrecht, 2008) in the Nordic automotive aftermarket as well (Interviewee 2). However, it has also been found that in comparison with previous theory (Sandberg et al., 2019; Wauters & Lambrecht, 2008), funding opportunities are not that limited as there is public funding available, as well as commercial loans (Interviewees 1 and 2), and it is also possible to start a business without having any prior certification in the area (Ihanainen-Alanko, 2022).

As for market-specific needs, this paper has only found evidence in one particular case (Ihanainen-Alanko, 2022); however, in that case, entrepreneurial journey has started with identifying this particular difference, making them a Type-2 refugee entrepreneur (Harima et al., 2021).

As a summary of this paper's findings Table 2 illustrates the connections between entrepreneurial barriers for refugees and entrepreneurial opportunities in the Nordic automotive aftermarket.

Table 2 Refugee entrepreneurial barriers and entrepreneurial opportunities in the Nordic automotive aftermarket

Barrier categories	Challenge	Opportunity in the Nordic automotive aftermarket
Market Opportunities	Lack of foundation for choice of sector (Wauters & Lambrecht, 2008)	Former network connections in the industry are likely in most cases either through friends or family, as the industry has wide social and economic impact (Interviewee 1 & 2)
	Entry barriers are difficult to overcome due to resource constraints such as lack of investment, resources, prior experience, and knowledge (Gold & Kibria, 1993; Wauters & Lambrecht, 2008)	Industry has low entry barriers, as funding is accessible through financial institutions, family, and friends, and initial investment is considered smaller relative to other industries (Hakkarainen, 2018, Interviewee 1 & 2)
	Market-specific needs could require new product and service design solutions (Harima et al., 2021)	The industry is considered global and homogeneous; therefore, market-specific needs are limited (Interviewee 1 & 2)
Social Integration	Language knowledge might be required to fully integrate into COR (Harima et al., 2019, Shneikat & Alrawadieh, 2019, Wauters & Lambrecht, 2008)	The industry is already international; therefore, compatriots can support each other in language skills (Saksa, 2016) Additionally, language learning courses are available on distinct platforms with financial aid available (Interviewee 2)
	Potential discrimination or negative public perception (Kupferberg, 2003)	As the industry is a customer-facing one, it is characterised with significant network embeddedness (Interviewee 2). Empirical evidence showed that face-to-face interactions with customers could improve social integration and public perception of refugee entrepreneurs (KirkkonummenSanomat, 2020; Prokirkkonummi, 2022)
Institutions	Ability to prove skills could be limited due to regulatory requirements for certifications (Wauters & Lambrecht, 2008)	As the industry is practical, demonstration of skills might also be acceptable in many cases (Interviewee 1 & 2). Empirical evidence also showed that previous experience and skills are not necessarily required to achieve success (Ihanainen-Alanko, 2022)
	Lack of knowledge about local customs, laws, and institutions (Harima et al., 2019)	Knowledge about local customs, laws, and institutions is easily accessible through industry networks and through compatriots or fellow refugees (Interviewee 1 & 2)

Source: Own design based on relevant studies and collected empirical data

6 Conclusion

Migration has always been part of humans' lives; however, during the last couple of years forced migration has gotten significantly more attention from researchers, policymakers, and businesses (UNHCR, 2022). Refugees represent almost 30 million people today (UNHCR, 2023), with an expected continuous growth. One option for refugees to get back the control over their lives is initiating entrepreneurial activities in their host countries (Pieloch et al., 2016; Shepherd et al., 2020).

6.1 Research Contributions

The aim of this paper was to provide an insight into the opportunities that lie ahead in the automotive aftermarket for refugee entrepreneurs through qualitative research. The research question this paper aimed to answer was the following:

> *""How can the Nordic automotive aftermarket provide opportunities for the challenges of disadvantaged entrepreneurs?"*

This paper has found that the automotive aftermarket can indeed provide an entry point for refugee entrepreneurs in the Nordics. The paper provides the following research contributions:

1. Support for language learning is widely accessible and affordable in the Nordics.
2. Most refugees can choose this industry as the place for their entrepreneurial activities, since many of them tend to have former network connections in the industry, or perhaps previous knowledge and experience. Even if a refugee entrepreneur doesn't have any of the following, low entry barriers can be attractive to start their journey.
3. By joining this industry, refugee entrepreneurs can further their social integration efforts. One reason for this is that the industry is international in the Nordics, enabling them to learn about cultural differences from fellow compatriots and get support for language learning. Secondly, in this industry most jobs are customer-facing, thus new network connections can be easily created.
4. Market-specific needs are limited, as the industry is considered global and homogenous.
5. Ability to prove skills tend to be difficult for newcomers; however, due to the practicality of the sector, certificates are not required to start an entrepreneurial journey.
6. Knowledge about local customs, laws, and institutions is easily accessible through industry networks.

6.2 Practical Implications

This paper provides practical implications for newcomers, e.g., refugees and immigrants, managers, and policymakers.

For refugees the main implication of this paper is that the automotive aftermarket provides a suitable entry point for initiating their entrepreneurial journey. As the industry is rapidly growing and it has significant social and economic global impact with low entry barriers, it might be an opportunity worth taken into consideration.

This paper suggests that policymakers assess the current processes for the required paperwork and skill certification for incoming refugees, as this paper's empirical data suggests that most refugees are finding institutional processes and bureaucracy some of the highest barriers.

The paper recommends to the sector and businesses in the Nordic automotive aftermarket to create frameworks and support such positive developments by initiating projects and cooperating with refugee entrepreneurs (cf. Wu, 2018). If successful, not only would these projects contribute impactfully to Sustainable Development Goals but would also help refugees increase their resilience and overcome their refugee traumas.

6.3 Limitations and Research Outlook

This study has several limitations. Firstly, since most of the qualitative data came from secondary empirical material, the authors did not have the chance to ask follow-up questions or to navigate the conversation. As a result, the cases derived from the secondary data sources might not be fully comparable. Secondly, the study has only been focusing on refugee entrepreneurs in Finland, and there was lack of empirical data from other Nordic countries. This could alter the results, as the research was aimed for the entire Northern-European region, not just for Finland. Lastly, as a limitation, the sample size of this study was relatively small, that limited the generalisability of the results and focused mainly on the domestic business operations. As mentioned in Sect. 3, collecting data for this research has proven to be a significant challenge. Additionally, the sample consists mainly of refugee entrepreneurs who have fled from emerging countries to developed ones. Barriers of entrepreneurship for refugees might be different in emerging countries than developed ones, therefore, this study has only focused on entrepreneurial opportunity creation in the Nordic automotive aftermarket sector. Future studies should address the potential of disadvantaged entrepreneurs in creating transnational export businesses, like the case of Dillin Kone and E-P:n Kone ja Tarvike in Finland, where one newcomer from Iraq has been able to generate exports of over 1000 agricultural vehicles and has generated several jobs while doing so (Latva-Mantila, 2023).

As for the academic implications and research outlook, this paper found two other potential research topics that would require further investigation. Firstly, empirical data showed that the choice of sector in the automotive aftermarket varied a lot. Consequently, this paper suggests another study focusing solely on the choice of sector market condition in the case of refugee entrepreneurs in the Nordic automotive aftermarket. Secondly, this paper found that geography has high influence on market conditions; however, this has not been extended on in the context of refugee entrepreneurs. Therefore, another research is recommended to explore what affects refugee entrepreneurs from a geographical point of view to initiate their journeys.

Acknowledgement The authors acknowledge the financial support by NOS HS grant No. 329071 for the development of this study. Maria Elo gratefully acknowledges financial support for finalizing this research by the Research Council of Finland, grant No. 348534.

References

Adam, H., Obodaru, O., Lu, J. G., Maddux, W. W., & Galinsky, A. D. (2018). The shortest path to oneself leads around the world: Living abroad increases self-concept clarity. *Organizational Behavior and Human Decision Processes, 145*, 16–29. https://doi.org/10.1016/j.obhdp.2018.01.002

Alexandre, L., Salloum, C., & Alalam, A. (2019). An investigation of migrant entrepreneurs: the case of Syrian refugees in Lebanon. *International Journal of Entrepreneurial Behavior & Research, 25*(5), 1147–1164. https://doi.org/10.1108/IJEBR-03-2018-0171

Alrawadieh, Z., Cetin, G., & Karayılan, E. (2018). Understanding the challenges of refugee entrepreneurship in tourism and hospitality. *Service Industries Journal, 39*. https://doi.org/10.1080/02642069.2018.1440550

Barberis, E. (2017). Diversity, entrepreneurship and local embeddedness. A case study in Milan, Italy. *Argomenti: Rivista di Economia, Cultura e Ricerca Sociale, 7*. https://doi.org/10.14276/1971-8357.1052

Bizri, R. M. (2017). Refugee-entrepreneurship: a social capital perspective [Article]. *Entrepreneurship and Regional Development, 29*(9–10), 847–868. https://doi.org/10.1080/08985626.2017.1364787

Bonacich, E. (1973). A Theory of Middleman Minorities. *American Sociological Review, 38*. https://doi.org/10.2307/2094409

Bowring, A. G. P. (1990). The U.S. automotive aftermarket : opportunities and constraints for developing country suppliers. In *Vol. Industry and Energy Department working paper. Industry series paper*. World Bank Group.

Braun, V., & Clarke, V. (2006). Using thematic analysis in psychology. *Qualitative Research in Psychology, 3*(2), 77–101. https://doi.org/10.1191/1478088706qp063oa

Dana, L. P., & Dana, T. E. (2005). Expanding the scope of methodologies used in entrepreneurship research. *International Journal of Entrepreneurship and Small Business, 2*(1), 79–88.

Dustmann, C. (1996). Return migration: the European experience. *Economic Policy, 11*(22), 213–250. https://doi.org/10.2307/1344525

Eisenhardt, K. M. (1989). Building theories from case study research. *The Academy of Management Review, 14*(4), 523–550.

Elo, M., & Hieta, H. (2017). From ethnic enclaves to transnational entrepreneurs: the American dream of the Finns in Oregon, USA. *International Journal of Entrepreneurship and Small Business, 31*(2), 204–226.

Elo, M., & Vincze, Z. (2019). Transnational intrapreneurship: Opportunity development in transnational teams in the Nordic periphery. *International Journal of Entrepreneurship and Small Business, 36*(1/2), 103–125.

Elo, M., Täube, F., & Volovelsky, E. K. (2019). Migration 'against the tide': location and Jewish diaspora entrepreneurs. *Regional Studies, 53*(1), 95–106. https://doi.org/10.1080/00343404.2018.1549359

Elo, M., Zubair, M., & Zhang, X. (2022). Disadvantaged Migrant Entrepreneurs and Their Selection of Location: Entrepreneurial Settlement and Making a "Home" Abroad. In *Disadvantaged Minorities in Business* (pp. 35–72). Springer International Publishing.

Elo, M., Täube, F., & Servais, P. (2022). Who is doing "transnational diaspora entrepreneurship"? Understanding formal identity and status. *Journal of World Business*. https://doi.org/10.1016/j.jwb.2021.101240

Gericke, D., Burmeister, A., Löwe, J., Deller, J., & Pundt, L. (2018). How do refugees use their social capital for successful labor market integration? An exploratory analysis in Germany. *Journal of Vocational Behavior, 105*, 46–61. https://doi.org/10.1016/j.jvb.2017.12.002

Gold, S. J. (1988). Refugees and small business: The case of Soviet Jews and Vietnamese. *Ethnic and Racial Studies, 11*(4), 411–438. https://doi.org/10.1080/01419870.1988.9993613

Gold, S., & Kibria, N. (1993). Vietnamese Refugees and Blocked Mobility. *Asian and Pacific Migration Journal, 2*(1), 27–56. https://doi.org/10.1177/011719689300200102

Goodman, R. D., Vesely, C. K., Letiecq, B., & Cleaveland, C. L. (2017). Trauma and resilience among refugee and undocumented immigrant women. *Journal of Counseling & Development, 95*(3), 309–321. https://doi.org/10.1002/jcad.12145

Hakkarainen, T. (2018). *Uussuomalainen autotaivas—irakilainen Alan pyörittää monikansallista korjaamoa*. APU. Retrieved 2022-12-10 from https://www.apu.fi/artikkelit/uussuomalainen-autotaivas-irakilainen-alan-pyorittaa-monikansallista-korjaamoa

Harima, A., Freudenberg, J., & Halberstadt, J. (2019). Functional domains of business incubators for refugee entrepreneurs. *Journal of Enterprising Communities: People and Places in the Global Economy, 14*(5), 687–711. https://EconPapers.repec.org/RePEc:eme:jecpps:jec-11-2018-0081

Harima, A., Periac, F., Murphy, T., & Picard, S. (2021). Entrepreneurial Opportunities of Refugees in Germany, France, and Ireland: Multiple Embeddedness Framework. *International Entrepreneurship and Management Journal, 17*(2), 625–663. https://doi.org/10.1007/s11365-020-00707-5

Ihanainen-Alanko, S. (2022). *Suomessa ei kannusteta yrittäjyyteen, sanoo 21-vuotias irakilainen autokorjaamoyrittäjä*. Retrieved 11.13.2022 from https://www.kauppalehti.fi/uutiset/suomessa-ei-kannusteta-yrittajyyteen-sanoo-21-vuotias-irakilainen-autokorjaamoyrittaja/a5765f9d-c393-4750-9464-fdb30b2a8231

IOM. (2022). *What is migration?* https://www.iom.int/what-migration

IOM (n.d.). source: https://weblog.iom.int/false-dichotomy-betweeneconomic-migrants-and-refugees

KirkkonummenSanomat. (2020). *Lulin Autohuollosta osa Mekonomen-ketjua*. Retrieved 24.10.2022 from https://www.kirkkonummensanomat.fi/neo/?app=NeoDirect&com=6%2F159%2F56149%2Fe03c758bed

Kiss, A. N., Danis, W. M., Nair, S., & Suddaby, R. (2020). Accidental tourists? A cognitive exploration of serendipitous internationalisation. *International Small Business Journal, 38*(2), 65–89. https://doi.org/10.1177/0266242619884032

Kupferberg, F. (2003). The Established and the newcomers: What makes immigrant and women entrepreneurs so special? *International Review of Sociology, 13*(1), 89–104. https://doi.org/10.1080/0390670032000087005

Kvale, S., & Brinkmann, S. (2009). *Interviews: Learning the craft of qualitative research interviewing*. Sage.

Latva-Mantila, P. (2023). Irakista paennut konemies Dilli haluaa "elää ja tulla haudatuksi" Seinäjoelle—yksi asia Suomessa harmittaa. *Taloussanomat, 7*(10), 2023. https://www.is.fi/taloussanomat/art-2000009883518.html

Lee, S. K. (2018). From political activists to social entrepreneurs: Burmese refugees in South Korea. *Journal of Refugee Studies, 31*(3), 371–389. https://doi.org/10.1093/jrs/fey011

Massey, D. S., Alarcón, R., Durand, J., & González, H. (1987). *Return to Aztlan the social process of international migration from Western Mexico*. University of California Press. http://www.jstor.org/stable/10.1525/j.ctt1ppp3j

Mathivathanan, D., Kannan, D., & Haq, A. N. (2018). Sustainable supply chain management practices in Indian automotive industry: A multi-stakeholder view. *Resources, Conservation and Recycling, 128*, 284–305.

Mawson, S., & Kasem, L. (2019). Exploring the entrepreneurial intentions of Syrian refugees in the UK. *International Journal of Entrepreneurial Behavior & Research, 25*(5), 1128–1146. https://doi.org/10.1108/IJEBR-02-2018-0103

Merriam, S. B. (2009). *Qualitative research: A guide to design and implementation* (2nd ed.). Jossey-Bass.

Mohammed, M., Omar, S. N., & Kayadibi, S. (2016). Refugee MICROENTERPRISES. *Prospects and Challenges, 2*, 55–69.

Mueller, E. (2014). Entrepreneurs from low-skilled immigrant groups in knowledge-intensive industries: company characteristics, survival and innovative performance. *Small Business Economics, 42*, 871–889.

Mulholland, J., & Ryan, L. (2014). Doing the business: Variegation, opportunity and intercultural experience among intra-EU highly-skilled migrants. *International Migration, 52*(3), 55–68.
Mulvey, G. (2015). Refugee integration policy: The effects of UK policy-making on refugees in Scotland. *Journal of Social Policy, 44*(2), 357–375. https://doi.org/10.1017/S004727941500001X
Nanda, R., & Khanna, T. (2010). Diasporas and domestic entrepreneurs: Evidence from the Indian Software Industry. *Journal of Economics & Management Strategy, 19*(4), 991–1012. https://doi.org/10.1111/j.1530-9134.2010.00275.x
Pedrini, M., Bramanti, V., & Cannatelli, B. (2016). The impact of national culture and social capital on corporate social responsibility attitude among immigrants entrepreneurs. *Journal of Management & Governance, 20*, 759–787.
Pieloch, K., McCullough, M. B., & Marks, A. (2016). Resilience of children with refugee statuses: A research review. *Canadian Psychology/Psychologie Canadienne, 57*, 330–339. https://doi.org/10.1037/cap0000073
Portes, A., & Manning, R. D. (2008). The immigrant enclave: Theory and empirical examples.
Portes, A., & Shafer, S. (2006). Revisiting the enclave hypothesis: Miami twenty-five years later. *Princeton University, Woodrow Wilson School of Public and International Affairs, Center for Migration and Development, Working Papers, 25*. https://doi.org/10.1016/S0733-558X(06)25005-5
Prokirkkonummi. (2022). *Kuntavaaliehdokas - Lulzim Luli Verbovci*. Retrieved 24.11.2022 from https://www.prokirkkonummi.fi/kuntavaaliehdokas-lulzim-luli-verbovci/
Pugliese, J. (2009). Civil modalities of refugee trauma, death and necrological transport. *Social Identities, 15*(1), 149–165. https://doi.org/10.1080/13504630802693687
Rath, J. (2003). Undressing the garment industry. Immigrant entrepreneurship in seven cities. *Nederlands Tijdschrift Voor Geneeskunde - NED TIJDSCHR GENEESKD*, 253–285.
Rodgers, G. (2004). Hanging out' with forced migrants: Methodological and ethical challenges. *Forced Migration Review, 21*, 48–49.
Saksa, M. (2016). *Irakista paennut automekaanikko Razak lähti etsimään Suomesta töitä—vastassa karanteeni ja byrokratia*. Retrieved 2022-09-29 from https://seura.fi/asiat/tutkivat/irakista-paennut-automekaanikko-razak-lahti-etsimaan-suomesta-toita-vastassa-karanteeni-byrokratia/
Sandberg, S., Immonen, R., & Kok, S. (2019). Refugee entrepreneurship: taking a social network view on immigrants with refugee backgrounds starting transnational businesses in Sweden. *International Journal of Entrepreneurship and Small Business, 36*(1–2), 216–241. https://doi.org/10.1504/IJESB.2019.096967
Scherschel, K. (2011). Who is a refugee? Reflections on social classifications and individual consequences. *Migration Letters, 8*(1), 67–76.
Schönwälder, J., & Weber, A. (2023). Maturity levels of sustainable corporate entrepreneurship: The role of collaboration between a firm's corporate venture and corporate sustainability departments. *Business Strategy and the Environment, 32*(2), 976–990. https://doi.org/10.1002/bse.3085
Shepherd, D. A., Saade, F. P., & Wincent, J. (2020). How to circumvent adversity? Refugee-entrepreneurs' resilience in the face of substantial and persistent adversity. *Journal of Business Venturing, 35*(4), 105940. https://doi.org/10.1016/j.jbusvent.2019.06.001
Shneikat, B., & Alrawadieh, Z. (2019). Unraveling refugee entrepreneurship and its role in integration: empirical evidence from the hospitality industry. *The Service Industries Journal, 39*(9–10), 741–761. https://doi.org/10.1080/02642069.2019.1571046
Tavakoli, J. (2020). Cultural entrepreneurship of Sahrawi refugees. *African Identities, 18*(3), 279–294. https://doi.org/10.1080/14725843.2020.1777086
Triandafyllidou, A. (2018). Globalisation and migration: An introduction. In *Handbook of migration and globalisation* (pp. 1–14). Edward Elgar Publishing.
Turner, L. (2020). '#Refugees can be entrepreneurs too!' Humanitarianism, race, and the marketing of Syrian refugees. *Review of International Studies, 46*(1), 137–155. https://doi.org/10.1017/S0260210519000342

UNHCR. (2022). *What is a Refugee?* Retrieved 14.12.2022 from https://www.unhcr.org/what-is-a-refugee.html
UNHCR. (2023). *Data Finder: Population figures - UNHCR data on displacement.* https://www.unhcr.org/refugee-statistics/download/?url=2bxU2f
Vinogradov, E., & Isaksen, E. (2008). Survival of new firms owned by natives and immigrants in Norway. *Journal of Developmental Entrepreneurship (JDE), 13*, 21–38. https://doi.org/10.1142/S1084946708000831
Vinokurov, A., Trickett, E. J., & Birman, D. (2017). Community context matters: Acculturation and underemployment of Russian-speaking refugees. *International Journal of Intercultural Relations, 57*, 42–56. https://doi.org/10.1016/j.ijintrel.2017.02.002
Waldinger, R. (1989). Structural opportunity or ethnic advantage? Immigrant business development in New York. *International Migration Review, 23*(1), 48–72. https://doi.org/10.1177/019791838902300103
Wauters, B., & Lambrecht, J. (2008). Barriers to refugee entrepreneurship in Belgium: Towards an explanatory model. *Journal of Ethnic and Migration Studies, 34*(6), 895–915. https://doi.org/10.1080/13691830802211190
Williams, T. A., & Shepherd, D. A. (2016). Victim entrepreneurs doing well by doing good: Venture creation and well-being in the aftermath of a resource shock. *Journal of Business Venturing, 31*(4), 365–387. https://doi.org/10.1016/j.jbusvent.2016.04.002
Wu, G. (2018). *Ready for inspection—the automotive aftermarket in 2030.* https://www.mckinsey.com/industries/automotive-and-assembly/our-insights/ready-for-inspection-the-automotive-aftermarket-in-2030
Zhang, Z., & Chun, D. (2018). Becoming entrepreneurs: How immigrants developed entrepreneurial identities. *International Journal of Entrepreneurial Behavior & Research, 24*(5), 947–970. https://doi.org/10.1108/IJEBR-07-2016-0214
Zolin, R., Chang, A., Yang, X., & Ho, E. Y. H. (2016). Social capital or ethnic enclave location? A multilevel explanation of immigrant business growth. *Thunderbird international business review, 58*(5), 453–463.

David Breitenbach Research Assistant specializing in entrepreneurship, particularly focusing on migrant entrepreneurship, migration, and work in international contexts. Harnessing experience as a CMO and Partner in a global startup, his research navigates the complexities of entrepreneurial ecosystems, underscoring the influential roles and challenges encountered by migrant entrepreneurs in the evolving business landscapes.

Susanne Sandberg is an Associate Professor at the School of Business and Economics, Linnaeus University, Sweden. Her research spans within international marketing and entrepreneurship. Focus has been given to internationalisation processes of small and medium sized enterprises entering and taking off from emerging markets while current research covers transnational and refugee entrepreneurship as well as emerging market exit and resilience of SMEs. Publications have been made in e.g. Journal of Business Research, International Business Review, International Marketing Review, International Journal of Entrepreneurship and Small Business, and Journal of International Entrepreneurship.

Maria Elo full professor of international business and entrepreneurship at the University of Southern Denmark, professor at the BRIIB Institute of International Business at Shanghai University, Senior Research Fellow, University of Turku, Adjunct professor at Åbo Akademi University and Research Fellow at Institut für Arbeit und Technik works in the area of international business, -entrepreneurship and -migration with topics such as sustainability, business ecosystems, policy frameworks, networks, internationalization, mobility forms, human stickiness, migrant entrepreneurship, transnational and family businesses, diaspora investment and remittances. She has published books and articles, for example, in Journal of World Business, Journal of International Management, Journal of International Business Policy, Industrial Marketing Management,

Regional Studies, Journal of International Entrepreneurship and International Journal of Entrepreneurship and Small Business. Maria is one of the founders of Talents4AA and a Talents4AA ambassador leading the collaboration with the education sector.

Automotive Remanufacturing: The Ultimate Form of Circular Economy

Daniel C. F. Köhler

Abstract

In this chapter a brief overview of Remanufacturing in the context of circular economy and the automotive aftermarket will be provided. The multiple benefits that a company can derive out of a Remanufacturing process and a Remanufacturing business model will be explained. They reach from, e.g. Core-Control over reduction of quality costs to an improved environmental footprint. This chapter is the first publication that comprehensively describes the big management picture of Remanufacturing, mainly based on the role of an Original Equipment Supplier company in the automotive aftermarket. It also provides thoughts on the right identification of products that are suitable for Remanufacturing as well as some Remanufacturing specific key performance indicators. A special focus will be given to the Core, the lifeblood of Remanufacturing, as well as the necessary Core return system. The declaration of Remanufacturing as the ultimate form of recycling and circular economy will be underlined by a discussion on the sustainability dimensions of Remanufacturing, where examples of CO_2 savings of a remanufactured heavy duty trailer air disc brake are given.

D. C. F. Köhler (✉)
Business Unit Truck and Trailer Components, BPW Bergische Achsen KG, Wiehl, Germany
e-mail: KoehlerD@bpw.de

M. Elo, F. Katsardis (eds.), *Automotive Aftermarket*, Management for Professionals, https://doi.org/10.1007/978-3-031-62419-3_16

1 Introduction

Remanufacturing uses the old product (Core)[1] (see e.g. APRA Europe, 2016, p. 2) as an essential supply, material that would otherwise be handed over to the waste treatment,[2] but is here industrially processed. "Remanufacturing is the ultimate form of recycling" (Steinhilper, 1998) is a famous statement and book title of Rolf Steinhilper, retired Remanufacturing researcher, professor at the University of Bayreuth and member of the Remanufacturing hall of fame of the APRA Europe[3] (Automotive Parts Remanufacturers Association). It was already published in 1998. Thus, it is surprising that scientific as well as business publications about Remanufacturing are very rare and hard to find. If found at all, most of the publications focus on logistics or production process challenges.[4]

Therefore, this chapter tries to close the gap by providing a comprehensive overview of Remanufacturing in the context of the automotive aftermarket and the circular economy. In addition, a general introduction to Remanufacturing and its specialties will be given.

Competition between companies is continuously growing in a globalized world, especially in the automotive aftermarket. In the last decades, the aftermarket was a high revenue business where a lot of actors could easily make money. In our days, this is changing. The European parliament and court define the aftermarket as a level playing field (The European Parliament and the European Council, 2018) and a lot of companies rediscovered Remanufacturing as a very powerful weapon in the battle for the aftermarket.[5] Thus, it is more than overdue to describe Remanufacturing as a holistic business model and explain its multiple business benefits.

After providing all insights, it can easily be understood why the most sustainable business model—Remanufacturing—is not only the ultimate form of recycling (Steinhilper, 1998; Koehler, 2011, pp. 108, 114) but also the ultimate form of circular economy.

2 Remanufacturing and the Circular Economy

2.1 Circular Economy

Remanufacturing as a fully industrial process is an integral part of the circular economy. In Remanufacturing an old and used product, the so-called Core is processed and a new remanufactured product is produced. In the practical life, many similar

[1] For a more detailed definition see (APRA Europe, 2016, p. 2; Steinhilper and Hudelmaier 1993, p. 24; Dunkel, 2008, p. 26; Martens, 2011, p. 3f).

[2] The term waste treatment is defined as the recovery or disposal operations of waste. For more detailed definition see 2008/98/EG and Martens (2011, p. 3f).

[3] APRA Europe is an international European non-profit organization representing the whole Remanufacturing business field. APRA Europe honors outstanding personalities for their achievements in and for Remanufacturing with a hall of fame.

[4] Personal perception of the author.

[5] Personal perception of the author.

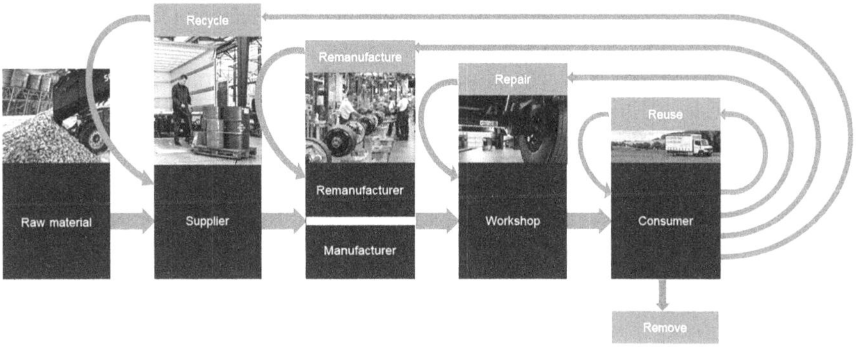

Fig. 1 Circular economy and Remanufacturing (based on Ellen Mac Arthur Foundation, 2013, p. 24, *source*: BPW, used with permission)

expressions are used for Remanufacturing. It is often called—only to mention some examples—overhaul, refurbishing, reconditioning, exchanging or regeneration. All of these terms and processes are not exactly defined and specified, while Remanufacturing represents the most comprehensive form with the highest degree of quality and product specification. The Remanufacturing process can be extended to a Regenerative Supply Chain under the special emphasis of the regeneration (Koehler, 2011, p. 114). Such closed-loop supply chains are part of a circular economy (Fig. 1).[6]

In Remanufacturing, most of the value adding processes of the new production e.g. primary shaping by forging processes, material altering like hardening and final fine shaping by milling are avoided and therefore regenerated (Koehler, 2011, p. 39). Thus, Remanufacturing is really on the forefront of the circular economy and can easily be distinguished from the classical recycling that works only on material level and can also be considered as downcycling (Ellen Mac Arthur Foundation, 2013, p. 25). As the output of the Remanufacturing process is a product which is in line with technical specifications, it can easily be distinguished from other end-of-life options like reuse or repair. Remanufacturing creates a completely new life cycle, while repair only fixes a specific defect in a product and makes it work again, thus only prolonging the existing life cycle.

3 Remanufacturing Process and Supply Chain

The Remanufacturing process typically consists of five different manufacturing steps: disassembly, cleaning, reconditioning, 100% checking and reassembly (see, e.g. Steinhilper, 1988, p. 47, 1999, p. 40; Koehler, 2011, p. 15). Figure 3 illustrates the Remanufacturing process of the pneumatic air disc brake TSB[7] of BPW Bergische Achsen KG (BPW), the German market leader for running gears for trailers. The first step is the disassembly of the Core. During disassembly, all

[6] For further information about circular economy see Ellen Mac Arthur Foundation (2013), p. 7.

[7] TSB is the abbreviation for the English-German term Trailer-Scheibenbremse.

components that are not intended to be reconditioned and regenerated are automatically scrapped. Figure 2 shows the Remanufacturing bill of materials of the BPW TSB Caliper. The dark grey components are high durable components that are forged (e.g. the lever, upper right component) or casted (e.g. caliper housing, upper left), while the light grey components are components that are directly scrapped in the disassembly, e.g. sealings, bellows, bushings and screws.

The disassembly is followed by the second step, the thorough cleaning of all components. When required, reconditioning processes are executed on the components. Then, all components that are intended to be regenerated are 100% checked. This process step is also a significant difference to a new serial production, where usually only capable processes exist and control sampling as well as statistics are applied. As the input quality and the wear and tear in the field operation of a Core are very individual, every reconditioned component, meaning 100%, must be checked on every relevant characteristic, meaning 100%. Finally, all components are reassembled to the remanufactured product using also new components, e.g.

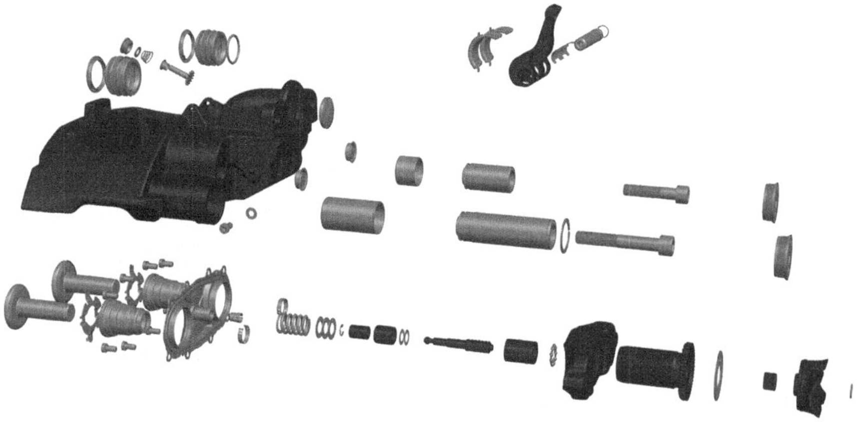

Fig. 2 Explosive drawing of the pneumatic air disc brake of BPW TSB (*Source*: BPW, used with permission)

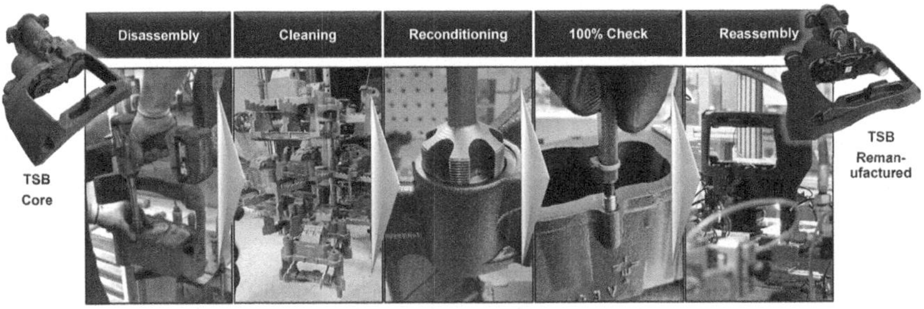

Fig. 3 Remanufacturing process of BPW TSB (Illustration of BPW.)(*Source*: BPW, used with permission)

sealings, bushings and screws. Sometimes the sorting of the Cores, before disassembly, is also considered to be part of the Remanufacturing process. In some cases, it can be smart to execute a basic inspection or functional test of the Core during the sorting process.

Besides the five process steps, a more comprehensive definition of the Remanufacturing is required. Consequently, in 2016 multiple international associations that are active in the automotive aftermarket, Remanufacturing or automotive aligned the following definitions under the lead of APRA Europe.[8]

> Remanufacturing is a standardized industrial process by which Cores are returned to same-as-new, or better, condition and performance. The process is in line with specific technical specifications, including engineering, quality and testing standards. The process yields fully warranted products. An industrial process is an established process, which is fully documented, and capable of fulfilling the requirements established by the remanufacturer. (APRA Europe, 2016, p. 2)

The Core that is also understood as the lifeblood of Remanufacturing is defined as "a previously sold, worn or non-functional product or part, intended for the Remanufacturing process. During reverse logistics, a Core is protected, handled and identified for Remanufacturing to avoid damage and to preserve its value. A Core is not waste nor scrap and is not intended to be reused before Remanufacturing." (APRA Europe, 2016, p. 2)

Furthermore, the output of the Remanufacturing process, the remanufactured part, "is defined as a part that fulfils a function which is at least equivalent compared to the original part. It is restored from an existing part (Core), using standardized industrial processes in line with specific technical specifications. A remanufactured part is given the same warranty as a new part, and it clearly identifies the part as a remanufactured part and states the remanufacturer." (APRA Europe, 2016, p. 1)

Not all products are suitable for Remanufacturing. The products need to have a certain basic value so that Core return and reman process efforts can be afforded. Products also need to have a certain design that they can be remanufactured, not mentioning that due to serial production cost optimization not a single product is designed to be remanufactured. At this point, it also must be mentioned that technically everything can be remanufactured in our days. Thus, Remanufacturing is not a technical question. It is rather a technical-economical question, as technical aspects determine the majority of the costs. In automotive there are a lot of products that typically are remanufactured. To mention only some: engines, transmissions, starters, alternators, turbochargers, brake calipers, gearboxes, water pumps, air compressors, A/C compressors and electronic control units.

---

[8] For more details see APRA Europe (2016), p. 1.

3.1 Remanufacturing in the Automotive Aftermarket

As described before Remanufacturing requires the Core as an essential input. As the Core is a previously sold, used, worn or non-functional product remanufacturing publicly only exists in the automotive aftermarket. Figure 4 shows a simplified overview of the European automotive aftermarket. On a very aggregated level, the automotive aftermarket operates as described in the following paragraph.

The Original Equipment Supplier (OES), also known as first-Tier Supplier, produces the new product, e.g. a turbocharger. The vehicle Original Equipment Manufacturer (OEM) assembles that product usually in serial production in his own products, e.g. luxury passenger car with a turbocharged combustion engine. Now, some years later, after the serial production of the luxury vehicle, a fleet or vehicle operator has a service event and needs to exchange a worn out or defective turbocharger in that vehicle. Therefore, the engaged workshop will order the service product from a wholesaler or distributor. Workshops usually only cooperate with some wholesalers and distributors. They fulfil the important functions of bundling demands and creating availability of spare parts and service products over hundreds of brands and part numbers. The wholesaler or distributor can usually order the turbocharger at the OES or the OEM. In some cases, the complete product flow can happen within the company group of the luxury car manufacturer, e.g. part of the organization can be a service or spare part organization that delivers the products to the luxury car workshops that are also owned by that same luxury car company group. In general, there are a lot of different constellations that make the aftermarket very complex.

As the commercial relation of OEM and OES is usually based on long-term serial delivery agreements, the OEM has quite competitive conditions versus the

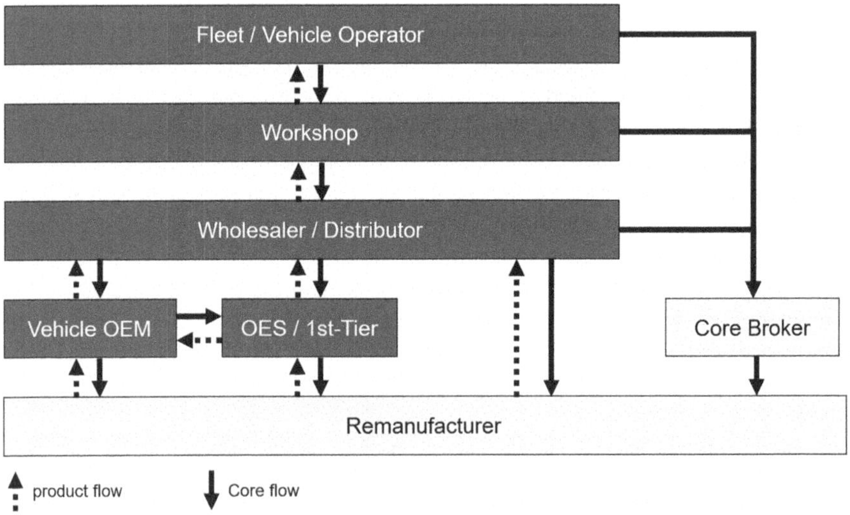

Fig. 4 Product and Core flow in the automotive aftermarket

OES. In addition, sometimes the OEM also forces a dual sourcing in serial production, resulting in two competing OESs in the aftermarket. This aspect is important to underline the role of Remanufacturing and competition in the aftermarket. In a lot of cases, Remanufacturing is the element that enables real competition in the aftermarket. Not every product is suitable for Remanufacturing, but products relevant for Remanufacturing are usually difficult to copy or to reverse engineer. Therefore, without Remanufacturing the automotive aftermarket would only have this small competition between OES and OEM. In the moment where a Core return takes place, new competitors can enter the automotive aftermarket for this specific product—Remanufacturers. In our example, the turbocharger may have a too demanding design to be simply copied, but Remanufacturing can easily be executed with the only need to reverse engineer, e.g. the sealings, bushing, bearings and screws and the overall product functions.

Figure 4 also illustrates the Core flow that can be initiated by all market participants. The Core flow basically follows the preliminary product flow but in an inverted direction. The Cores will enable Remanufacturers approach the market and intensify competition. Remanufacturers can be independent and operate for their own brands or can also work for the OEM or OES or for some wholesalers and dealers. Some OEM or OES also run their own Remanufacturing business units. With the Core flow, some experts for Cores also enter the market, the so-called Core Brokers or Core Dealers. They are specialized in finding and bundling Cores. They find available Cores in the market and even in, e.g. broken down vehicles on scrap yards. In addition, spare part suppliers enter the market that are specialized in, e.g. reverse engineering and manufacturing of sealings and supply the required spare parts and components to the remanufacturers. Remanufacturing suddenly enables a very competitive aftermarket.

4 Remanufacturing Business Model

4.1 Advantages and Benefits of a Remanufacturing Business

Remanufacturing is often only respected as a process that can create cheap service products for the aftermarket, and by doing so, it also saves some environmental impacts, e.g. kg CO_2[9] and kg new material, compared to the new production. Unfortunately, this view is much too narrow. Remanufacturing as a business model that is properly operated by a company can provide much more benefits. This is at the author's best knowledge the first time that a brief but comprehensive description of the benefits of the Remanufacturing business model is provided. As illustrated in Fig. 5, the benefits out of the view of an OES or OEM can be broken down into six main aspects.

[9] If not otherwise specified in this chapter CO_2 is understood as kg CO_2 equivalents measured with the Global Warming Potential over 100 years, kg CO_2eq_{100a}.

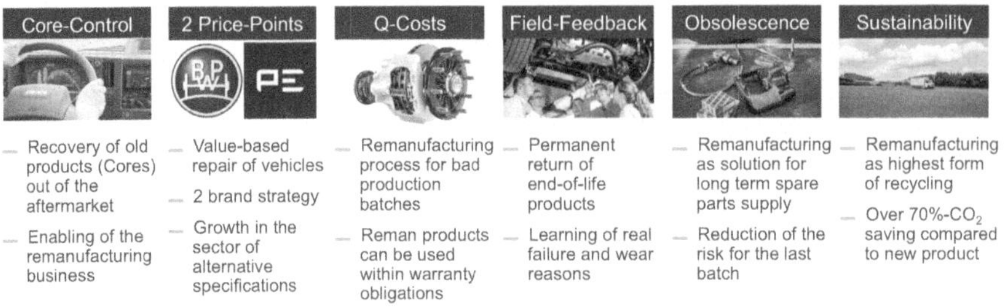

Fig. 5 Benefits of the Remanufacturing business model (*Source*: BPW, used with permission)

A first very important benefit is that with the Remanufacturing a Core return system will be established that will lead to a better Core-Control of the respective OE products in the market. This will drain the Cores from the market and provide a solid foundation for starting the Remanufacturing business. The customers, in that case the wholesalers and distributors, only accept participating in the Core return system when they can see also the opportunity of getting a remanufactured product offer.

The second possible benefit is that with a remanufactured product it is easy to establish a second quality or a second specification for the specific product, which is usually cheaper as the service new product. Usually, the design, the suppliers and the basic supply chain do not need to be changed. Therefore, it can be easily implemented. With that optional second brand or the reman line product, the sales department is enabled to better target customer groups. Typically, new and very valuable vehicles target for the OE service new product, while older and less valuable vehicles target for the remanufactured product.

A Remanufacturing process can even contribute to reduce quality and warranty costs (Q-costs). The Remanufacturing process provides a highly industrialized disassembly and reassembly process. In a lot of quality cases this process can be used to, e.g. efficiently rework bad production batches, or, e.g. to remanufacture products that have been recalled from the field. In some cases, it is possible to use remanufactured products for, e.g. a preventive exchange in the scope of warranty obligations or field risk avoidance.

Extremely valuable for the research and development and engineering department is to get the real feedback from the field. If a Remanufacturing and Core return system exists, the real failure modes and the real wear and tear of products can easily be identified and analyzed. In many cases, the technical departments only have products available out of long-term stationary tests or some long-term vehicle tests and of course from early field failures or warranty claims. In some cases, especially mechatronic products, even the memory chips can be read out to get better and detailed field statistics. All that helps to better understand the product and optimize and improve it or its next generation in the future.

An improved obsolescence management can be a fifth benefit of a Remanufacturing business model. Often, products that are close to the end of their life cycle become obsolete. It is often very difficult to maintain for only small aftermarket quantities a full assembly line or reinvest into expensive casting tools, especially when the sales quantity outlook for the next years is very small and foggy. Electronics and mechatronics can become even sooner obsolete as the general life cycle of electronic components is much shorter than a regular automotive product's life cycle. Remanufacturing can be a solution here or even lower the quantity risk of so-called last batch orders. In some cases, Remanufacturing can keep the complete product available in the aftermarket even if some components already became obsolete in new production, as these components can easily be reconditioned.

The last but not least benefit of Remanufacturing is the fact that Remanufacturing saves a lot of natural resources and also environmental impacts, e.g. more than 70% kg CO_2-equivalents (CO_2eq) compared to the new production. It helps promote sustainability and can be used in the sustainability and environmental reporting of the company. There even might be some customers that prefer to buy the greener Remanufacturing product.

4.2 Core Return System

As described before, the Core is the lifeblood of Remanufacturing and by far the most rare and crucial resource. Consequently, the Core return system a company has implemented is very important for the success of the Remanufacturing business. A Core return system can be on a very first view and basic approach compared with buying a crate of beer in a German supermarket. In Germany you have to pay a deposit for the beer crate and the beer bottles themselves in addition to the price for the fluid beer. Once the very tasty German fluid beer left the bottles, the crate and the empty bottles can be returned to the supermarket and the deposit will be refunded by the supermarket. In this comparison, the fluid beer would be the service product, that can be a new product or a remanufactured product, and the beer crate and the beer bottles would be the Core.

Based on the beer example we will describe the state-of-the-art Core return system out of the perspective of an OEM or OES company. In the European automotive aftermarket, the new or remanufactured service products are usually sold with a so-called upfront surcharge that is usually directly invoiced with the sales price of the product, therefore the term upfront. Depending on the conception of the Core return system the upfront surcharge might be defined as the price for a resale option. This means that the customer automatically also buys a resale option that entitles him to resell the Core to the supplier and certain conditions to be met. The higher the surcharge or the price for the resale option in comparison to the sales price, the higher the probability that the Core is returned. Usually, it is 30–50% of the sales price if the Core return system should be effective. It is very important that the quality and the status of the Core are specified, so that the customer is clearly aware under which conditions the surcharge is refunded or the Core can be resold to the supplier under the resale option. In addition, the surcharge or the resale option are timewise limited. Usually, they expire after 12–24 months.

If the OEM or OES operate a Core return system as described above, the wholesalers and dealers forward the surcharge or resale option in a similar way to the workshop and the workshop to the fleet or vehicle operator. It is also common that a physical one-to-one exchange is directly done. Independent remanufacturers operate in a similar manner as the OEM or OES. It was experimented with other methods, e.g. buyback programs or delayed surcharges, all of them did not turn out to be effective.

The Core return system leads to a significant logistics effort in the reverse supply chain. The wholesalers and dealers need to sort the Cores according to the different suppliers and need to organize shipments and transports. All of this only works as it is somehow common in Europe and the market participants recognize the benefit of having a remanufactured product offer with a reasonable price. In some European countries, it is—due to legal limitations—still quite difficult to establish a cross-border Remanufacturing business. An example would be Turkey. The resale option as well as the surcharge are of course recognized somehow as a part of the price or the price point. Therefore, the effects of surcharge-less or resale-option-less reimports, e.g. from Africa or Middle East, must be respected when drafting the Core return system.

As mentioned above, the returned Core needs to fulfil certain quality requirements so that the surcharge or resale option can be refunded. This Core quality is usually checked and confirmed as well as the basic Core identity by a Core sorting and inspection process. The operation of a Core return system is highly complex. Usually, a so-called Core bank account is required per customer where the resale options are managed. ERP systems usually struggle to provide the proper setup to manage this. Therefore, a lot of companies have own programmed IT solutions. There are also companies that provide this Core return system operation as a service. They offer the complete Core return system operation from the physical logistics up to the Core bank customer accounting of the resale options as a service. They sometimes even extend their services, e.g. also to warranty and claim process management, as this also heavily involves reverse logistics and a Core.

The relevant key performance indicators (KPI) that measure the quality and performance of the Core return system are the Regeneration Rate Sorting (RR_{Sor}) and the Core Return Rate (CRR).

4.3 Key Performance Indicators in Remanufacturing

The Remanufacturing process and business model is a very special and not so common model. Thus, its management also requires a special approach. Of course, the very basic management figures apply also for the Remanufacturing business in a company. These include for example, the sales figures, the profit and loss statements, the productivity of the reman process and so on. The first very important Remanufacturing KPIs are the Regeneration Rates. As scrap is a common element in the Remanufacturing process and the circular economy, it needs a special attention. Compared to serial new production it has a significant level, thus its

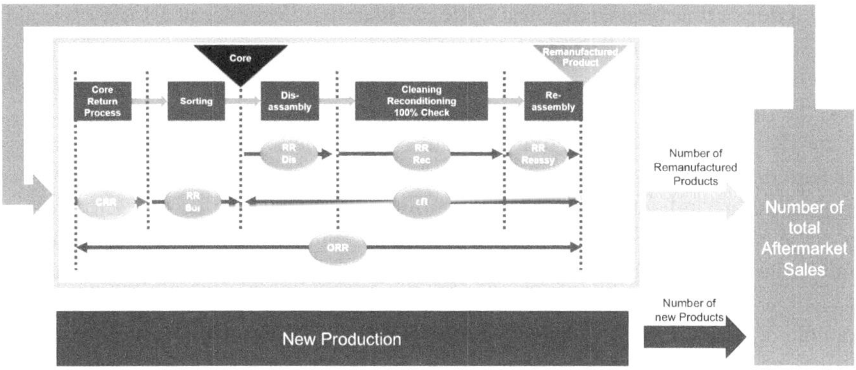

Fig. 6 Remanufacturing KPI process overview

management is more important than regular approaches to process optimization where, e.g. process times are optimized. The Regeneration Rates (RR) exist for every relevant process step (Koehler, 2011, p. 41). It is important that the complete Remanufacturing supply chain is covered by the RR. This is necessary to have no blind spots in the end. An RR can never exceed 100% and is basically defined as the process output divided by the process input. Usually, the Remanufacturing supply chain is split into the following RR[10]:

1. Core sorting process: RR_{Sor}
2. Disassembly process: RR_{Dis}
3. Reconditioning process (including cleaning and 100% Check): RR_{Rec}
4. Reassembly process (including end of line functional testing): RR_{Reassy}

A very important KPI can in the end be calculated, the so-called external RR (eR) (Koehler, 2011, p. 42). Figure 6 illustrates the processes that eR covers as well as the general overview of the KPIs in this paragraph. eR is the product of RR_{Dis} RR_{Rec} and RR_{Reassy}. It is the KPI that describes the general capability and performance of the Remanufacturing Process. RR_{Sor} on the other hand describes mainly the quality of the returned Cores. The RR are not so easy to derive out of the common ERP systems, but it is possible. The aggregation level in the RR is switching from the complete Core or product to the components. This becomes especially challenging when the RR are combined to the eR. Here, every company needs to define for its own business what is most suitable. The eR and RR can, e.g. be defined by the bottleneck component, the component with the lowest RR or the most valuable component.

Another important figure is the CRR. The CRR describes the general capability and performance of the Core return system. CRR is defined by the realized return options that were due in a time frame, divided by the return options that were in

[10] For more details about Regeneration Rates see Koehler (2011), p. 41.

general due in that time frame. The CRR can usually never exceed 100%.[11] A good Core return system usually operates at a CRR above 90%.

In the context of circularity, the RR and CRR can be combined to the Overall Regeneration Rate (ORR). The ORR can be compared to the Remanufacturing Share (RS). The RS is defined as the sold Remanufacturing products, divided by the sum of the sold Remanufacturing products and the sold equivalent service new products. This indicator provides interesting information. If the ORR is greater than the RS, then the Remanufacturing process and the Core return system can supply the market with Remanufactured products. A circular economy is achieved. If ORR is lower than the RS, the Remanufacturing and Core return system is not capable of supplying the market with remanufactured products. This will lead to a situation where orders cannot be fulfilled, or the company decides to fill in new components or products. This is to a certain degree common in the Remanufacturing business and depending on the products. These products are then called Infill or Infill New.

To better illustrate the described KPIs we will draft an example. An undefined OES company sells per year 60.000 units of a product family X. As they did not invest a lot into sales and marketing of product family X they sell 40.200 units of remanufactured products from X. This results in an RS of 67%. The Core return system is very customer friendly and well established, thus from the 60.000 due resale options for the customers realized 54.600 per year, resulting in a CRR of 91%. Unfortunately, in the Remanufacturing workshop 13% of all Cores in the warehouse inbound and sorting process of X must be scrapped due to heavy corrosion, what results in an RR_{Sor} of 87%. The Remanufacturing process is quite effective but since X had its start of serial production already 20 years ago, a lot of the Cores are quite old and worn out, so that in the 100% quality check a lot of the bearing tolerances are not okay and cannot be reconditioned. This results in an eR of 73%. The overall process of product family X led to an ORR of 57,8% (= 91% X 87% X 73%) that is lower than the RS of 67%. Consequently, not all customer orders for remanufactured products can be supplied with really remanufactured products, which would be 34.676 remanufactured products per year in our example, leading finally to the case of Infill products.

4.4 Remanufacturing Product Portfolio

It is often unclear which products are suitable and which products are not suitable for Remanufacturing. To answer the question, which products can be part of the Remanufacturing product portfolio, a comprehensive analysis of the technical characteristics as well as some market characteristics must be executed. As described

[11] In rare cases the CRR can be higher than 100%. This requires in a first step that the Core return system allows so-called overreturns. Second the Cores then overreturned either origin from already in the past expired resale options or come from sources outside the Core return system, e.g. from Core Brokers.

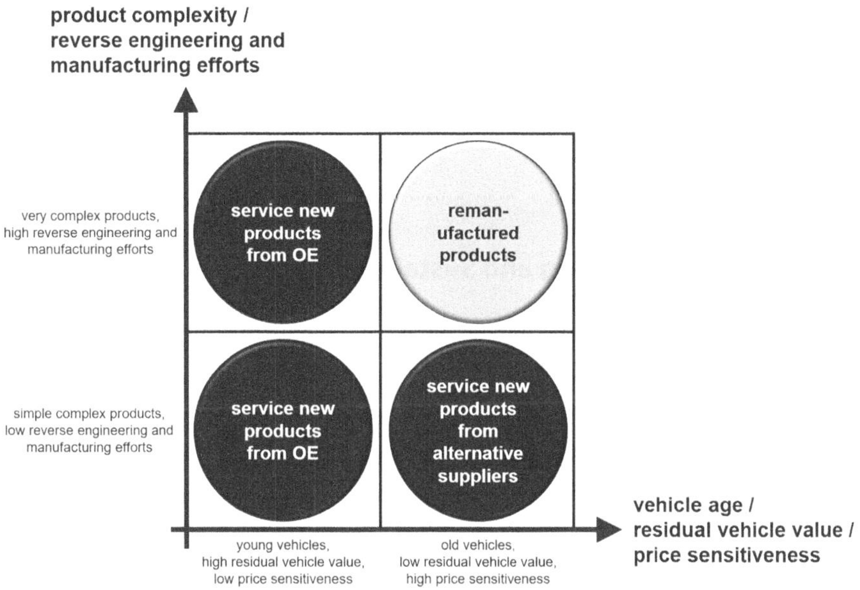

Fig. 7 Automotive aftermarket sectors with preferred service solution product

above, on the one hand, products for Remanufacturing need to fulfil certain criteria as, e.g. minimum value, technical design and wear and tear behaviour. On the other hand, the products should also have a market and based on the market it needs to be possible to organize a proper Core return system. As a result, the automotive aftermarket can be roughly divided into four sectors that define the preferred service solution product as described in Fig. 7.

On one axis, the market is divided into the young and old aftermarket. For young vehicles that require a service, owners tend to purchase a service new OE product, being supplied from the OEM or the OES. This is similarly valid for vehicles that have a high residual value, that can be, e.g. in heavy duty some dangerous goods transporters like tankers. Besides this, there can also be just very price-sensitive markets or customers. On the other end of the axis, there are older vehicles or vehicles that have a low residual value. Owners of such vehicles tend to purchase more alternative and pricewise more sensitive service products, e.g. remanufactured products or products from alternative specification manufacturers. The second axis that divides the automotive aftermarket is a more technical axis. We can distinguish here complex from less complex products. Complex products are products that are in practical business also called captive products. These are products that cannot easily be reverse engineered and manufactured by alternative non-OE players. These are products that are very complex as they, e.g. combine mechanical and electronical parts and software. These can also be products that require a very expensive and special tooling and therefore there is no business case for alternative manufacturers to step into that market. Less complex products are market-wise not suitable for

Remanufacturing, even if technically a Remanufacturing would be easily possible, as usually alternative manufacturers and suppliers will approach the market. When these alternative manufacturers approach, they will sell their reverse engineered new products without a resale option or surcharge. As this is recognized as a part of the price and leads to additional efforts, the market will hardly buy and ask for remanufactured products.

5 Remanufacturing and Sustainability

Sustainability usually consists of three dimensions, the economical, the social and the ecological dimension. In the public awareness, the ecological dimension of sustainability is the most dominant during the transformation discussions to a circular economy. To measure the ecological dimension, the Life Cycle Assessment (LCA) is used. It is a method in the field of environmental management, defined in the standards DIN EN ISO 14040 and 14,044 and used for the measurement of the environmental impact of systems.[12]

An LCA regards all inputs and outputs in the form of energies and materials of any system and evaluates their environmental impact, thus it can be very well applied for a supply chain, which is identical also with the production phase of a life cycle (LC). Interactions occurring between a system and the environment, so-called environmental impacts have a vast spectrum and therefore are also quantified and measured in a lot of different ways in LCAs.[13] The actual interest of the public awareness of summing up all environmental issues to CO_2-aspects is based on enormous preliminary scientific results. These results manifest themselves in a well-proven scientific method to measure environmental impacts, the Global Warming Potential (GWP), which was developed by the Intergovernmental Panel on Climate Change (IPCC)[14] (IPCC, 2007, p. 20; Jungbluth, 2007, p. 120; Klöpffer & Grahl, 2009, p. 252). A result of the IPCC's investigations is the responsibility of greenhouse gases (GHGs)[15] for climate change and global warming. The developed method is an accepted scientific method for the aggregation of the climatic effects of greenhouse gas emissions (IPCC, 2007, p. 23; Klöpffer & Grahl, 2009, p. 252). The aligned target values of the Kyoto Protocol are based on the GWP and it is also

[12] See DIN EN ISO 14040 (2006a, p. 15ff.); DIN EN ISO 14044 (2006b, p. 15ff). and as representatives Bundesumweltministerium & Umweltbundesamt (2001, p. 299ff.), Hermann (2010, p. 152ff) and Klöpffer and Grahl (2009, p. 12ff).

[13] Other major characterization categories are fossil resource consumption, eutrophication, acidification, terrestrial and aquatic ecotoxicity, acidification, terrestrial eutrophication and human toxicity.

[14] The IPCC is playing a very important role in the science of climate since 1989. For further information see: www.ipcc.ch.

[15] The term greenhouse gases are defined as the tracer gases in the troposphere, which are causative for the anthropogenic greenhouse effect. They are partly identical with the natural greenhouse gases. See ISO 14064-1 p. 1, IPCC (2007, p. 23ff) and Klöpffer and Grahl (2009, p. 253).

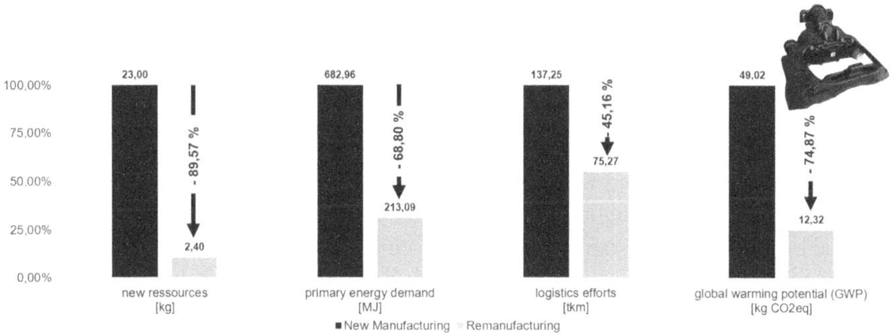

Fig. 8 Environmental savings of Remanufacturing compared to new manufacturing of the BPW TSB (Internal study of BPW Bergische Achsen KG in 2022. Life cycle assessment according to DIN EN ISO 14040 and 14,044. Cradle-to-gate study, calculation of the emissions according to IPCC 2013 (GWP100). Modelled with openLCA 1.10.3 and Ecoinvent 3.7.1. Ready-to-ship air disc brake TSB as a functional unit. Environmental impact of Core considered, not given for free. Based on the published LCA methodology for the comparison of new production and Remanufacturing of Koehler in 2011) (*Source*: BPW, used with permission)

preferably used for the impact assessment of LCAs.[16] Since different greenhouse gases have different environmental impacts in the atmosphere, because their radiation absorption behaviour is different (radiative forcing), CO_2eq are used as an entity, thus the GWP of different GHGs is normalized with the GWP of CO_2 (IPCC, 2007, p. 20). Furthermore, it has to be taken into account that GHGs degrade at different rates in the atmosphere (Jungbluth, 2007, p. 120; IPCC, 2007, p. 20). This fact forces a time component in the conversion of GHGs into CO_2eq. The SETAC (Federal Environment Agency) recommends an application of the 100-year-baseline (100a), which is also applied in the Kyoto Protocol.[17] The method is described as GWP 100a (IPCC, 2007, p. 33).[18]

The BPW Bergische Achsen KG executed an LCA study for their TSB Remanufacturing in 2022 according to DIN EN ISO 14040 and 14,044 as well as according to the published LCA methodology for the comparison of new production and Remanufacturing of Koehler in 2011. The impact assessment of the Remanufacturing is remarkably positive as Fig. 8 demonstrates, compared to the SCs of new production. Figure 8 provides a detailed view of the results of the life cycle impact assessment.

With the Remanufacturing of the BPW TSB, 89,57% of the product's mass is regenerated, which is identical with the avoidance of 74,87% of the CO_2eq_{100a} as

[16] See UNFCCC (1997, p. 34) of the Kyoto Protocol in the appendix "Beschluss 2/CP.3 Methodische Frage hinsichtlich des Protokolls von Kyoto", paragraph 3.

[17] Recommendation of Federal Environmental Agency of Germany, cited from Hermann (2010, p. 158) and UNFCCC (1997, p. 34) of the Kyoto Protocol in the appendix "Beschluss 2/CP.3 Methodische Frage hinsichtlich des Protokolls von Kyoto", paragraph 3.

[18] Following this method sulfurhexafluorid (SF6), the strongest greenhouse gas, has a 22.800-times higher GWP than CO_2 (GWP_{100a}(1 kg SF6) = 22.800 kg CO_2eq_{100a}) and therefore a correlating high contribution to the global warming than an identical mass of CO_2.

well as 68,80% of the MJ-equivalents (MJeq).[19] The required transfer processes of the TUL-logistics[20] measured in logistics efforts of tkm, mainly transports, show a reduction of 45,16%, despite the great efforts that are required for the Core return system. Other companies as well as Koehler, 2011 published similar convincing figures that demonstrate the great ecological advantage Remanufacturing can bring (Bosch, 2010, pp. 3, 59 and Koehler, 2011, p. 104).

A comprehensive evaluation of Remanufacturing as a very sustainable business also needs to address the social and the economical dimension. For the economical dimension it can be easily stated that there are a lot of professional OEM and OES companies that have a dedicated Remanufacturing product line and brand as well as own business units or plants. This facts are indicating that Remanufacturing definitely also has a positive economical dimension. Not mentioning companies which only exist due to the possibility of Remanufacturing like the remanufacturers and Core Brokers.

A way to measure the social dimension of a business could be the number of jobs that are created and supported by this business. The Remanufacturing processes are commonly very manual processes. Due to the non-defined quality and damages of the Core as well as the high variety of part numbers and variants, the Core sorting as well as the disassembly process are still highly manual. The majority of all costs in the production costs of a Remanufactured product are therefore labour costs. When this is compared with newly produced products in serial production where the major share of all costs for automotive parts are material costs, driven by raw material prices and energy, it can be easily stated that Remanufacturing is much more labour-intensive as new production. This leads also to a positive social dimension. Per value added or production costs of a Remanufacturing product, much more jobs are created than of a newly produced product. Remanufacturing shows a positive performance in all three sustainability dimensions when compared to new production.

6 Conclusion and Outlook

It could be demonstrated that Remanufacturing can save more than 74% of CO_2eq_{100a} compared to new production while still creating a product with a new life cycle that is in its functions at least equivalent to a new product. Reflecting this, together with the aspect that Remanufacturing also has a tremendous positive contribution to all three dimensions of sustainability, it can be concluded that Remanufacturing is not only the most sustainable business model but also the ultimate form of circular economy.

Looking into the future, the automotive aftermarket will see more and more companies that develop their Remanufacturing business to secure their aftermarket

[19] The MJeq-data was derived from the identical systems as the CO_2eq-data. The cumulative energy demand method was used.

[20] The processes of TUL-logistics are defined as the common transfer processes of transportation (German: Transport), handling (German: Umschlagen) and storing (German: Lager).

business and competitiveness. These initiatives will be supported by the positive ecological and sustainable aura of Remanufacturing. More publications and research on the comprehensive business model of Remanufacturing would be helpful to support that transition to more circularity in the automotive aftermarket. Remanufacturing is more than an industrial process—it is a beneficial business model and a sharp weapon in the aftermarket.

References

APRA Europe. (2016). Remanufacturing Associations agree on international industry definition. www.apraeurope.org

Bundesumweltministerium & Umweltbundesamt. (2001). *Handbuch Umweltcontrolling*. Franz-Vahlen-Verlag.

DIN Deutsches Institut für Normung e.V. (2006a). *DIN EN ISO 14040. Umweltmanagement—Ökobilanz—Grundsätze und Rahmenbedingungen (ISO 14040:2006). Deutsche und Englische Fassung EN ISO 14040:2006*. Beuth-Verlag.

DIN Deutsches Institut für Normung e.V. (2006b). *DIN EN ISO 14044. Umweltmanagement—Ökobilanz—Anforderungen und Anleitungen (ISO 14044:2006). Deutsche und Englische Fassung EN ISO 14044:2006*. Beuth-Verlag.

Dunkel, M. (2008). *Methodenentwicklung für Lean Remanufacturing*. Shaker-Verlag.

Ellen MacArthur Foundation. (2013). Towards the circular economy Vol.1. Economic and business rationale for an accelerated transition. https://ellenmacarthurfoundation.org

Hermann, C. (2010). *Ganzheitliches life cycle management*. Springer-Verlag.

IPCC. (2007). Climate change 2007: The physical basis. Contribution of Working Group I. 2007. In S. Solomon, D. Qin, M. Manning, C. Zhenlin, M. Marquis, K. Averyt, M. Tignor, & H. Miller (Eds.), *Fourth Assessment Report of the Intergovernmental Panel on climate change* (p. I–989). Cambridge University Press.

Jungbluth, N. (2007). IPCC 2001 (climate change). In R. Frischknecht & N. Jungbluth (Eds.), *Implementation of life cycle assessment methods* (Ecoinvent report No 3, v2.0) (pp. 120–129). Swiss Center for Life Cycle Inventories.

Klöpffer, W., & Grahl, B. (2009). *Ökobilanz (LCA)*. Wiley-Verlag.

Koehler, D. (2011). *Regenerative supply chains*. Shaker-Verlag.

Martens, H. (2011). *Recyclingtechnik*. Spektrum Akademischer-Verlag.

Robert Bosch GmbH. (2010). *Geschäftsbericht 2009* (pp. 1–130). Robert Bosch GmbH Unternehmenskommunikation (C/CC). http://www.bosch.com/content/language1/downloads/GB2009_De.pdf

Steinhilper, R. (1988). *Produktrecycling im Maschinenbau*. Springer-Verlag.

Steinhilper, R. (1998). *Remanufacturing: The Ultimate Form of Recycling*. Fraunhofer IRB-Verlag.

Steinhilper, R. (1999). *Produktrecycling*. Fraunhofer IRB-Verlag.

Steinhilper, R., & Hudelmaier, U. (1993). *Erfolgreiches Produktrecycling zur erneuten Verwendung oder Verwertung*. Druck Partner Rübelmann-Verlag.

The European Parliament And The Council Of The European Union. (2018). Regulation (EU) 2018/858 of the European Parliament and of the Council of 30 May 2018 on the approval and market surveillance of motor vehicles and their trailers, and of systems, components and separate technical units intended for such vehicles, amending Regulations (EC) No 715/2007 and (EC) No 595/2009 and repealing Directive 2007/46/EC. Official Journal of the European Union.

UNFCCC. (1997). United Nations Framework Convention on Climate Change. Das Protokoll von Kyoto zum Rahmenübereinkommen der Vereinten Nationen über Klimaänderungen. Sekretariat der Klimarahmenkonvention (Hrsg.). http://unfccc.int/resource/docs/convkp/kpger.pdf

Dr.-Ing Daniel C. F. Koehler is a German business engineer and holds a doctorate degree in engineering. Since 2021, he is head of the research and development department of the business unit Truck and Trailer Components of the BPW Bergische Achsen KG. For today, he has more than 10 years of experience in higher technical management positions, e.g., he was 7 years in charge of the Remanufacturing business of Knorr-Bremse Systeme für Nutzfahrzeuge GmbH. For more than 15 years, Dr. Koehler is involved in the Remanufacturing business as well as the automotive industry. Since 2017, he also has the honour to be the elected Chairman of the international non-profit association Automotive Parts Remanufactures Association Europe AISBL (APRA Europe). While he is regularly invited for holding speeches and presentations with a focus on technical management, business development, and evolution, Dr. Koehler also irregularly publishes expert articles.

Leveraging Car Connectivity in the Automotive Aftermarket and Beyond

Felix Sterk

Abstract

The research objective of this article is to explore and enhance the general understanding of how automotive companies operating in the automotive aftermarket and beyond conceptualize and design business models and leverage platform ecosystems to capture value from connected cars. Since players in the independent aftermarket do not enjoy exclusive in-vehicle data access like original equipment manufacturers (OEMs), they have to rely on alternative technical gateways to access this valuable data source. Our article investigates the various archetypal business models that can be formulated using vehicle data. Subsequently, we focus on one particular archetype among these models. Finally, we adopt the perspective of OEMs and analyze Android Automotive, an operating system developed by Google, which is adopted by multiple manufacturers and holds significant potential as an enabler for future use cases in the independent aftermarket.

1 Introduction

The global automotive industry is experiencing a significant shift towards connectivity, which is turning modern vehicles into highly intelligent computers on wheels (Häberle et al., 2015; Kaiser et al., 2018). These vehicles are now equipped with advanced sensor technology that allows them to collect and share an ever-increasing

F. Sterk (✉)
Robert Bosch GmbH, Automotive Aftermarket, Karlsruhe Institute of Technology, Institute of Information Systems and Marketing, Institute of Information Systems and Marketing (IISM), Karlsruhe, Germany
e-mail: felix.sterk@kit.edu

M. Elo, F. Katsardis (eds.), *Automotive Aftermarket*, Management for Professionals, https://doi.org/10.1007/978-3-031-62419-3_17

volume of data, such as geolocation, fuel consumption, vehicle performance, and driver condition (Hood et al., 2019; Soley et al., 2018; Winkler et al., 2020).

In the present day, a connected vehicle generates a substantial 25 GB of data per hour, making big data and analytics vital for creating new value propositions (Heid et al., 2018). Looking ahead to 2030, McKinsey & Company projects that the annual incremental value from car data monetization could range from \$250 billion to \$400 billion (Martens & Schneiderbauer, 2021). To capitalize on the transformative potential of connected cars, various players in the automotive aftermarket, including original equipment manufacturers (OEMs), insurers, rental companies and repair shops, increasingly explore data-driven services (Sterk et al., 2022a). The connected cars perfectly illustrate the Internet of Things (IoT) in our study since they collect valuable data from a wide range of sensors, making them highly profitable and draw in a continuously expanding ecosystem of industry participants (Cichy et al., 2021).

The main aim of this article is to delve into how automotive companies—operating in the automotive aftermarket and beyond—can conceptualize and design business models while leveraging platform ecosystems to extract value from connected cars. Our article is organized as follows: First, we explore a range of archetypal business models that can be created using vehicle data. Second, we focus on one design of a business model based on a specific archetype, namely fleet management. Finally, we examine Android Automotive, an operating system developed by Google and adopted by multiple manufacturers. This system holds substantial potential as an enabler for future use cases in the independent aftermarket.

2 Business Model Archetypes for the Independent Aftermarket and Beyond[1]

Despite the growing importance of car connectivity, the existing literature lacks a comprehensive analysis that explicitly examines the structure, including stereotypical patterns (archetypes), of business models for connected car companies (Sterk et al., 2022c). To address similar issues, scholars have repeatedly explored the application of stereotypical business model patterns or archetypes, which represent typical combinations of characteristics, as a promising approach for strategic decision-making (e.g., Gimpel et al., 2018; Hunke et al., 2021; Weking et al., 2020). Interestingly, it has been found that as much as 90% of all business model innovations can be categorized as recombinations of preexisting patterns (Gassmann et al., 2014). Our objective is to make a valuable contribution by examining higher-level business model configurations for connected cars through empirical analysis of emerging archetypes. We adopt a comprehensive approach to study the connected car ecosystem, ensuring that our archetypes not only explore potential business models of OEMs but also encompass the broader ecosystem, including aftermarket players, such as garages, insurance companies, and fleet management system

[1] This subsection is based on the published article (Sterk et al., 2022c) and the working paper (Sterk et al., 2023c).

providers. By taking this holistic perspective, our analysis is more scientifically rigorous and reflective of the interconnected nature of the ecosystem. The archetypes illuminate established innovation paths that executives can adopt to digitalize their business models and effectively capitalize on car data monetization. Consequently, our research revolves around the following central question:

RQ1: What are the archetypal patterns of data-driven business models in the connected car domain?

To address this question, we follow a sequential mixed methods design (Venkatesh et al., 2013, 2016), comprising two iterations. In the initial iteration, we constructed a business model taxonomy based on existing literature and analyzed 154 connected car business models sourced from various reputable outlets. The sample of business models was queried from the current literature, practitioner-oriented business reports published by leading consulting firms, and Crunchbase, the world's largest startup database. The taxonomy design undergoes a comprehensive evaluation, combining quantitative and qualitative methods, including 12 expert interviews and its application to a set of 154 real-world business models. The outcome of this process reveals ten key dimensions and 48 corresponding characteristics, providing a holistic description of the connected car business models (Table 1).

In the second iteration, we use the taxonomy classified with the 154 real-world business models and perform a cluster analysis (Kaufman & Rousseeuw, 1990) to derive cluster groups of business models that share similar characteristics across the taxonomy dimensions. By comparing the respective cases within each cluster, we derived seven archetypes as qualitative interpretations that describe and distinguish

Table 1 Taxonomy of data-driven business models in the connected car domain

	Dimension	Characteristic						
Value Proposition	Value for car owner or driver	Safety & security	Cost reduction	Traffic efficiency	Infotainment	Environmental sustainability	Convenience	Indirect value
	Car data impact on value	Car data core business model	Car data-enabled business model					
	Car autonomy impact on value	Enhanced value by autonomy	Reduced value by autonomy	Autonomy not relevant				
Value Architecture	Data category	PII	Contextual data	Diagnostic data	Usage data	ADAS data	Application data	
	Data access	OEM proprietary access	OEM specific cloud or neutral server	OBD2-dongle	Other retrofit devices	Smartphone or other non-in-vehicle sources		
	Enabler technology	Blockchain	Augmented reality	Over-the-air architectures	ADAS Technology	Artificial intelligence	Cellular networks	
Value Network	Role in ecosystem	End-customer solution provider	Platform provider	Technology provider				
	Customer segment	Private individuals (B2C)	Fleet providers (B2B)	OEMs (B2B)	Third-party providers (B2B)	Governments (B2G)		
Value Finance	Data monetization strategy	Data source & data provision	Data aggregation & data exchange	Data analysis & data insights	Data application & data service			
	Revenue model	One-time payment	Pay-per-use	Subscription fee	Licensing fee	Commission fee	On-demand	Open source

ideal configurations of connected car business models. By analyzing the companies in the seven different cluster groups, we develop the following interpretive labels for the archetypes: (A1) data platforms, (A2) location-based services, (A3) fleet management, (A4) diagnostics and maintenance, (A5) driving analytics, (A6) cyber-physical protection, and (A7) connected infotainment. Table 2 presents a comprehensive summary of the seven archetypes, outlining their unique features and offering illustrative examples of typical applications associated with each archetype.

Four of the identified archetypes are essential for the aftermarket in the automotive industry due to their potential to extend firms business models from the pre-digital age with data-driven features:

- *(A1) Data platforms* function as marketplaces for trading vehicle data between companies. These platforms serve as neutral intermediaries, enabling OEMs to monetize collected vehicle data by selling it to independent aftermarket players. These aftermarket companies, in turn, use the data to develop data-driven services. This business model archetype's primary value proposition is to offer a centralized access point for vehicle data and essential functionalities like consent management and secure data exchange between parties.
- *(A3) Fleet management* plays a crucial role in aiding fleet managers to oversee and minimize the overall cost of ownership for different types of fleets, encompassing corporate fleets, logistics, and mobility service providers. Beyond cost management, these services also have the potential to directly or indirectly enhance vehicle safety, optimize transportation efficiency, and contribute to environmental sustainability efforts.
- *(A4) Diagnostics and maintenance* services focus on monitoring and enhancing vehicle health, while managing maintenance activities between vehicle owners and related businesses, such as repair shops. The value proposition for customers lies in increased vehicle uptime, convenience, and cost reduction through remote services and proactive maintenance enabled by continuous vehicle monitoring. Players in the independent aftermarket achieve this by leveraging vehicle usage and diagnostic data from various sources, including neutral servers, OBD2-dongles, or other retrofitted devices.
- *(A5) Driving analytics* seeks to decrease end-customer usage costs by monitoring and profiling their driving behaviors. Insurance companies have been quick to adopt this archetype, introducing usage-based insurance (UBI) programs that utilize dynamic behavioral data collected through OBD2-dongles, other retrofit devices (e.g., black boxes), or modern smartphones to calculate premiums.

Our taxonomy and archetypes are valuable strategic management tools for managers, identifying business opportunities and potential market entry points in the automotive aftermarket and beyond and evaluating their relevance within a company's unique context. Additionally, these artifacts help explain the current business model to stakeholders, improve specific operational aspects, and develop new business models aligned with the corporate strategy (Spieth et al., 2014). Utilizing a

Table 2 Summary of the identified business model archetypes

ID	Archetype	Distinguishing characteristics	Typical applications
A1	Data platforms	• Operating as a data platform provider in the automotive ecosystem • Making car data available for third-party service providers at scale • Accessing data directly from the OEM and acting as a neutral server	Car data marketplaces, contextual data providers
A2	Location-based services	• Delivering end-customer value through navigation or parking services enhancing overall transportation efficiency • Collecting data by mobile mapping vehicles equipped with remote sensing systems • Exploiting contextual data such as external road and environmental conditions to create real-time maps	Navigation systems, parking applications
A3	Fleet management	• Addressing fleet providers, logistics service providers and mobility service providers as customer segment • Delivering end-customer value primarily in the areas of driving safety, cost reduction, and traffic efficiency • Accessing data commonly by means of retrofitted OBD2-dongles	Fleet management systems, digital driver logbooks
A4	Diagnostic and maintenance	• Analyzing diagnostic data such as trouble codes to offer proactive and predictive maintenance • Providing value not only to end-customers but also to aftermarket players such as car dealers or workshops • Utilizing over-the-air architectures to do incremental remote updates and repairs in future scenarios	Remote diagnostics services, predictive maintenance services
A5	Driving analytics	• Analyzing car usage data for monitoring driving patters and behavior • Reducing cost through pay-as-you-drive tariffs or recommendations for fuel-efficient driving • Providing driving assistance and driver tutoring to increase road safety	Usage-based insurance tariffs, applications for driving style suggestions
A6	Cyber-physical protection	• Delivering end-customer value by ensuring occupant safety and cybersecurity • Enabling data-driven business models by providing crucial and secure connected car technologies • Harboring huge future potentials, for instance trough over-the-air driver-assistance system updates	Cybersecurity solutions, driver-assistance systems
A7	Connected infotainment	• Delivering end-customer value via in-car infotainment applications • Providing a standardized platform that enables in-car third-party applications on the head unit • Increasing value through car autonomy, e.g., by watching virtual reality films or playing video games	Digital cockpit solutions, infotainment operating systems

morphological analysis (Geum et al., 2016), our work supports the systematic development of innovative ideas

3 Utilizing Car Data from Data Platforms to Enhance Fleet Management from an Independent Aftermarket Perspective[2]

Connected cars have the potential to offer a unique customer experience and bring cost and revenue benefits to mobility enterprises (Coppola & Morisio, 2016; Stocker et al., 2021). Currently, OEMs monetize valuable car data by providing digital services like BMW ConnectedDrive or Mercedes me connect, which offer features such as concierge services, remote diagnostics, and on-street parking information (Kaiser et al., 2021). However, this data is not exclusively of interest to OEMs. Independent service providers, encompassing suppliers, workshops, and insurers, also seek access. These entities often use telematics-enabled dongles that are retrofitted to the on-board diagnostics (OBD) port to facilitate remote data collection (Bätz et al., 2020; Kaiser et al., 2019; Marabelli et al., 2017).

Although OBD dongles were initially expected to gain widespread adoption, the market remains fragmented, facing challenges such as time-consuming installations, expensive hardware purchases, limited car park coverage, and data point availability (Coppola & Morisio, 2016; Pütz et al., 2019; Stocker et al., 2021). To address these drawbacks, third-party data access directly from OEMs has emerged as an alternative solution. Aspiring car data platforms like Caruso Dataplace or Otonomo act as neutral intermediaries inspired by successful transaction platforms like eBay, Uber, and Airbnb. They facilitate the exchange of standardized data between OEMs and independent service providers (Martens & Müller-Langer, 2018; Sterk et al., 2022c). One major advantage is that these platforms offer a single point of entry to access data from multiple OEMs (Stocker et al., 2021). However, scaling up remains challenging as marketplaces depend on data access conditions set by OEMs, including data pricing, availability, and willingness to share (Sterk et al., 2023a). Nevertheless, in the digital era, where data is pivotal in enabling new business models, these data marketplaces are expected to become increasingly significant (Jung et al., 2021).

While the previous research question focused on the general conceptualization of connected car business models, our current objective is to conduct a more in-depth investigation into designing a specific car-data-based service. Leading consultancies have identified data-based fleet management as one of the most impactful use cases in car data monetization (Arif et al., 2019; Carter et al., 2018). The decline in private vehicle ownership has increased demand for professionally managed fleets (Pütz et al., 2019). As a result, McKinsey & Company predicts that the global connected fleet solutions market will grow at approximately 23% annually, reaching a value of $75.79 billion by 2025 (Carter et al., 2018).

[2] This subsection is based on the published article (Sterk et al., 2023a).

Since car data marketplaces are still in their early stages and offer limited data, fleet management emerges as a promising starting point for designing connected services due to its high utility and manageable data requirements (Arif et al., 2019). Additionally, fleet vehicles typically have a relatively low age, resulting in a significant number of connected cars in modern fleets capable of providing data. However, scholars have scarcely touched on designing connected car or fleet services incorporating the concept of data marketplaces (Sterk et al., 2022b). Hence, we pose the following research question:

RQ2: How to design a connected fleet management system in order to use car data from data marketplaces effectively?

To address the inquiry, we embarked on a design science research (DSR) project (Kuechler & Vaishnavi, 2008), utilizing insights from an extensive literature review and conducting practical interviews with domain experts. We conducted semi-structured interviews with 21 fleet domain experts who operate in five distinct areas: corporate fleet ($n = 11$), car subscription ($n = 4$), car sharing ($n = 2$), ride pooling ($n = 1$), and fleet service provider ($n = 3$). Through these interviews, the fleet experts highlighted several crucial issues they encounter in their daily work, which could potentially be addressed using vehicle data. These issues are then aggregated into three dimensions:

- *Economic Sustainability.* First, the total cost of ownership (TCO) is vital for identifying cost-saving opportunities and reducing operating costs stemming from fuel, maintenance, tires, or repairs (Fatin Amirah et al., 2013; López-Ibarra et al., 2020). One expert stressed that *"the topic of cost transparency is still in its infancy. Even the big fleet management companies still work with Excel."*
- *Environmental Sustainability.* Second, ambitious greenhouse gas reduction targets dominate current discussions about fleet management. Thereby, a sustainability strategy should also include appropriate measures to raise drivers' partially limited awareness of sustainable driving. To this end, one interviewee explicated that *"employees opted for PHEVs*[3] *primarily because of the tax advantage, never drove electric, and left the charging cable in its original packaging."*
- *Vehicle Health.* Another prominent concern fleet managers face is maintaining vehicle conditions to ensure long-lasting vehicle health and driver safety (Coppola & Morisio, 2016). Regarding this, one interviewed expert mentioned that they *"usually only find out too late when maintenance intervals are not adhered to, or vehicles run without oil for weeks, causing enormous costs."*

Drawing from the identified issues, we derive six design principles (DP) for the Connected Fleet Management System (CFMS) by adhering to established

[3] PHEV = plug-in hybrid electric vehicles.

guidelines (Gregor et al., 2020). These design principles are categorized into two key areas of fleet management, namely strategic and operational aspects:

- *Strategic Fleet Management.* The DPs around strategic fleet management include the following aspects: (DP1) essential KPIs and their visualization via comprehensive dashboards, (DP2) a reporting tool including essential KPIs and visualizations within, and (DP3) a planning tool for calculating expected or desired KPIs.
- *Operational Fleet Management.* The DPs around operational fleet management comprise the following aspects: (1) a fleet overview that can be filtered by vehicle specifications, (2) a vehicle-specific status overview, and (3) frequent status updates and proactive notifications to drivers.

To implement the design principles (DPs), we created a prototype that relies on in-vehicle data, including mileage, fuel level, state of charge, and tire pressure. The in-vehicle data used in the prototype was collected from 89 cars during a field test conducted by Caruso Dataplace (Mokeev et al., 2021). The final step involved developing a prototype of the Connected Fleet Management System (CFMS) using Microsoft Power BI. This prototype empowers fleet managers to make effective use of car data, as depicted in Fig. 1.

In conclusion, car data marketplaces present a fascinating array of data that enables the realization of initial fleet management use cases, leading to improved economic and environmental sustainability, as well as vehicle health. Nevertheless, ample potential exists for advancing research and development efforts to break the data monopoly held by automotive OEMs. This monopoly puts other market

Fig. 1 Instantiation of the connected fleet management system (CFMS) prototype

participants, such as suppliers, repair shops, and insurance companies, at a disadvantage, as they heavily rely on the data supply from OEMs (Martens & Mueller-Langer, 2020). To counteract this situation, car data marketplaces have emerged, accessing OEMs' cloud systems, harmonizing the data, and reselling it to independent service providers. However, prominent car data marketplaces like Otonomo and Wejo currently face financial challenges, as certain OEMs are hesitant to share extensive data or only offer data that may not hold inherent value for external buyers, unless it is shared across multiple brands. Consequently, there is a pressing need to explore ways to establish fair competition between OEMs and alternative service providers. One promising foundation for research endeavors could be the European Commission's Data Act, which regulates fair data access and utilization (European Commission, 2022). Scholars can delve into different data governance models and frameworks that effectively implement the provisions of the Data Act. Another approach worth considering is assessing the potential impact of the Data Act on the co-creation of value between OEMs, independent service providers, and data marketplaces. By doing so, researchers can gain valuable insights into fostering a more equitable and thriving automotive data ecosystem.

4 The Case of Google's Automotive Ecosystem Involvement and Its Implications on the Independent Aftermarket[4]

Not only service providers operating in the IAM but also OEMs face the emergence of sophisticated digital technologies that present disruptive opportunities. To date, incumbent OEMs have encountered challenges in embracing a digital mindset and implementing strategic management techniques required to achieve the vision of software-defined vehicles (Dremel et al., 2017; Svahn et al., 2017). Drivers now have a strong demand for integrated navigation and entertainment features, with nearly 40% even considering switching vehicle brands in search of better digital services (Heineke et al., 2020). This shift in consumer preferences gives an advantage to tech players with expertise in smartphones. They can leverage their proficiency by using the vehicle's infotainment system as a bridge to establish a digital connection between the driver and the vehicle (Schreieck et al., 2022; Weiss et al., 2021). One such example is Google, which entices users with its Android Automotive OS (AAOS) and includes Google Automotive Services (GAS) like Google Maps, Google Assistant, and Google Play Store in the vehicle. This level of integration grants Google significant control over the user interface, user data, and the app store within the vehicle. To delve into the complexities of OEMs' platform strategies and the consequences of various degrees of involvement by Google, we formulate the following research question:

RQ3: How and why do incumbent firms decide on a certain level of tech player involvement in their digital strategy?

---

[4]This subsection is based on a working paper (Sterk et al., 2023b).

To address this research question, we utilize a revelatory single case strategy following Yins (2014) approach. This enabled us to explore previously inaccessible dynamics of the phenomenon under investigation. For several reasons, we selected the automotive industry and focused on Google's AAOS and its underlying GAS (i.e., Google Maps, Google Assistant, and Google Play Store). First, the automotive industry is experiencing significant IT-driven innovation efforts from established players and external tech firms. Second, Google has established a dominant position in the market with its highly integrated AAOS. Third, unlike other automotive solutions like Android Auto or Apple CarPlay, AAOS is purpose-built for direct in-vehicle integration, providing greater capabilities to interact with the car's internal systems and deliver innovative features. Last, the extent of Google's access to individual vehicle functions and data (e.g., AAOS with or without GAS) has become a central question due to the increasing collaboration between OEMs and Google. Within the context of Google's involvement in the automotive ecosystem, we examine the strategic positioning of various incumbent firms concerning Google's AAOS and GAS offerings over a span of time from 2017 to 2023 (Table 3). Employing a sampling logic emphasizing subunit diversity (Yin, 2014), three distinct OEM actualization strategies are identified by comparing their strategic actions during the

Table 3 Three actualization strategies used by OEMs regarding Google involvement

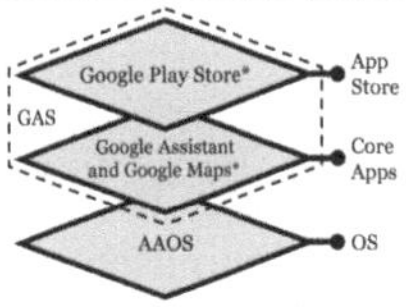 *Design and user data controlled by Google	*Volvo's Google Built-In Approach*: Swedish carmaker Volvo Cars and its subsidiary Polestar, both owned by Chinese carmaker Geely, have been offering cars with built-in AAOS since 2020. Volvo is fully committed to the partnership resulting in all new cars featuring the whole infotainment system supplied by Google, including the preinstalled GAS. Thereby, Volvo or Polestar car drivers are prompted to link their Google accounts. Furthermore, Google and Volvo are taking their partnership to the next level with the integration of "HD Maps," where Google Maps will use additional car sensor data in real time to provide highly detailed and up-to-date road information
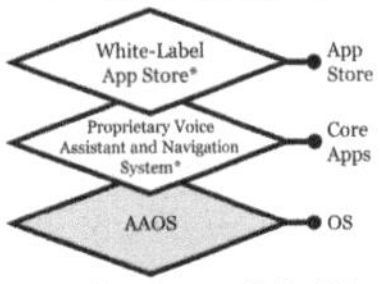 *Design and user data controlled by OEM	*BMW's Open-Source Approach*: Starting in 2023, the BMW Group will be the first German carmaker to launch an infotainment system based on the open-source variant of AAOS, called BMW OS 9. This approach excludes permanently installed GAS applications (e.g., Google Maps), as BMW wants to retain independence in these areas. BMW also does not use the Google Play Store and instead tries to build up its own Android-based commercial ecosystem supported by selected suppliers. Here, BMW integrates Faurecia Aptoide's white-label app store, with BMW developing the user interface to preserve its brand-specific design and experience
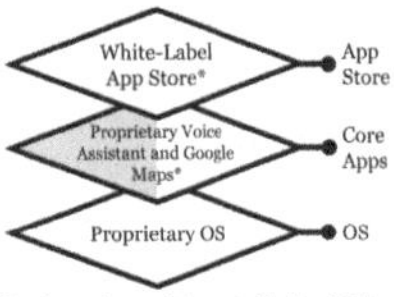 *Design and user data controlled by OEM	*Mercedes' Exclusive Approach*: Mercedes-Benz took a distinctive approach to its software strategy starting in 2023, opting against an off-the-shelf operating system like AAOS. Instead, Mercedes developed a proprietary infotainment system called MB.OS to retain control over customer relationships and data privacy, and to integrate unique car functions. Mercedes is using Faurecia Aptoide's white-label app store, but has also established a strategic, long-term partnership with Google to be the first OEM to build its own branded navigation system based on in-car data and Google Maps navigation capabilities

specified period. These diversified strategic approaches of traditional OEMs served as the foundation for extracting knowledge across multiple embedded units of analysis.

Although the three strategic orientations described above are clearly focused on the OEM side of the ecosystem, the emergence of AAOS and GAS also comes with decisive managerial implications for the IAM. Google offers with AAOS a standardized platform, on which services can be developed by service providers in the automotive aftermarket for a wide range of OEMs, brands, and models, ultimately enabling implementation of scalable multi-brand business models. However, it remains unclear which app categories the OEMs will allow in their white-label app stores and to what extent they will prevent the deployment of certain third-party services that could jeopardize their business potential, such as remote diagnostic services or workshop booking services. Moreover, app stores such as Google Play Store or white label app stores and the underlying AAOS will, for the first time, enable aftermarket players to find a direct channel into the vehicle and thus expand their business logic in the direction of the end user, for example, through driving statistics or workshop booking tools. Finally, it remains unclear whether and which vehicle data will be made available by OEMs to third-party developers and under which conditions. At least for the OEMs using GAS, Google could oblige them to share specific data points with third-party developers and thus also independent aftermarket players. In addition, Google could decide from 1 day to the next not to release future versions of Android Automotive as open source and instead only sell and support new versions under B2B licensing agreements that require the use of GAS.

5 Conclusion

This article investigates connected cars and their impressive capacity to exchange real-time data within their ecosystem. However, the automotive industry is on the brink of the next revolutionary leap, the "software-defined vehicle." This innovative concept prioritizes the vehicle's software over mechanical hardware, taking charge of and executing vehicle functions (Ohlsen, 2022; Windpassinger, 2022). This shift opens up research opportunities for studying user preferences, analyzing usage patterns, and creating personalized features that enhance user satisfaction and engagement. Car owners can now select their desired features and services, tailoring their driving experience to their needs.

The potential to shape this future presents a unique chance for researchers and innovators to explore novel approaches and technologies that boost safety, efficiency, and overall user experience. However, it is evident that incumbent OEMs and traditional players in the automotive aftermarket are not fully prepared for this paradigm shift and cannot navigate it successfully in isolation (Sterk et al., 2022b). Crucial elements of vehicle operating systems or cloud environments rely increasingly on frameworks from major hyper-scalers such as Amazon, Google, or Microsoft. This situation calls for further exploration of collaborative approaches and cooperation to help incumbents bridge their knowledge gaps and actively contribute to the value creation of software-defined vehicles beyond basic products.

In conclusion, we invite scholars to join us in exploring the vast possibilities of the connected car and the soon-to-be software-defined dream car. Together, we can delve into this exciting new territory and contribute towards further innovation in the automotive industry.

References

Arif, S., Kane, A., Yelamarthi, K., Walsh, F., & Abdelgawad, A. (2019). "Connected Vehicle Trend Radar," *Capgemini*, pp. 1–50.

Bätz, A., Heger, S., Gimpel, H., & Wöhl, M. (2020). "Driving Sustainably - The Influence of IoT-Based Eco-Feedback on Driving Behavior," in *HICSS 2020 Proceedings*, pp. 912–921. https://doi.org/10.24251/hicss.2020.114.

Carter, C., Gould, R., Tiwari, S., & Zafar, S. (2018). *Driving Value from Fleet Telematics* (pp. 1–4). McKinsey & Company. https://www.mckinsey.com/business-functions/operations/our-insights/driving-value-from-fleet-telematics

Cichy, P., Salge, T. O., & Kohli, R. (2021). Privacy concerns and data sharing in the internet of things: Mixed methods evidence from connected cars. *MIS Quarterly, 45*(4), 1863–1891. https://doi.org/10.25300/MISQ/2021/14165

Coppola, R., & Morisio, M. (2016). Connected car: Technologies, issues, future trends. *ACM Computing Surveys, 49*(3), 1–36. https://doi.org/10.1145/2971482

Dremel, C., Herterich, M. M., Wulf, J., & Brenner, W. (2017). How AUDI AG established big data analytics in its digital transformation. *MIS Quarterly Executive, 16*(2), 81–100.

European Commission. (2022). "Proposal for a regulation of the European Parliament and the Council on Harmonised Rules on Fair Access to and Use of Data (Data Act)." https://eur-lex.europa.eu/legal-content/EN/TXT/?uri=COM%3A2022%3A68%3AFIN

Fatin Amirah, A. S., Ruzanna, M. J., Aishah, R., & Mohamad Syazwan, M. A. (2013). An overview of fleet maintenance and operating cost: Key components and methods. *International Journal of Commerce, Business and Management, 2*(6), 443–452.

Gassmann, O., Frankenberger, K., & Csik, M. (2014). *The business model navigator: 55 models that will revolutionise your business*. Financial Times Publishing Pearson.

Geum, Y., Jeon, H., & Lee, H. (2016). Developing new smart services using integrated morphological analysis: Integration of the market-pull and technology-push approach. *Service Business, 10*(3), 531–555. https://doi.org/10.1007/s11628-015-0281-2

Gimpel, H., Rau, D., & Röglinger, M. (2018). Understanding FinTech Start-Ups—a Taxonomy of Consumer-Oriented Service Offerings. *Electronic Markets, 28*(3), 245–264. https://doi.org/10.1007/s12525-017-0275-0

Gregor, S., Chandra Kruse, L., & Seidel, S. (2020). Research perspectives: the anatomy of a design principle. *Journal of the Association for Information Systems*, 21(6), 1622–1652.

Häberle, T., Charissis, L., Fehling, C., Nahm, J., & Leymann, F. (2015). The Connected Car in the Cloud: A Plattform for Prototyping Telematics Services. *IEEE Software, 32*(6), 11–17.

Heid, B., Huth, C., Wu, G., & Kempf, S. (2018). Ready for Inspection: The Automotive Aftermarket in 2030. *McKinsey Insights*, 1–52. http://search.ebscohost.com/login.aspx?direct=true&db=bth&AN=133513620&site=ehost-live

Heineke, K., Kampshoff, P., Möller, T., & Wu, T. (2020). *From no mobility to future mobility: Where COVID-19 has accelerated change* (pp. 1–259). McKinsey & Company.

Hood, B. J., Hoda, O., & Robinson, R. (2019). Monetizing Data in the Age of Connected Vehicles. *Deloitte Insights*, 1–7.

Hunke, F., Heinz, D., & Satzger, G. (2021). Creating customer value from data: Foundations and archetypes of analytics-based services. *Electronic Markets, 32*(1), 503–521. https://doi.org/10.1007/s12525-021-00506-y

Jung, S., Wortmann, F., Bronner, W., & Gassmann, O. (2021). Platform economy: Converging IoT platforms and ecosystems. In O. Gassmann & F. Ferrandina (Eds.), *Connected business* (pp. 35–54). Springer.

Kaiser, C., Stocker, A., Festl, A., Lechner, G., & Fellmann, M. (2018). "A Research Agenda for Vehicle Information Systems," in *ECIS 2018 Proceedings*, pp. 1–12.

Kaiser, C., Stocker, A., & Fellmann, M. (2019). "Understanding Data-Driven Service Ecosystems in the Automotive Domain," in *AMCIS 2019 Proceedings*, pp. 1–10.

Kaiser, C., Stocker, A., Viscusi, G., Fellmann, M., & Richter, A. (2021). Conceptualising value creation in data-driven services: The case of vehicle data. *International Journal of Information Management, 59*(1), 1–15. https://doi.org/10.1016/j.ijinfomgt.2021.102335

Kaufman, L., & Rousseeuw, P. J. (1990). *Finding groups in data: An introduction to cluster analysis*. John Wiley & Sonns.

Kuechler, B., & Vaishnavi, V. (2008). On theory development in design science research: anatomy of a research project. *European Journal of Information Systems*, 17(5), 489–504.

López-Ibarra, J. A., Gaztañaga, H., Saez-de-Ibarra, A., & Camblong, H. (2020). Plug-in hybrid electric buses total cost of ownership optimization at fleet level based on battery aging. *Applied Energy, 280)*, Elsevier Ltd, 115887. https://doi.org/10.1016/j.apenergy.2020.115887

Marabelli, M., Hansen, S., Newell, S., & Frigerio, C. (2017). The light and dark side of the black box: Sensor-based technology in the automotive industry. *Communications of the Association for Information Systems, 40*(1), 351–374. https://doi.org/10.17705/1cais.04016

Martens, B., & Mueller-Langer, F. (2020). Access to digital car data and competition in aftermarket maintenance services. *Journal of Competition Law & Economics, 16*(1), 116–141. https://doi.org/10.1093/joclec/nhaa005

Martens, B., & Müller-Langer, F. (2018). "Access to digital car data and competition in aftersales services."

Martens, C., & Schneiderbauer, T. (2021). *Unlocking the full life-cycle value from connected-car data* (pp. 1–15). McKinsey & Company.

Mokeev, A., Webel, C., De Calan, G., Knodel, J., & Rauschenbach, T. (2021). Connected vehicle field test. *Caruso Dataplace*, 1–84.

Ohlsen, J. (2022). The software-defined vehicle is overwhelming the automotive industry. *ATZelectronics Worldwide, 17*(6), 56–56.

Pütz, F., Murphy, F., Mullins, M., & O'Malley, L. (2019). Connected automated vehicles and insurance: Analysing future market-structure from a business ecosystem perspective. *Technology in Society, 59*(1), 1–9. https://doi.org/10.1016/j.techsoc.2019.101182

Schreieck, M., Wiesche, M., & Krcmar, H. (2022). Governing innovation platforms in multi-business organisations. *European Journal of Information Systems*, 1–22.

Soley, A. M., Siegel, J. E., Suo, D., & Sarma, S. E. (2018). Value in vehicles: Economic assessment of automotive data. *Digital Policy, Regulation and Governance, 20*(6), 513–527. https://doi.org/10.1108/DPRG-05-2018-0025

Spieth, P., Schneckenberg, D., & Ricart, J. E. (2014). Business model innovation - state of the art and future challenges for the field. *R&D Management, 44*(3), 237–247. https://doi.org/10.1111/radm.12071

Sterk, F., Dann, D., & Weinhardt, C. (2022a). "Monetizing car data: A literature review on data-driven business models in the connected car domain," in *HICSS 2022 Proceedings*, pp. 1–10.

Sterk, F., Heinz, D., Peukert, C., Fleuchaus, F., Kölbel, T., & Weinhardt, C. (2022b). "Fostering value co-creation in incumbent firms: The case of Bosch's IoT ecosystem landscape," in *ICIS 2022 Proceedings*.

Sterk, F., Peukert, C., Hunke, F., & Weinhardt, C. (2022c). "Understanding car data monetization: A taxonomy of data-driven business models in the connected car domain," in *WI 2022 Proceedings*, pp. 1–16.

Sterk, F., Frank, S., Lauster, I., & Weinhardt, C. (2023a). "Utilizing fleet data: Towards designing a connected fleet management system for the effective use of multi-brand car data," in *HICSS 2023 Proceedings*, pp. 1–10.

Sterk, F., Heinz, D., Hengstler, P., & Weinhardt, C. (2023b). "Reallocating uncertainty in incumbent firms through digital platforms: The case of Google's Automotive ecosystem involvement," *Working Paper*.
Sterk, F., Stocker, A., Heinz, D., & Weinhardt, C. (2023c). "Unlocking the value from car data: A taxonomy and archetypes of connected car business models," *Working Paper*.
Stocker, A., Lechner, G., Kaiser, C., & Fellmann, M. (2021). "Digitalized Mobility," in *AMCIS 2021 Proceedings*, pp. 1–10.
Svahn, F., Mathiassen, L., & Lindgren, R. (2017). Embracing Digital Innovation in Incumbent Firms: How Volvo Cars Managed Competing Concerns. *MIS Quarterly, 41*(1), 239–253.
Venkatesh, V., Brown, S. A., & Bala, H. (2013). Bridging the Qualitative-Quantitative Divide: Guidelines for Conducting Mixed Methods Research in Information Systems. *MIS Quarterly, 37*(1), 21–54.
Venkatesh, V., Brown, S. A., & Sullivan, Y. W. (2016). Guidelines for conducting mixed-methods research: An extension and illustration. *Journal of the Association for Information Systems, 17*(7), 435–495. https://doi.org/10.17705/1jais.00433
Weiss, N., Schreieck, M., Wiesche, M., & Krcmar, H. (2021). From product to platform: How can BMW compete with platform giants? *Journal of Information Technology Teaching Cases, 11*(2), 90–100.
Weking, J., Mandalenakis, M., Hein, A., Hermes, S., Böhm, M., & Krcmar, H. (2020). The impact of blockchain technology on business models—A taxonomy and archetypal patterns. *Electronic Markets, 30*(2), 285–305. https://doi.org/10.1007/s12525-019-00386-3
Windpassinger, H. (2022). On the way to a software-defined vehicle. *ATZelectronics worldwide, 17*(7–8), 48–51.
Winkler, M., Mehl, R., Matthies, M., Monske, S., Kolhas, C., Kiefer, N., Purrucker, P., & Gräber, R. (2020). *Monetizing vehicle data how to fulfill the promise* (pp. 1–20). Capgemini.
Yin, R. K. (2014). *Case study research: Design and methods* (5th ed.). SAGE Publications.

Dr. Felix Sterk is a former doctoral student affiliated with Robert Bosch GmbH in the Automotive Aftermarket business unit in collaboration with the Karlsruhe Institute of Technology at the Institute of Information Systems and Marketing. His research centers on data-driven business models for connected cars, primarily focusing on monetizing vehicle data. Specifically, his work delves into business ecosystems and platform business models, including exploring car data marketplaces or infotainment platforms, both vital enablers for cross-manufacturer business models and developing data-based services.

Glossary

Glossary and Terms[1]

Independent Jobbers Jobbers that are not owned, in part or in full, by a warehouse.

Jobber/Retailer Wholesalers who also sell parts, chemicals, and accessories to retail customers. Retail sales account for more than 50% of a jobber/retailer's total sales.

Jobbers Firms that sell more than 50% of their products to repair shops and more than 50% of their purchases are from distributors.

OE Original Equipment. Parts and components supplied to manufacturers for motor vehicle production.

OEM Original Equipment Manufacturers. Companies that supply parts and components to produce motor vehicles.

OES Original Equipment Suppliers. Companies that supply parts and components to produce motor vehicles.

Service Stations Establishments, which may or may not sell products over the counter, for which gasoline accounts for more than 50% of total sales. Examples include BP, Shell, and Exxon.

Three-Step Distribution Traditional aftermarket distribution process where products flow from the manufacturer to the warehouse distributor to the jobber to the service outlet.

Tier 1 Suppliers Automotive parts manufacturers that supply final equipment directly to vehicle manufacturers. Increasingly, tier one suppliers are becoming producers of major subassemblies and modular components that can be installed into a vehicle as a unit, such as a complete drivetrain.

Tier 2 Suppliers Manufacturers that produce components for Tier One suppliers.

Tier 3 Suppliers Manufacturers that supply raw materials used in the production of components.

Tire Dealers Stores that generate more than 50% of their sales from automotive tires.

[1] Source: Glossary of Automotive Terms, Auto Care

M. Elo, F. Katsardis (eds.), *Automotive Aftermarket*, Management for Professionals, https://doi.org/10.1007/978-3-031-62419-3

Tool and Equipment Segment Specialists in providing the tools and equipment needed to perform repair and maintenance of motor vehicles.

Two-Step Distribution Distribution process where products flow from the manufacturer to the warehouse distributor to the service outlet directly, eliminating the jobber.

Two-Step Warehouse Distributors Firms that sell more than 50% of their products to repair shops and more than 50% of their purchases are from manufacturers.

Warehouse Distributors Firms that sell more than 50% of their products to automotive jobbers or retailers.

Relevant Concepts

After-Sales Services and Aftermarket Support Activities that aim to enhance customer satisfaction, loyalty, and retention after the purchase of a product. After-sales services include warranty, technical assistance, spare parts distribution, and customer care. Aftermarket support involves secondary market transactions for additional products and product recovery processes (see more in Durugbo, C. M. (2020). After-sales services and aftermarket support: a systematic review, theory and future research directions. *International Journal of Production Research, 58*(6), 1857–1892. https://doi.org/10.1080/00207543.2019.1693655)

Automotive Interaction Design The design of in-car user interfaces that enable drivers and passengers to interact with various functions and features of the vehicle. Automotive interaction design combines user-centred design principles with psychological theories of attention, perception, memory, emotion, and motivation (see more in Chen, F., & Terken, J. (2023). *Automotive Interaction Design: From Theory to Practice.* Springer Singapore. https://doi.org/10.1007/978-981-19-3448-3)

Circular Economy The concept of designing products and systems that minimize waste and resource consumption and maximize reuse and regeneration. Circular economy can improve environmental sustainability, economic efficiency, and social equity (see more in Ghisellini, P., Cialani, C., & Ulgiati, S. (2016). A review on circular economy: the expected transition to a balanced interplay of environmental and economic systems. *Journal of Cleaner Production, 114*, 11–32. https://doi.org/10.1016/j.jclepro.2015.09.007)

Customer Relationship Management The process of managing interactions with existing and potential customers to build long-term relationships and increase profitability. Customer relationship management involves collecting, analysing, and using customer data to provide personalized services and offers (see more in Nguyen B., Simkin L., Canhoto A.I., Dibb S. (2019) The state-of-the-art of CRM: A literature review 2008–2018. Journal of Strategic Marketing 29(1):1–30. https://www.tandfonline.com/doi/full/10.1080/0965254X.2019.1672639)

Market Disruption The phenomenon of new entrants or innovations that challenge the established market players or practices. Market disruption can create new value propositions, customer segments, or business models that alter the competitive landscape (see more in Heid, B., Huth, C., Kempf, S., & Wu, G. (2018, August 27). *Ready for inspection: The automotive aftermarket in 2030.* McKinsey & Company. https://www.mckinsey.com/industries/automotive-and-assembly/our-insights/ready-for-inspection-the-automotive-aftermarket-in-2030)

Remanufacturing The process of restoring used products to a like-new condition. Remanufacturing can reduce environmental impacts, save costs, and create value for customers and manufacturers (see more in Casper, R., & Sundin, E. (2021). Electrification in the automotive industry: effects in remanufacturing. *Journal of Remanufacturing, 11,* 121–136. https://doi.org/10.1007/s13243-020-00094-8)

Servitization The transformation of a product-oriented company into a service-oriented company. Servitization can enhance customer value, differentiation, and loyalty by offering integrated solutions that combine products and services (see more in Baines, T., Ziaee Bigdeli, A., Bustinza, O. F., Shi, V. G., Baldwin, J., & Ridgway, K. (2017). Servitization: revisiting the state-of- the-art and research priorities. *International Journal of Operations & Production Management, 37*(2), 256–278. https://doi.org/10.1108/IJOPM-06-2015-0312)

SPRINGER NATURE

GPSR Compliance

The European Union's (EU) General Product Safety Regulation (GPSR) is a set of rules that requires consumer products to be safe and our obligations to ensure this.

If you have any concerns about our products, you can contact us on ProductSafety@springernature.com

In case Publisher is established outside the EU, the EU authorized representative is:

Springer Nature Customer Service Center GmbH
Europaplatz 3
69115 Heidelberg, Germany

The manufacturer's authorised representative in the EU is Springer Nature Customer Service Centre GmbH, Europaplatz 3, 69115 Heidelberg, Germany. If you have any concerns regarding our products, please contact ProductSafety@springernature.com

Printed and bound by CPI Group (UK) Ltd, Croydon, CR0 4YY
01/10/2026
02225849-0011